**BIOS** INSTANT NOTES

# Immunology

THIRD EDITION

# BIOS INSTANT NOTES

# Immunology

## THIRD EDITION

### Peter Lydyard
Division of Infection and Immunity
University College Medical School
University College, London, UK

### Alex Whelan
Department of Immunology
Trinity College, Dublin, Ireland

### Michael Fanger
Department of Microbiology and Immunology
Dartmouth Medical School
Lebanon, New Hampshire, USA

Garland Science
Taylor & Francis Group

NEW YORK AND LONDON

*Garland Science*
Vice President: Denise Schanck
Editor: Elizabeth Owen
Editorial Assistant: Louise Dawnay
Production Editor: Georgina Lucas
Copyeditor: Sally Huish
Typesetting and illustrations: Phoenix Photosetting
Proofreader: Mac Clarke
Printed by: MPG Books Limited

ISBN 978-0-4156-0753-7

**Library of Congress Cataloging-in-Publication Data**

Lydyard, Peter M.
  Immunology / Peter Lydyard, Alex Whelan, Michael Fenger. — 3rd ed.
    p. cm. — (BIOS instant notes)
  Rev. ed. of: Instant notes in immunology. 2003.
  ISBN 978-0-415-60753-7 (pbk.)
  1. Immunology—Outlines, syllabi, etc. 2. Immunity—Outlines, syllabi,
  etc. I. Whelan, A. II. Fanger, Michael W. III. Lydyard, Peter M. Instant
  notes in immunology. IV. Title.
  QR182.55.L93 2011
  571.9'.6—dc22                                              2011004856

Published by Garland Science, Taylor & Francis Group, LLC, an informa business,
270 Madison Avenue, New York NY 10016, USA, and 2 Park Square, Milton Park, Abingdon,
OX14 4RN, UK.

15 14 13 12 11 10 9 8 7 6 5 4 3 2 1

Garland Science
Taylor & Francis Group

Visit our web site at http://www.garlandscience.com

# Contents

# Preface

Immunology as a science probably began with the observations by Metchnikoff in 1882 that starfish when pierced by a foreign object (a rose thorn) responded by coating it with cells (later identified as phagocytes). Immunology—the study of the way in which the body defends itself against invading organisms or internal invaders (tumors)—has developed rapidly over the last 50 years, and more recently with the advent of molecular techniques has contributed critical tools for research and diagnosis, and therapeutics for treatment of a wide range of human diseases.

Immunology has thus become an integral part of human and life science courses as well as medical studies. In this third edition, we have included some of the recent important aspects of immunology and have (1) updated all of the material presented, including adding more figures and tables; (2) emphasized the important role of pattern recognition in immune responses; (3) described critical aspects of antigen processing and presentation to T cells in more detail; (4) described the recent advances in our understanding of how immune responses are regulated, particularly through regulatory T cells and the neuroendocrine system; (5) described important microbial infections in some detail from entry of the microbe to mechanisms of immunity; (6) made more definitive the material on the role of age and gender on immune responses; (7) added a new chapter on immunotherapy for the treatment of a number different diseases.

Overall, we believe that this third edition of *Instant Notes in Immunology* will provide a firm basis for the understanding of contemporary human immunology and its relevance to the mechanisms and treatments of human diseases.

For ease of understanding, we have divided the subject matter in this book into seven main areas:

1. Cellular and molecular components of the immune system (Sections A–D).

2. Mechanisms involved in the development of immunity (Sections E–G), including antibody and cellular responses, and their regulation.

3. The immune system in action (Sections H and I), including immunity to infection and vaccination.

4. Diseases and deficiencies of the immune system (Sections J–L), including allergy, autoimmunity, and congenital and acquired immune deficiency.

5. The immune response to tumors and transplants (Sections M and N).

6. The influence of gender and aging on the immune response (Sections O and P).

7. A new chapter on immunotherapy including monoclonal antibodies, cytokines, and cellular therapy (Section Q).

We are grateful for the helpful discussions with Dr Michael Cole, Professor Peter Delves, Dr Derek Doherty, Professor Paul Guyre, Dr Aideen Long, Professor Randy Noelle, and Dr Mark Smith.

We would also like to thank our wives, Meriel, Annette, and Sharon for longstanding support and understanding during preparation of the third edition.

PML, AW, and MWF

# A1 The need

| Key Note | |
|---|---|
| The ubiquitous enemy | Infectious microbes and larger organisms such as worms are present in our environment. They range from being helpful (e.g., *E. coli*) to being major pathogens which can be fatal (e.g., HIV). |
| Related topic | (H1) The microbial cosmos |

## The ubiquitous enemy

Microbes are able to survive on animal and plant products by releasing digestive enzymes directly and absorbing the nutrients, and/or by growth on living tissues (extracellular), in which case they are simply bathed in nutrients. Other microbes infect (invade and live within) animal/human cells (intracellular), where they not only survive, but also replicate utilizing host-cell energy sources. Both extracellular and intracellular microbes can grow, reproduce, and infect other individuals. There are many different species of microbes and larger organisms (such as worms) that invade humans, some of which are relatively harmless and some even helpful (e.g., *E. coli* in the intestines). Many others cause disease (human pathogens). There is a constant battle between invading microbes and the immune system (Section H2). Although most pathogenic microbes do not cause mortality, they do produce global morbidity. Table 1 shows the range of organisms that can infect humans.

## Table 1. Range of infectious organisms

| | |
|---|---|
| Worms (helminths) | e.g., tapeworms, filaria |
| Protozoans | e.g., trypanosomes, leishmania, malaria |
| Fungi | e.g., *Candida*, *Aspergillus* |
| Bacteria | e.g., *Bacteroides*, *Staphylococcus*, *Streptococcus*, mycobacteria |
| Viruses | e.g., polio, pox viruses, influenza, hepatitis B, human immunodeficiency virus (HIV) |
| Prions | e.g., Creutzfeldt–Jakob disease (CJD) |

# A2 External defenses

## Key Notes

| | |
|---|---|
| **Physical barriers to entry of microbes** | Microbes gain entrance into the body actively (penetration of the skin), or passively (ingestion of food and inhalation). They have to pass across physical barriers such as the skin or epithelial cells which line the mucosal surfaces of the respiratory, gastrointestinal, and genitourinary tracts. |
| **Secretions** | Secretions from epithelial surfaces at external sites of the body are important for protection against entry of microbes. Sweat, tears, saliva, and gastric juices all contain antimicrobial substances such as enzymes, peptides (e.g., defensins), fatty acids, and secreted antibodies. |
| **Microbial products and competition** | Nonpathogenic bacteria (commensals) colonize epithelial surfaces, and by releasing substances toxic to other microbes, for example colicins, competing for essential nutrients required by pathogens, and occupying the microenvironment, they prevent invasion by pathogenic bacteria. |
| **Related topics** | (C3) Mucosa-associated lymphoid tissues (H1) The microbial cosmos |

## Physical barriers to entry of microbes

Before a microbe or parasite can invade the host and cause infection, it must first attach to and penetrate the surface epithelial layers of the body. Organisms gain entrance into the body by active or passive means. For example, they might burrow through the skin, be ingested in food, be inhaled into the respiratory tract, enter via the genitourinary tract, or penetrate through an open wound. In practice, most microbes take advantage of the fact that we have to breathe and eat, and therefore enter the body through the respiratory and gastrointestinal tracts. Whatever their point of entry, they have to pass across physical barriers such as the dead layers of the skin or living epithelial cell layers which line the cavities in contact with the exterior—the respiratory, genitourinary, and gastrointestinal tracts. These are the main routes of entry of microbes into the body.

Many of the cells at the interface with the outside world are mucosal epithelial cells, which secrete mucus. In addition to providing a physical barrier, these cells have other properties useful in minimizing infection. For example, epithelial cells of the nasal passages and bronchi of the respiratory system have **cilia** (small hair-like structures) that beat in an upward direction to help remove microorganisms that enter during breathing. This is the **mucociliary escalator** (Figure 1).

## Secretions

A variety of secretions at epithelial surfaces are important in defense (Table 1), as they help to create a hostile environment for microbial habitation. Some substances are known to directly kill microbes, for example lysozyme digests proteoglycans in bacterial cell walls;

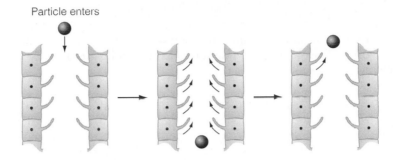

Figure 1. The mucociliary escalator. When a particle is inhaled, it comes into contact with cilia of the bronchial or nasal epithelia, which beat in an upwards direction to a position where the particle can be coughed up or sneezed out.

**Table 1. Secretions at epithelial surfaces**

| Site | Source | Specific substances secreted |
|------|--------|------------------------------|
| Eyes | Lacrimal glands (tears) | Lysozyme, IgA and IgG, antimicrobial peptides* |
| Ears | Sebaceous glands | Oily waxy secretion, fatty acids |
| Mouth | Salivary glands (saliva) | Digestive enzymes, lysozyme, IgA, IgG, lactoferrin |
| Skin | Sweat glands (sweat) | Lysozyme, high NaCl, short-chain fatty acids |
| | Sebaceous glands | Oily secretion and fatty acids (sebum) |
| Stomach | Gastric juices | Digestive enzymes (pepsin, rennin), acid (low pH, 1–2) |

* Antimicrobial peptides are found in all of these secretions.

others, for example transferrin, compete for nutrients (i.e., Fe), and others interfere with ion transport (e.g., NaCl). Mucus (containing mucin) secreted by the mucosal epithelial cells coats cell surfaces and makes it difficult for microbes to contact and bind to these cells—a prerequisite for entry into the body.

The washing action of tears, saliva, and urine also helps to prevent attachment of microbes to the epithelial surfaces. In addition, IgA **antibodies** in tears and saliva prevent the attachment of microbes. These antibodies are also secreted across epithelial cells into the respiratory, gastrointestinal, and genitourinary tracts (Sections C3 and O1).

A large number of different antimicrobial peptides, produced by epithelial cells in the respiratory, gastrointestinal, and genitourinary tracts, are found in body secretions (Table 1). These peptides, which are also produced by phagocytic cells, have potent antibacterial properties and include cecropins, magainins, and defensins (Section B2).

## Microbial products and competition

Normal commensals (**nonpathogenic bacteria**) are also important in protection from infection. These nonpathogenic microorganisms are found on the skin, in the mouth, and in the reproductive and gastrointestinal tracts. The gastrointestinal tract contains many billions of bacteria that have a symbiotic relationship with the host. These bacteria help to prevent pathogens from colonizing the site by preventing attachment, competing for essential nutrients, and releasing antibacterial substances such as **colicins** (antibacterial proteins) and short-chain fatty acids. Gut flora also perform such house-keeping duties as further degrading waste matter and helping gut motility. Normal microbial flora occupying the site of entry (e.g., throat and nasal passages) of other microbes probably function in a similar manner. Some commensal bacteria such as lactobacilli, which inhabit the vagina, cause their environment to become acidic (pH 4.0–4.5), discouraging the growth of many microbes (Section O2).

# A3 Immune defense

## Key Notes

**The immune system**

The immune system protects us from attack by microbes and worms. It uses specialized organs designed to filter out and respond to microbes entering the body's tissues and a mobile force of molecules and cells in the bloodstream to respond rapidly to attack. The system can fail, giving rise to immunodeficiency, or "over-react" against foreign microbes, giving rise to tissue damage (immunopathology). It has complex and sophisticated mechanisms to regulate it.

**Innate versus adaptive immunity**

The innate immune system is the first line of defense against infection. It works rapidly, gives rise to the acute inflammatory response, and has some specificity for microbes, but has no memory. In contrast, the adaptive immune system takes longer to develop, is very highly specific, and responds more quickly to a microbe that it has encountered previously (i.e., shows memory).

**Interaction between innate and adaptive immunity**

The innate and adaptive immune systems work together through direct cell contact and through interactions involving chemical mediators, cytokines, chemokines, and antibodies. Moreover, many of the cells and molecules of the innate immune system are also used by the adaptive immune system.

**Adaptive immunity and clonal selection**

All immunocompetent individuals have many distinct lymphocytes, each of which is specific for a different foreign substance (antigen). When an antigen is introduced into an individual, lymphocytes with receptors for this antigen seek out and bind it and are triggered to proliferate and differentiate, giving rise to clones of cells specific for the antigen. These cells or their products specifically react with the antigen to neutralize or eliminate it. The much larger number of antigen-specific cells late in the immune response is responsible for the "memory" involved in adaptive immunity.

**T and B cells and cell cooperation**

There are two major types of lymphocytes: B cells and T cells. T cells mature under the influence of the thymus and, on stimulation by antigen, give rise to cellular immunity. B cells mature mainly under the influence of bone marrow and give rise to humoral immunity, immunity that involves production of soluble molecules—**antibodies (immunoglobulins)**. Interactions between T and B cells, as well as between T cells and antigen-presenting cells, are critical to the development of specific immunity.

| Related topics | (C) The adaptive immune system | (F) The T-cell response – cell-mediated immunity |
|---|---|---|
| | (D) Antibodies | (G) Regulation of the immune response |
| | (E3) The cellular basis of the antibody response | (H) Immunity to infection |

## The immune system

The immune system is composed of many cell types, the majority of which are organized into separate lymphoid tissues and organs (Section C2). Because attack from microbes can come at many different sites of the body, the immune system has a mobile force of cells in the bloodstream that are ready to attack the invading microbe wherever it enters the body. Although many of the cells of the immune system are separate from each other, they maintain communication through cell contact and molecules secreted by them (such as cytokines and chemokines). Like the other body systems, the immune system is only apparent when it goes wrong. This can lead to severe, sometimes overwhelming, infections and even death. One form of dysfunction is **immunodeficiency**, which can, for example, result from infection with the human immunodeficiency virus (HIV) causing acquired immune deficiency syndrome (AIDS). On the other hand, the immune system can be "hypersensitive" to a microbe (or even to "an inert" substance such as pollen) and this itself can cause severe tissue damage sometimes leading to death. Thus, the immune system must strike a balance between producing a life-saving response and a response that causes severe tissue damage. This regulation (Section G) is maintained by cells and molecules of the immune system (e.g., regulatory T cells-Tregs, and cytokines) and from without by nonimmune cells and tissues and their products (e.g., the neuroendocrine system).

## Innate versus adaptive immunity

Having penetrated the external defenses, microbes come into contact with cells and products of the immune system and the battle commences. A number of cell types and defense molecules are usually present at the site of invasion, or migrate (**home**) to the site. This "first line of defense" is the "**innate immune system**." It is present at birth and changes little throughout the life of the individual. The cells and molecules of this innate system are mainly responsible for the first stages of expulsion of the microbe and may give rise to inflammation (Section B4). Some of the most important cells in the innate immune system are phagocytes, because they are able to ingest and kill microbes.

The second line of defense is the "**adaptive immune system**," which is brought into action even as the innate immune system is dealing with the invading microbe, and especially if it is unable to remove it. The key difference between the two systems is that the adaptive system shows far more specificity and remembers that a particular microbe has previously invaded the body. This leads to a more rapid expulsion of the microbe on its second or third time of entry. The cells, molecules, and characteristics of innate and adaptive immune systems are shown in Table 1.

## Interaction between innate and adaptive immunity

Although innate and adaptive immunity are often considered separately for convenience and to facilitate their understanding, it is important to recognize that they work together. For example, macrophages are phagocytic but produce important **cytokines** (Section B2)

**Table 1. The innate and adaptive immune systems**

| Characteristics | Cells | Molecules |
| --- | --- | --- |
| **Innate immunity** | | |
| Responds rapidly | Phagocytes (PMNs and macrophages) | Antimicrobial peptides |
| | | Complement |
| Has some specificity | Natural killer cells | Cytokines |
| But no memory | Mast cells | Chemokines |
| | Dendritic cells | Acute phase proteins |
| **Adaptive immunity** | | |
| Slow to start | T and B cells | Antibodies |
| Highly specific | | Cytokines |
| Memory | | Chemokines |

that help to induce the adaptive immune response. Cytotoxic T cells kill virus-infected body cells. These have to be cleared from the body by phagocytic cells. Complement components of the innate immune system can be activated directly by microbes, but can also be activated by antibodies, molecules of the adaptive system. The various cells of both systems work together through direct contact with each other, and through interactions with chemical mediators, the cytokines and chemokines (Section B2). These chemical mediators can be either cell-bound or released as localized **hormones**, acting over short distances. Cells of both systems have a large number of surface receptors: some are involved in adhesion of the cells to blood endothelial walls (e.g., leukocyte function antigen, LFA-1), some recognize chemicals released by cells (e.g., complement, cytokine, and chemokine receptors), and others trigger the function of the cell such as activation of the phagocytic process.

## Adaptive immunity and clonal selection

All immunocompetent individuals have many distinct lymphocytes. Each of these cells is specific for a different foreign substance (**antigen**). Specificity results from the fact that each lymphocyte possesses cell-surface receptors all of which are specific for a particular antigen. When antigen is introduced into an individual, lymphocytes with appropriate receptors seek out and bind the antigen and are triggered to proliferate and differentiate into the effector cells of immunity (i.e., they give rise through division to large numbers of cells). All members of this **clone** of cells are specific for the antigen initially triggering the response and they, or their products, are capable of specifically reacting with the antigen or the cells that produce it and can mediate its elimination. In addition, there are a much larger number of cells specific for the immunizing antigen late in the immune response. These cells are able to respond faster to antigen challenge, giving rise to the "**memory**" involved in immunity. That is, individuals do not usually become infected by the same organism twice, as their immune system remembers the first encounter and protects against a second infection by the same organism. Of particular importance, all immunocompetent individuals have produced enough different specific lymphocytes to react with virtually every antigen with which an individual may potentially come in contact. How this diversity is developed is considered in Section D3.

Clonal selection as it applies to the B-cell system is shown in Figure 1 and is presented in more detail in Section E3. In particular, on encounter with antigen, B cells with receptors for that antigen bind and internalize it and receive help from T cells (Section F5). These B cells are triggered to proliferate, giving rise to clones of daughter cells. Some of these cells serve as memory cells, others differentiate and mature into **plasma cells**, which produce and secrete large quantities of specific antibody (Section C1).

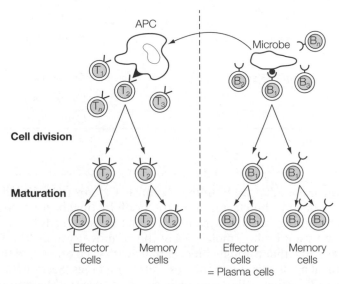

Figure 1. Clonal selection. From a large pool of B and T cells ($T_n$ and $B_n$), antigen selects those which have receptors for it (e.g., $T_2$ and $B_1$) and stimulates their expansion and differentiation into memory and effector cells. Although B cells can recognize and bind native antigen, T cells only see antigen associated with MHC molecules on antigen-presenting cells (APC).

## T and B cells and cell cooperation

The lymphocytes selected for clonal expansion are of two major types, B cells and T cells, each giving rise to different kinds of immunity. T lymphocytes mature under the influence of the thymus and, on stimulation by antigen, give rise to cellular immunity. B lymphocytes mature mainly under the influence of bone marrow and give rise to lymphoid populations which, on contact with antigen, proliferate and differentiate into **plasma cells**. These plasma cells make antibody (immunoglobulin) which is specific for the antigen and able to neutralize and/or eliminate it (humoral immunity).

The development of the immune response to an antigen also requires cell cooperation between cells of the immune system. T cells need to interact with both B cells and other antigen-presenting cells (APC) for the development of specific immune responses. Subpopulations of T cells help (Th cells) or suppress (Treg cells) antibody and cellular immune responses. Although immune responses to most antigens (especially proteins) require cell cooperation, some antigens (**T-independent**) are able to initiate an immune response in the absence of T lymphocytes.

# **A4** Antigens

---

**Key Notes**

| | |
|---|---|
| **Range of antigens** | Antigens are defined as substances which induce an immune response. They include proteins, carbohydrates, lipids, and nucleic acids. Microbes have many different components which can be recognized by the innate and adaptive immune systems. |
| **Antigen structures recognized by T and B lymphocytes** | Antigens may contain a number of different antigenic determinants (epitopes) to which individual antibodies or T-cell responses are made. The smallest unit (antigenic determinant) to which an antibody can be made is about three to six amino acids or about five to six sugar residues. All large molecules have many determinants. Antibodies bind to conformational antigenic determinants (dependent on folding of the molecule) while T-cell receptors recognize linear amino acid sequences. Molecules which can stimulate an immune response ("immunogens") can be distinguished from those that react with antibodies but cannot initiate an immune response (haptens or individual antigenic determinants). |
| **Related topics** | (E1) The B-cell receptor complex, co-receptors, and signaling      (F2) T-cell recognition of antigen      (M2) Transplantation antigens |

---

## Range of antigens

The first stage of removing an invading organism is to recognize it as being foreign, that is, not "self" (Sections E and F). The immune system sees the invader as having a number of antigens. In the broad sense, an antigen is any substance which induces an immune response. Antigens recognized by cells of the innate immune system are called "pathogen-associated molecular patterns" (PAMPs) and initiate production of cytokines, chemokines, and antimicrobial peptides by these cells (Section B2). Recognition of antigens by lymphocytes results in proliferation and production of cytokines and/or antibodies. Antigens recognized by lymphocytes include proteins, carbohydrates, lipids, and nucleic acids. Responses can be made to virtually anything. Even self molecules or cells can act as antigens under appropriate conditions, although this is quite well regulated in normal healthy individuals (Section G).

## Antigen structures recognized by T and B lymphocytes

It is usual that an antigen, a molecule which is capable of initiating an adaptive immune response, possesses several unique molecular structures, each of which can elicit an adaptive immune response. Thus, antibodies or cells recognizing an antigen are not directed against the whole molecule but against different parts of the molecule. These "antigenic determinants" or "epitopes" (Figure 1) are the smallest unit of an antigen to which an antibody or cell can bind. For a protein, an antibody binds to a unit which is about three to six amino acids while for a carbohydrate it is about five to six sugar residues. Therefore, most large molecules possess many antigenic determinants per molecule, that is, they are "multideterminant." However, these determinants may be identical or different from each other on the same molecule. For example, a carbohydrate with repeating sugar units will have several identical determinants, while a large single chain protein will usually not have repeating 3–5 amino acid sequences, and will thus have many different antigenic determinants.

Although the linear sequence of the residues in a molecule has been equated with an antigenic determinant, the physical structures to which antibodies bind are primarily the result of the conformation of the molecule. As a result of folding, residues at different parts of the molecule may be close together and may be recognized by a B-cell receptor or an antibody as part of the same determinant (Figure 1). Thus, antibodies made against the native (natural) conformation of a molecule will not, in most instances, react with the denatured molecule even though the primary sequence has not changed. This is in contrast to the way in which T-cell receptors recognize antigenic determinants—in the form of linear amino acid sequences (Section F2), which have to be presented by major histocompatibility complex (MHC) molecules (Figure 2). The anchor amino acid residues of the linear peptide are important for attachment to the MHC molecule (Section F2).

In practical terms, microbes have a large number of different molecules and therefore potentially many different antigenic determinants, all of which could stimulate an immune response. However, all antigenic determinants are not equal—some may elicit strong and others weak responses. This is determined by the health, age, and genetics of the individual (Section G1).

Very small molecules, which can be viewed as single antigenic determinants, are incapable of eliciting an antibody response by themselves. These **haptens**, as they are called,

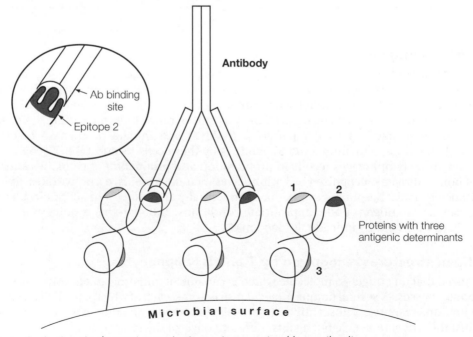

Figure 1. Antigenic determinants (epitopes) recognized by antibodies.

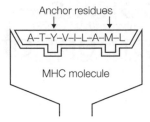

Figure 2. Linear sequence of peptides recognized by T cells.

can be attached covalently to larger molecules (**carriers**) and in this physical form can, with the help of T cells, induce the formation of antibodies. Therefore, one can distinguish between molecules which can stimulate an immune response (**immunogens**) and those which react with antibodies but cannot initiate an immune response (haptens or individual antigenic determinants).

# A5 Hemopoiesis – development of blood cells

---

**Key Notes**

| | |
|---|---|
| **A common stem cell** | The majority of the cell types involved in the immune system are produced from a common hemopoietic stem cell (HSC). HSCs are found in the fetal liver, fetal spleen, and in the neonate and adult bone marrow. They differentiate into functionally mature cells of all blood lineages. |
| **Stromal cells** | Direct contact with stromal cells (including epithelial cells, fibroblasts, and macrophages) is required for the differentiation of a particular lineage, as are adhesion molecules and cytokines. |
| **Role of cytokines** | Stromal cells produce many cytokines, including stem cell factor (SCF), monocyte colony-stimulating factor (M-CSF) and granulocyte colony-stimulating factor (G-CSF). Interaction of stem cells with stromal cells and M-CSF or G-CSF results in the development of monocytes and granulocytes, respectively. |
| **Related topic** | (B2) Molecules of the innate immune system |

---

## A common stem cell

The majority of cell types involved in the immune system are produced from a common hemopoietic stem cell (HSC) and develop through the process of differentiation into functionally mature blood cells of different lineages, for example monocytes, platelets, lymphocytes, and so forth (hemopoiesis: Figure 1). These stem cells are replicating self-renewing cells, which in early embryonic life are found in the yolk sac and then in the fetal liver, spleen, and bone marrow. After birth, the bone marrow contains the HSCs.

The lineage of cells differentiating from the HSC is determined by the microenvironment of the HSC and requires contact with stromal cells and interaction with particular cytokines. These interactions are responsible for switching on specific genes coding for molecules required for the function of the different cell types, for example those used for phagocytosis in macrophages and neutrophils, and the receptors on lymphocytes which determine specificity for antigens. This is, broadly speaking, the process of differentiation.

## Stromal cells

Stromal cells, including epithelial cells and macrophages, are necessary for the differentiation of stem cells to cells of a particular lineage, for example lymphocytes. Direct

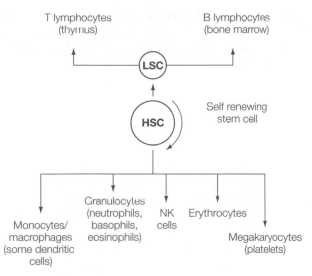

Figure 1. Origin of blood cells (hemopoiesis); LSC, lymphoid stem cell; HSC, hemopoietic stem cell; NK, natural killer.

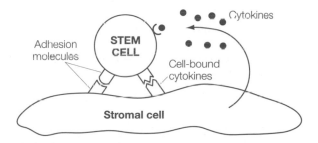

Figure 2. Role of stromal cells in hemopoiesis. Stromal-cell-bound cytokines (e.g., stem cell factor) and released cytokines (e.g., IL-7) determine the differentiation pathway of the stem cell attached through specific adhesion molecules (e.g., CD44) on the stem cell attached to hyaluronic acid molecules on the stromal cell.

contact of the stromal cell with the stem cell is required. Within the fetal liver, and in the thymus and bone marrow, different stromal cells (including macrophages, endothelial cells, epithelial cells, fibroblasts, and adipocytes) create discrete foci where different cell types develop. Thus, different foci will contain developing granulocytes, monocytes, or B cells. Cytokines are essential for this process, and it is thought that adhesion molecules also play an important role (Figure 2).

## Role of cytokines

Different cytokines are important for renewal of HSC and their differentiation into the different functionally mature blood cell types. Although an oversimplification, the processes related to HSC regeneration depend largely on stem cell factor (SCF), interleukin-1 (IL-1), and IL-3. The development of granulocytes and monocytes require, among other cytokines, monocyte colony-stimulating factor (M-CSF) and granulocyte colony-stimulating factor (G-CSF), both of which are produced by stromal cells. Thus, interaction of stem cells with stromal cells and with M-CSF or G-CSF results in the development of monocytes and granulocytes, respectively (Figure 3). Other cytokines are important for the early differentiation of T cells in the thymus and B cells in particular locations within the bone marrow.

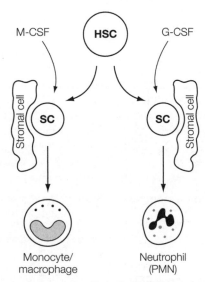

Figure 3. Different cytokines and stromal cells induce different pathways of differentiation. PMN, polymorphonuclear cell. SC, stem cell.

# B1 Cells of the innate immune system

## Key Notes

**Phagocytes**

Most white blood cells are mobile phagocytes (or eating cells), called neutrophils or polymorphonuclear cells (PMNs), that patrol the blood in search of invading microbes. Other primary phagocytic cells are part of the mononuclear phagocyte system, and include monocytes and macrophages. Monocytes are present in the blood and settle in the tissues as macrophages (MØ). These phagocytes are attracted to sites of infection (chemotaxis), bind to the microbe (adhere), and ingest (phagocytose) and kill the microbe. Molecules coating a microbe, such as complement or antibody, enhance contact and ingestion (opsonization) of the microbe.

**Natural killer (NK) cells**

NK cells are found throughout the tissues of the body but mainly in the circulation, and are important for protection against viruses and some tumors. Changes in the surface molecules of cells as the result of virus infection allow NK cells to bind to and kill infected cells by releasing perforins and inducing apoptosis. In addition, on binding to virus-infected cells, NK cells secrete interferon gamma (IFNγ), which protects adjacent cells from infection by viruses and helps to activate T-cell-mediated immunity.

**Mast cells and basophils**

Mast cells (in connective tissues) and basophils (in the circulation) are produced in the bone marrow and have similar morphology and functions. When activated, these cells degranulate, releasing pharmacological mediators resulting in vasodilation, increased vascular permeability, and leukocyte migration.

**Dendritic cells (DCs)**

DCs represent an important link between the innate and adaptive immune systems. Most interact with T cells through their expression of MHC class II molecules while others (follicular dendritic cells) interact with B cells in the B-cell follicles of the lymphoid tissues. Those that interact with T cells are either myeloid DCs or plasmacytoid DCs, so called because they look like plasmacytes. Immature myeloid DCs pick up antigen and transport it to the lymphoid tissues. The main property of plasmacytoid DCs is their ability to produce large amounts of type I interferons following activation by viruses. Both types of DCs are involved in inducing and regulating immune responses.

| | |
|---|---|
| **NKT cells** | Although these cells have many of the properties of NK cells they also have some T-cell characteristics, making it difficult to classify them as innate immune cells. Expressing CD3 and using a limited T-cell receptor gene repertoire, they respond to lipids and glycolipids but not peptides, which are recognized by most T cells. They recognize antigens displayed on CD1d (not conventional MHC molecules) expressed by antigen-presenting cells. Like NK cells, they can be cytotoxic, but produce a wider spectrum of cytokines than NK cells that function to regulate immune responses. |
| **Other cells playing a role in innate immunity** | A variety of other cells, including epithelial cells, eosinophils, platelets, and erythrocytes play a role in immune defense. Epithelial cells produce antimicrobial peptides. Eosinophils are granular leukocytes that attack and kill parasites by releasing the toxin major basic protein. Platelets, on activation, release mediators that activate complement leading to attraction of leukocytes. Erythrocytes bind and remove small immune complexes. |
| **Related topics** | (D2) Antibody classes  (H1) The microbial cosmos<br>(D8) Antibody functions  (K2) IgE-mediated (type I)<br>(F2) T-cell recognition of         hypersensitivity: allergy<br>         antigen |

## Phagocytes

Phagocytes are specialized "eating" cells (*phagein*: Greek, "to eat"), of which there are two main types, neutrophils and macrophages. **Neutrophils**, often called polymorpho-nuclear cells (PMNs) because of the multilobed nature of their nuclei (Figure 1), are mobile phagocytes that comprise the majority of blood leukocytes (about $8 \times 10^6$ per ml of blood). They have a very short half-life (days) and die in the bloodstream by apoptosis

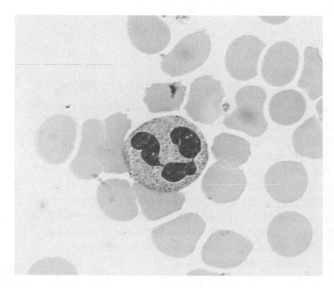

Figure 1. A polymorphonuclear cell (neutrophil) in the blood. From Male D, Brostoff J, Roth D & Roitt I (2006) Immunology, 7th ed. With permission from Elsevier.

(programmed cell death). They have granules that contain peroxidase, alkaline and acid phosphatases, and defensins (Section B2), which are involved in microbial killing. These granulocytes stain with neutral dyes and have a different function from granulocytes that stain with eosin (eosinophils) or basic dyes (basophils). PMNs have receptors for chemotactic factors released from microbes, for example muramyl dipeptide (MDP), and for complement components activated by microbes (Table 1). Their main function is to patrol the body via the bloodstream in search of invading microbes. As such they are pivotal cells in acute inflammation (Section B4). Like the majority of cells involved in the immune system, these phagocytes are produced in the bone marrow (Section A5).

The **mononuclear phagocyte system** (previously called the reticuloendothelial system) is a widely distributed tissue-bound phagocytic system whose major function is to dispose

## Table 1. Surface receptors on polymorphonuclear cells (PMNs)

| Surface molecules | Function |
| --- | --- |
| **Pattern recognition receptors** | |
| TLRs*, NLRs, and MR | Recognize PAMPs; trigger cytokine production and cell activation |
| **Complement receptors** | |
| CR1 (CD35) | Binds to C3b, iC3b, C4b, and mannose binding ligand (opsonization) |
| CR3 (CDIIb/CD18) | Binds to C3b, iC3b; permits removal (opsonization) of complement coated antigens and microbes |
| **Fc receptors for IgG** | |
| CD16 (FcγRIII), CD32 (FcγRII) | Binds to IgG–antigen complexes (opsonization) |
| CD64 (FcγRI) | Expressed on activated PMNs (opsonization) |
| **Chemokine receptors** | |
| CXCR1 and CXCR2 | Binds IL-8 produced by macrophages and epithelial cells |
| **Chemoattractant receptors** | |
| C5aR | Binds to C5a for attraction towards microbes after C activation |
| Leukotriene B4 receptors | LTB4 produced from leukocytes and necrotic cells |
| Receptors for FMLP | FMLP produced by breakdown of some bacterial proteins |
| Receptor for MDP (NOD2) | MDP is a peptidoglycan constituent of both Gram-positive and Gram-negative bacteria (see above and Sections B2 and B3) |
| **Adhesion receptors** | |
| LFA-1 | Binds to ICAM-1 on endothelium for extravasation |
| VLA-4 | Binds to VCAM-1 on endothelium for extravasation |

TLR ,Toll-like receptor; NLR, nucleotide-binding oligomerization domain-like receptor; MR, mannose receptor; PAMPs, pattern associated molecular pattern; FMLP, formyl-methionyl-leucyl-phenylalanine; MDP, muramyl dipeptide; LFA-1, leukocyte function antigen-1; ICAM-1, intercellular adhesion molecule-1; VLA-4, very late activation antigen-4; VCAM-1, vascular cell adhesion molecule-1. * Some TLRs and all NLRs are intracellular receptors. NOD2 is one of the NLRs.

of microbes and dead body cells through the process of phagocytosis. Monocytes (Figure 2) are blood-borne precursors of the major tissue phagocytes, macrophages. Different organs/tissues each have their versions of monocyte-derived phagocytic cells (Table 2).

Phagocytosis is a multistep process (Table 3) and the major mechanism by which microbes are removed from the body. It is especially important for defense against extracellular microbes (Section H2).

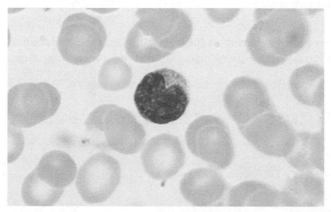

Figure 2. A monocyte in the blood. From Roitt I, Brostoff J & Male D (1998) Immunology, 5th ed. With permission from Elsevier.

**Table 2. Cells of the mononuclear phagocyte system**

| Cells | Location |
| --- | --- |
| Monocytes | Bloodstream |
| Kupffer cells | Liver |
| Mesangial cells | Kidney |
| Alveolar macrophages | Lungs |
| Microglial cells | Brain |
| Sinus macrophages | Spleen, lymph nodes |
| Serosal macrophages | Peritoneal cavity |

**Table 3. Stages in phagocytosis**

| Stage | Mechanism |
| --- | --- |
| 1 Movement of phagocyte towards the microbe | Chemotactic signals, e.g., MDP, complement (C5a) |
| 2 Attachment of microbe to the phagocyte surface | Binding to mannose, complement and/or Fc receptors |
| 3 Endocytosis of microbe leads to the formation of a phagosome | Invagination of surface membrane |
| 4 Fusion of phagosome with lysosome | Microtubules involved |
| 5 Killing of microbe | Oxygen-dependent killing, e.g., $O_2$ radicals, nitric oxide; oxygen-independent, e.g., myeloperoxidase |

Opsonization is the process of making a microbe easier to phagocytose. A number of molecules called "opsonins" (*opsonein*: Greek, "to buy provisions") do this by coating the microbe. They aid attachment of the microbe to the phagocyte and also trigger activation of phagocytosis. Opsonins include the complement component C3b and antibody itself, the latter acting as a bridge between the innate and adaptive immune systems (Sections B2 and D8). Mononuclear phagocytes use their surface receptors (Table 4), which bind to C3b or to the Fc region of IgG antibody (Fc receptors, FcγR), to attach to C3b or IgG coating the microbes, respectively.

Killing by mononuclear phagocytes is generally very efficient, as there are many cytotoxic mechanisms available to these cells. In particular, these cells contain many different

## Table 4. Surface receptors on monocytes/macrophages

| Surface molecules | Function |
| --- | --- |
| **Pattern recognition receptors** | |
| TLRs*, NLRs, MRs, and others | Mediate cytokine production and induce adaptive immune responses. Some mediate phagocytosis |
| **Fc receptors for IgG** | |
| CD16 (FcγRIII), CD32 (FcγRII), CD64 (FcγR1) | Binds to IgG–antigen complexes and IgG-coated target cells, mediating phagocytosis and cytokine production |
| **Fc receptor for IgA** | |
| CD89 (FcαR) | Binds to IgA–antigen complexes, mediating phagocytosis and cytokine production |
| **Complement receptors** | |
| CD35 (CR1) | Involved in enhancing phagocytosis of IgM/IgG-coated microbes on which complement has been activated (opsonization) |
| CR3 (CDIIb/CD18) | Binds to C3b, iC3b; permits removal of complement-coated antigens and microbes (opsonization) |
| **Adhesion receptors** | |
| CD18/11a,b,c (LFA-1, CR3, CR4) | Adhesion molecules facilitating interactions with other cells, including binding to endothelial cells for extravasion (monocytes) |
| VLA-4 | Binds to VCAM-1 on endothelial cells for extravasion (monocytes) |
| **Antigen presentation molecules (HLA)** | |
| MHC class I (HLA A,B,C) | Presentation of peptides to Tc cells |
| MHC class II (HLA D) | Presentation of peptides to Th cells |
| (MHC-like) CD-1 | Presentation of lipids to NKT cells** |

TLR, Toll-like receptor; NLR, nucleotide-binding oligomerization domain-like receptor; MR, mannose receptor; PAMP, pattern associated molecular pattern; FMLP, formyl-methionyl-leucyl-phenylalanine; MDP, muramyl dipeptide; LFA-1, leukocyte function antigen-1; VLA-4, very late activation antigen-4; VCAM-4, vascular cell adhesion molecule-4.
*Some TLRs and all NLRs are intracellular.
**To be dealt with in Section C.

enzymes, cationic proteins and antimicrobial peptides (defensins: Section B2) that in concert can mediate killing and digestion of the microbe. In addition, on activation, these mononuclear phagocytes can kill intracellular pathogens via oxygen-dependent mechanisms through oxygen metabolites, including superoxide and nitric oxide.

## Natural killer (NK) cells

Natural killer (NK) cells differ from classical lymphocytes in that they are larger, contain more cytoplasm, and have (electron) dense granules (Figure 3). Produced in the bone marrow, they are found throughout the tissues of the body, but mainly in the circulation where they comprise 5–15% of the total lymphocyte fraction (Section C1). Their cell surface receptors (Table 5) include killer activation receptors (KARs) and killer inhibitory receptors (KIRs). These interact with molecules on body cells that activate or inhibit NK cell activity, respectively. FcγRIII on these cells mediates antibody-dependent cellular cytotoxicity (ADCC) of antibody-coated microbes/tumor cells. There are also some cell adhesion molecules involved in activation and traffic of NK cells. Toll-like receptors (TLRs: Section B3) on NK cells contribute to activation of cytotoxicity and cytokine production.

The main function of NK cells is to kill virus-infected self cells, as well as some tumor cells. When NK cells bind to uninfected self cells, their KIRs provide a negative signal to the NK cell, preventing it from killing the self cell. This is because KIRs recognize MHC class I (Section F2) leader peptides presented in an MHC-like molecule, human leukocyte antigen-E (HLA-E). However, infection of cells by some viruses reduces the expression of MHC molecules, and therefore decreases the loading of class I peptides in HLA-E, thus allowing the activation through KARs to induce NK-cell killing of the infected cell. This is an important mechanism, allowing NK cells to recognize normal self cells and ignore them, while killing infected or malignant self cells.

The mechanisms by which NK cells mediate killing are identical to those used by cytotoxic T cells (Section F5) and involve release of granule contents (perforins and granzymes) onto the surface of the infected cell. Perforin has a structure similar to that of C9, a component of complement which can create pores in the cell membrane (Sections B2 and D8), allowing the passage of the granzymes (proteolytic enzymes) into the cell to induce apoptosis. NK cells, like cytotoxic T cells, are also able to induce target cell

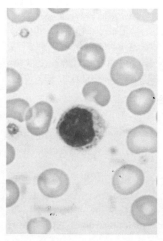

Figure 3. An NK cell in the blood. From Male D, Brostoff J, Roth D & Roitt I (2006) Immunology, 7th ed. With permission from Elsevier.

**Table 5. Surface receptors on natural killer cells**

| Molecules | Function |
|---|---|
| **Activation/inhibitory receptors** | |
| KIRs | Contain ITIMs and bind to MHC class-I-like molecules associated with self peptide and prevent NK cells from killing |
| KARs | Bind to self antigens (e.g., carbohydrate on self cells) and are associated with other molecules that contain ITAMs. On activation by KAR binding (in the absence of simultaneous engagement of KIRs) they initiate release of cytotoxic molecules from the NK cells |
| **Fc receptors** | |
| CD16 (FcγRIII) | Binds to IgG-coated target cells and mediates ADCC |
| **Adhesion/accessory molecules** | |
| CD2 | Binds to LFA-3 |
| CD56 (NCAM) | Binds to FGFR-1 that is bound to and secreted by fibroblasts |
| LFA-1 | Binds to ICAM-1 |

KIR, killer inhibitory receptor; KAR, killer activation receptor; ADCC, antibody-dependent cellular cytotoxicity; ITIM, immunoreceptor tyrosine-based inhibitory motif; ITAM, immunoreceptor tyrosine-based activation motif; NCAM, neural cell adhesion molecule; LFA-3, leukocyte function antigen-3; FGFR-1, fibroblast growth factor receptor-1; ICAM-1, intercellular adhesion molecule-1.

apoptosis through binding of their surface FasL molecules to Fas molecules on the surface of the virus-infected cell (Section F5).

IL-2 activates NK cells to kill, and these lymphokine-activated killer (LAK) cells have been used in clinical trials to treat tumors (Section N5). When NK cells are "activated" by recognizing a virus-infected cell they secrete IFNγ. This helps to protect surrounding cells from virus infection, although IFNα and IFNβ are probably more important in this role (Section B2). In addition, IFNγ can also enhance the development of specific T-cell responses directed to virus-infected cells (Sections F4 and F5).

## Mast cells and basophils

Mast cells (Figure 4) are found throughout the body in connective tissues close to blood vessels and particularly in the subepithelial areas of the respiratory, genitourinary, and gastrointestinal tracts. Basophils are granulocytes which stain with basic dyes and are present in very low numbers in the circulation (<0.2% of the granular leukocytes). Basophils and mast cells are very similar in morphology. Both have large characteristic electron-dense granules in their cytoplasm which are very important for their function. Like all the granulocytes, basophils and mast cells are produced from stem cells in the bone marrow.

Mast cells/basophils can be stimulated to release their granules as a result of:

- Their binding to the anaphylatoxins C3a and C5a

- Binding of allergens to antiallergen IgE bound to their cell surface Fc receptor (FcεR) and the resulting crosslinking of FcεR (Section K2)

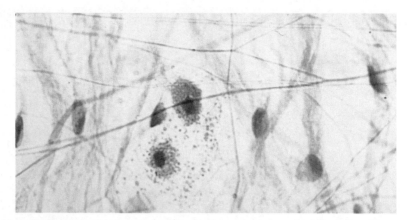

Figure 4. Mast cells. Note the large granules in the cytoplasm which contain pharmacological mediators. From Leboffe MJ & Pierce BE (1996) A Photographic Atlas for the Microbiology Laboratory. With permission from Morton Publishing.

- Binding to lectins (molecules that bind carbohydrates)
- Some drugs (e.g., opioids, succinylcholine, vancomycin) as well as radiocontrast agents

Stimulation results in the fusion of the intracellular granules with the surface membrane and the release of their contents to the exterior by the process of exocytosis. This release is almost instantaneous and is essential in the development of the acute inflammatory response (Section B4). Granule contents include a variety of preformed pharmacological mediators. Other pharmacological mediators are produced *de novo* when the cells are stimulated (Table 6). Production *de novo* of pro-inflammatory cytokines such as TNFα and IL-8 can be induced through binding of microbial products to TLRs expressed on mast cells (Section B3). When large numbers of mast cells/basophils are stimulated to degranulate, severe anaphylactic responses can occur, which in their mildest form give rise to the allergic symptoms seen in type I hypersensitivity (Section K2).

## Dendritic cells (DCs)

These are cells that link the innate and adaptive immune systems. Dendritic cells (DCs) are so called because of their many surface membrane folds that are similar in appearance to dendrites of the nervous·system (Figure 5). These folds allow maximum interaction with other cells of the immune system. Dendritic cells can be divided into two major groups: (i) those expressing high levels of MHC class II for antigen presentation to T cells;

## Table 6. Main mediators released and their effects

|  | Mediators | Effect |
|---|---|---|
| Preformed | Histamine | Vasodilation, vascular permeability |
| *De novo* synthesized | Cytokines, e.g. | |
|  | TNFα, IL-8, IL-5 | Attract neutrophils and eosinophils |
|  | PAF | Attracts basophils |

TNFα, tumor necrosis factor α; IL, interleukin; PAF, platelet-activating factor.

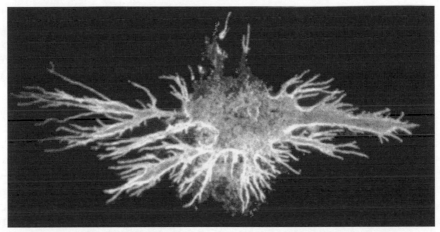

Figure 5. Dendritic cell. Note the many membrane processes that allow interactions with lymphocytes. Surface stained with a fluorescent antibody to CD44 (shown white). CD44 is an adhesion molecule that allows the dendritic cell to attach to connective tissue and other cells. (Figure courtesy of M. Binks.)

(ii) those in the follicles of lymphoid tissues (follicular dendritic cells, FDC) that do not express MHC class II but have other markers able to interact with B cells (Section C2).

The DCs that present antigen to T cells ((i) above) can be further divided into myeloid-derived DCs or plasmacytoid DCs, so called because they look like plasmacytes. The main characteristics of these DC populations are shown in Table 7.

DCs represent a primary interface between the innate and adaptive immune systems in that they recognize microbial antigens through pattern recognition receptors (PRRs), and through cytokine production and antigen processing and presentation they are able to initiate adaptive immune responses by presenting peptide antigens to T-helper (CD4$^+$) cells.

Since the T-cell antigen receptor can only recognize "pieces" of proteins in association with MHC molecules, proteins first need to be "processed" (cut up into short peptides). These peptides are then attached to MHC molecules (Section F2) for display on the

## Table 7. Dendritic cells

| Classification | Characteristics | Function |
|---|---|---|
| Myeloid* | Localized in blood and tissues. Relatively short lived and replenished by bone marrow precursors | Phagocytic when immature; migrate to lymphoid organs and tissues; main antigen presenting cells to prime CD4$^+$ T cells; regulate immune responses |
| Plasmacytoid | Present in bone marrow and peripheral organs: usually long lived | Produce large amounts of type I interferons following exposure to viruses; can present antigen to CD4$^+$ T cells and regulate immune responses |

*Langerhans cells (LH) are classified as myeloid-cell-derived DCs, but are found in the epidermis of the skin. Interdigitating cells of the T-cell areas of the secondary lymphoid tissues are also of this type (Section C2).

surface of myeloid and plasmacytoid DCs. Although macrophages can process and present antigen to T cells, the conventional DCs are much more efficient in carrying out this function. Some DCs can also act to induce T-cell tolerance to antigens (especially self), a very important mechanism leading to peripheral tolerance to self antigens (Section G3).

## NKT cells

These cells are difficult to classify as innate immune cells because although they have many of the properties of NK cells they also have some T-cell characteristics. They express CD3, but unlike conventional T cells that recognize peptides, NKT cells recognize and respond to lipid and glycolipid antigens. These are presented to them, not in a conventional MHC context (i.e., MHC class I and II molecules) but rather a nonpolymorphic MHC-like molecule, CD1d. They also possess a relatively invariant $\alpha\beta$ T-cell receptor using biased rearranged T-cell receptor genes (Section F3). Like NK cells, they can be cytotoxic, but they produce a wider spectrum of cytokines than NK cells, which function to regulate immune responses.

## Other cells playing a role in innate immunity

**Epithelial cells** lining the digestive, respiratory, and genitourinary tracts produce and secrete a variety of antimicrobial peptides (Section B2).

**Eosinophils** are granular leukocytes which stain with eosin. They are present at low levels in the circulation (2–5% of blood leukocytes), have some phagocytic activity, but are primarily responsible for extracellular killing of large parasites (e.g., schistosome worms) that cannot be phagocytosed (Section H2). They usually bind to an antibody-coated parasite through surface Fc receptors and release the contents of their granules (degranulate) onto the parasite surface. The granules contain peroxides and a toxin, major basic protein, which kill the parasite. Histaminase is also present in the granules. This anti-inflammatory substance dampens the effects of histamine released by mast cells earlier in the response.

**Platelets**, as well as having a major role in blood clotting, contain important mediators that are released when they are activated at the site of a damaged blood vessel. Parasites coated with IgG and/or IgE antibodies are also thought to activate platelets through surface Fc receptors for these antibody classes. Released mediators activate complement, which in turn attracts leukocytes to the site of tissue damage caused by trauma or infection by a parasite (Section B4).

**Erythrocytes** have surface complement receptors that bind to complement attached to small circulating immune complexes. They carry these complexes to the liver, where they are released to Kupffer cells which phagocytose them. Thus, erythrocytes play an important immunological role in clearing immune complexes from the circulation in persistent infections and in some autoimmune diseases.

# B2 Molecules of the innate immune system

---

## Key Notes

**Innate molecular immune defense**

A variety of molecules mediate protection against microbes during the period before adaptive immunity develops. These include antimicrobial peptides, complement, acute phase proteins, and cytokines. Many of these components are also required for the development of a functional adaptive immune system and highlight the interdependence of the two systems.

**Antimicrobial peptides (AMPs)**

AMPs are small peptides important to innate immune defense against a wide spectrum of microbes (Section C3). They are produced by epithelial cells that line the respiratory, gastrointestinal, and genitourinary tracts of the body, and by phagocytic cells once the microbes have breached the epithelial cell barrier.

**The complement system**

The complement system consists of over 20 interdependent proteins, which on sequential activation mediate protection against infection by some microbes. Synthesized by hepatocytes and monocytes, these proteins can be activated directly by microbes through the **alternative pathway** and thus have a pivotal role in innate immunity. This system can also be activated through the **classical pathway** by antibodies (adaptive immunity) and by lectin-like molecules bound to a microbe, the **lectin pathway**. On activation, the complement system can (i) initiate (acute) inflammation; (ii) attract neutrophils to the site of microbial attack (chemotaxis); (iii) enhance attachment of the microbe to the phagocyte (opsonization); (iv) kill the microbe.

**Acute phase proteins**

Acute phase proteins are a heterogeneous group of plasma proteins important in innate defense against microbes (mostly bacteria) and in limiting tissue damage caused by infection, trauma, malignancy, and other diseases. They include C-reactive protein (CRP), serum amyloid protein A (SAA), and mannose-binding protein (MBP). Acute phase proteins are mainly produced in the liver, usually as the result of a microbial stimulus, or in response to the cytokines IL-1, IL-6, TNFα, and IFNγ that are released by activated macrophages and NK cells. These proteins maximize activation of the complement system and opsonization of invading microbes.

| | |
|---|---|
| **Overview of cytokines** | Cytokines are small molecules that signal between cells, inducing growth, chemotaxis, activation, enhanced cytotoxicity, and/or regulation of immunity. Many of them are classified as interleukins as they are primarily produced by, and communicate with, leukocytes. Other groups include the chemokines that direct cell migration and the interferons that protect against viral infection, activate cells, and modulate immunity. |
| **Interleukins** | The interleukins, originally categorized by their ability to communicate between leukocytes, are growth and differentiation factors for immune cells. Other interleukins have pro-inflammatory activities that together with other cytokines are critical to immune defense and inflammation. |
| **Chemokines** | Chemokines are small cytokines produced by many cell types in response to infection or physical damage. They activate and direct effector cells expressing appropriate chemokine receptors to migrate to sites of tissue damage and regulate leukocyte migration into tissues. CC chemokines are chemotactic for monocytes, while CXC chemokines are chemotactic for PMNs. |
| **Interferons** | Interferons (IFNs) are produced in response to viral infection and inhibit protein synthesis. Type I IFNs (IFNα and IFNβ) are produced by many different cells. Type II interferon (IFNγ), mainly produced by Th1 cells and NK cells, induces Th1 responses, increases antigen presentation, and activates phagocytic and NK cells for enhanced killing. |
| **Other cytokines** | Other cytokines include colony-stimulating factors (CSFs) that drive development, differentiation, and expansion of cells of the myeloid series. GM-CSF induces commitment of progenitor cells to the monocyte/granulocyte lineage, and G-CSF and M-CSF commitment to the granulocyte and monocyte lineages, respectively. Transforming growth factor β (TGFβ) inhibits activation of macrophages and growth of B and T cells. Tumor necrosis factor β (TNFβ) is cytotoxic. |

## Innate molecular immune defense

There are many molecules of the innate immune system which are important in mediating protection against microbes during the period before the development of adaptive immunity. Although some of these molecules react with particular microbial structures, they are nonspecific in that they can react with many different microbes that express these structures. These molecules include the antimicrobial peptides, complement proteins, acute phase proteins, and interferons. Many of the molecules involved in innate immunity are also associated with adaptive immunity. Thus, the complement system can be activated through antibodies, and cytokines are involved in activation of antigen-presenting cells critical to triggering T-lymphocyte responses. Cytokines released by macrophages also play a role in acute inflammation. Thus, the immune response to microbes is continuous with both systems being intimately involved and synergistic.

## Antimicrobial peptides (AMPs)

There are a large number of antimicrobial peptides (AMPs) that are important in innate immunity. They are produced by epithelial cells that line the respiratory, gastrointestinal, and genitourinary tracts of the body (Section C3) to prevent entry of microbes. AMPs are also produced by phagocytic cells once the microbes have breached the epithelial cell barrier. These peptides are small (less than 60 amino acids), many have an overall positive charge, and they have a wide spectrum of antimicrobial activity. They are active against Gram-positive and Gram-negative bacteria, viruses, and fungi, and are either constitutively produced or induced by microbes through pattern recognition receptors (Section B3) or through the effects of cytokines. Their *modus operandi* is thought to be mainly through altering the membrane permeability of the pathogen but other mechanisms have also been suggested. Some AMPs also have chemotactic properties and attract T cells and dendritic cells, thus playing an important role in linking innate with adaptive immunity. The main human AMPs are the defensins and the cathelicidins (Table 1) but others are important and it is certain that many new AMPs will be described in the future. The magainins (originally isolated from *Xenopus* skin) and granulysin produced by NK cells (and cytotoxic T cells) are also antimicrobial peptides.

## The complement system

The complement system is a protective system common to all vertebrates (Section D8). In man it consists of 20 soluble glycoproteins (usually designated as C1, C2, etc., or as factors, e.g., factor B), many of which are produced by hepatocytes and monocytes. They are constitutively present in blood and other body fluids. On appropriate triggering, these components interact sequentially with each other (i.e., in a domino-like fashion). This "cascade"

## Table 1. Antimicrobial peptides

| Antimicrobial peptide | Cell origin |
| --- | --- |
| Defensins | |
| α defensins | Mainly neutrophils but some epithelial cells |
| β defensins | Mostly epithelial cells and keratinocytes |
| Cathelicidins | Macrophages, polymorphonuclear cells (PMNs), and some epithelial cells |
| Psoriasin | Skin (keratinocytes) |
| Calprotectin | PMNs |

of molecular events involves cleavage of some complement components into active fragments (e.g., C3 is cleaved to C3a and C3b) which contribute to activation of the next component, ultimately leading to lysis of, and/or protection against, a variety of microbes. This system can be "activated" (Figure 1) directly through the **alternative pathway** by certain molecules associated with microbes, or through the **classical pathway** by antibodies bound to a microbe or other antigen (Section D8). There is also a **lectin pathway** in which soluble mannose-binding lectin (MBL), a host protein, binds to mannose structures on microbes and activates the complement cascade beginning with C4. Ficolins (proteins quite similar to MBL) can also activate the complement cascade through the lectin pathway, by binding to *N*-acetylglucosamine on microbes (see Sections B3 and D8).

The alternative pathway is activated by interaction of C3 with certain types of molecules on microbes or by self molecules (e.g., CRP, see below) which react with these microbes. Complement component C3 is critical to this interaction and its cleavage into C3a and C3b is the single most important event in the activation of the complement system. More specifically, the alternative pathway depends on the normal continuous low-level breakdown of C3 (Table 2). In particular, cleavage of C3 to C3a and C3b activates a thioester bond in C3b, which is very reactive and can covalently bind to virtually any molecule or cell. If C3b binds to a self cell, regulatory molecules associated with this cell (Section D8) inactivate it, protecting the cell from complement-mediated damage. However, if C3b binds to a microbe, factor B is activated and its cleavage product Bb binds to C3b on the microbe. This C3bBb complex (C3 convertase) is enzymatically active and amplifies the breakdown of additional C3 to C3b. Equally importantly, the resulting enzyme cleaves

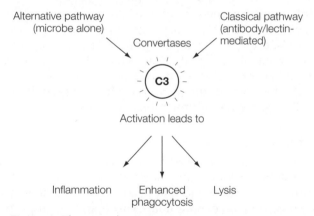

Figure 1. The complement system.

**Table 2. Sequence of complement activation by the alternative pathway leading to cell lysis**

| | |
|---|---|
| Microbe (M) + C3b | M-C3b |
| MC3b + factor B | M-C3b-Bb |
| M-C3b-Bb + C3b | M-C3b-Bb-C3b |
| M-C3b-Bb-C3b + C5 | M-C3b-Bb-C3b-C5b + C5a |
| M-C3b-Bb-C3b-C5b + C6 + C7 | M-C3b-Bb-C3b-C5b-C6-C7 |
| M-C3b-Bb-C3b-C5b-C6-C7 + C8 | M-C3b-Bb-C3b-C5b-C6-C7-C8 |
| M-C3b-Bb-C3b-C5b-C6-C7-C8 + C9 | M-C3b-Bb-C3b-C5b-C6-C7-C8-C9 Lysis of M |

C5 into C5a and C5b, both of which have critical protective functions. C5b is crucial to formation of the "membrane attack complex" (MAC), C5b–C6–C7–C8–C9, which mediates lysis of the microbe. This alternative pathway is important for control of infection in the absence of specific immunity. Thus, many different organisms are handled and eliminated as a result of their activation of the alternative pathway.

The major functions of the complement system are:

● Initiation, by C3a and C5a, of (acute) inflammation by direct activation of mast cells

● Attraction, by C5a, of neutrophils (chemotaxis) to the site of microbial attack

● Enhancement, by C3b, of attachment of the microbe to the phagocyte (opsonization)

● Killing, by C9, of the microbe activating the membrane attack complex (lysis)

## Acute phase proteins

Acute phase proteins are important in innate defense against microbes (mostly bacteria and protozoa) and in limiting tissue damage caused by microbial infection, trauma, malignancy, and other diseases, for example rheumatoid arthritis. They are also important in tissue repair. These molecules include C-reactive protein (CRP), complement components, opsonic proteins such as mannose-binding lectin (MBL: see above), metal-binding proteins, and protease inhibitors. The major acute phase proteins, CRP and serum amyloid protein A (SAA), have similar structures and are termed pentraxins, based on the pentagonal association of their subunits. CRP, which was named based on its ability to react with the C-protein of pneumococcus, is composed of five identical polypeptides associated by noncovalent interactions. MBL binds residues of mannose on glycoproteins or glycolipids expressed by microbes in a form different from that on mammalian cells. Thus, MBL is a soluble pattern recognition receptor (PRR) that binds to pathogen-associated molecular patterns (PAMPs) on a variety of microbes (Section B3).

Acute phase proteins, mainly produced by the liver, can either be produced *de novo* (e.g., CRP levels are increased by as much as 1000-fold within a few hours), or, as in the case of fibrinogen, are present at low levels and rapidly increase following infection. They are produced by hepatocytes in response to the cytokines IL-1, IL-6, TNFα and IFNγ released by activated macrophages and NK cells. IL-6 is important in enhancing production of acute phase proteins.

Acute phase proteins have several functions (Table 3), the most important being to maximize activation of the complement system and opsonization of invading microbes, and to limit tissue damage caused by these microbes. CRP binds to a wide variety of microbes and on binding activates complement through the alternative pathway, causing C3b deposition on the microbe (opsonization) and thus ultimately its phagocytosis by phagocytes expressing receptors for C3b. MBL binding to microbes also initiates complement activation and subsequent opsonization mediated by C3b, and in addition directly opsonizes these organisms for phagocytosis. Metal-binding proteins inhibit microbial growth, and protease inhibitors limit tissue damage by neutralizing lysosomal enzymes released by phagocytes.

Both CRP and SAA, as well as having complement activation properties, bind to DNA and other nuclear material from cells, helping in their clearance from the host. Quantitation of CRP in the serum of patients with inflammatory diseases (e.g., rheumatoid arthritis) is used as a way to assess the inflammatory activity of the disease. High levels of CRP signify a high level of disease activity.

**Table 3. Acute phase proteins and their functions**

| Protein | Function |
| --- | --- |
| C-reactive protein (CRP) | Binds to bacterial phosphocholine, activates complement through C1q, acts as an opsonin |
| Serum amyloid A (SAA) | Activates complement (through C1q), acts as an opsonin |
| Mannose-binding lectin (MBL) | Binds to mannose on bacteria, attaches to phagocyte; activates complement via classical pathway (Section D8) |
| Complement components | Chemotaxis, opsonization, and lysis (Section D8) |
| Metal-binding proteins | Removal of essential metal ions required for bacterial growth |
| Fibrinogen | Coagulation factor |
| $\alpha_1$-antitrypsin, $\alpha_1$-anti-chymotrypsin | Protease inhibitors |

## Overview of cytokines

Cytokines are low-molecular-weight molecules secreted by both immune and some nonimmune cells in response to a stimulus. They may have an effect on the cell that produces them (i.e., have an autocrine effect) and/or on other cells, with each cytokine often inducing several different biological effects. These signaling molecules may induce growth, differentiation, chemotaxis, activation, and/or enhanced cytotoxicity. Moreover, it is not uncommon for different cytokines to have similar activities and for many cytokines, some with opposing activities, to be released by a particular stimulus. Thus, the resulting biological effect is the sum of all of these activities.

There are several ways of classifying cytokines, including based on their structure or function. Originally they were defined by the type of cell that produced them, for example **monokines** produced by monocytes and macrophages and **lymphokines** by lymphocytes. Because some cytokines are produced by both cell types, we will use for these the term **interleukins** (IL), originally coined to identify those cytokines produced by leukocytes as communication molecules between leukocytes (*inter*: Latin, "between"). Even so, this term also has some problems because of overlap with other groups such as **chemokines** and **interferons**. Chemokines are small heparin-binding proteins that direct cell migration, and may also activate cells in response to infectious agents or tissue damage. **Interferons** are produced by a variety of cells in response to viral infection. Other cytokines are mainly growth factors responsible for controlling the differentiation of different cell populations.

It is important to note that the same cytokine can be made by several different cell populations. For example, IFNα is made by most if not all nucleated cells in response to viral infection. IFNγ is produced both by NK cells and Th1 cells. IL-1 is produced by macrophages, B cells, and nonimmune keratinocytes. Many different cell types make IL-6, several make IL-4, and so forth. Moreover, the same cytokine can induce different functions in different cell types. For example, IFNγ activates macrophages to kill intracellular microbes, induces B cells to switch their antibody class to IgG, and induces endothelial cells to increase expression of MHC class II molecules. TNFα can promote the proliferation of B cells but activate killing mechanisms in other cell populations.

## Interleukins

To date, 35 interleukins have been described and a few of the better characterized ones are shown in Table 4, together with other cytokines. Many are growth factors for lymphocytes

**Table 4. Some interleukins and other important cytokines**

| Cytokine | Produced by | Activity |
| --- | --- | --- |
| IL-1 | MØ, epithelial cells | Activates vascular endothelium, tissue destruction, increased effector cell access, fever, lymphocyte activation, mobilization of PMNs, and induces acute phase proteins (CRP, MBL) |
| IL-2 | T cells | Induces proliferation of T and NK cells |
| IL-3 | T cells, thymic cells | Induces proliferation and differentiation of hemopoietic cells |
| IL-4 | Th2 cells, mast cells | Induces B-cell activation and proliferation and Th2 IgE responses and inhibits Th1 responses |
| IL-5 | Th2 cells, mast cells | Induces eosinophil growth, differentiation, B-cell activation, and IgA responses |
| IL-6 | T cells, MØ | Induces lymphocyte activation, fever, and acute phase proteins |
| IL-8 | Mo, MØ, Fb, Kr | Increases tissue access for, and chemotaxis of, PMNs (also classified as a chemokine) |
| IL-10 | Th2 cells, MØ | Induces B-cell activation, suppression of MØ activity, and Th2 production, and inhibits Th1 responses |
| IL-12 | B cells, MØ | Induces Th1 and activates NK cells but inhibits Th2 responses |
| IFNγ | Th1 cells, Tc, NK cells | Activates MØ and PMNs; induces Th1 and inhibits Th2 responses |
| TNFα | MØ, T cells | Activates vascular endothelium; increases vascular permeability; induces fever, shock and mobilization of metabolites |

Mo, Monocytes; MØ, macrophages; En, endothelial cells; Fb, fibroblasts; Kr, keratinocytes; PMNs, neutrophils; Co; chondrocytes.

and/or influence the nature of the immune response. For example, IL-2 is made by T cells as a critical autocrine growth factor required for proliferation of T cells, especially Th0 and Th1 cells and cytotoxic T lymphocytes (CTLs). On activation (as a result of the interaction of their antigen receptor complexes with antigenic peptide in MHC molecules on antigen-presenting cells (APCs)), these T cells make IL-2 for secretion, and at the same time IL-2 receptors with which to bind and be stimulated by the secreted IL-2. In the absence of IL-2 and/or its receptor, many antigen-specific T cells do not expand, severely compromising immune responses.

IL-3 is involved in the growth and differentiation of a variety of cell types as a result of its synergistic activity with other cytokines in hemopoiesis. IL-4, produced by Th2 cells and mast cells, is a growth and differentiation factor for Th2 cells and B cells, and can induce B-cell class switch to IgE antibodies. IL-4 is important in influencing the nature of the immune response, as it can induce the development of Th2 cells from Th0 cells and can inhibit the development of Th1 responses (Table 4 and Section F5). Thus, IL-4 is not only involved in B-cell growth, but can also influence the B cell and its subsequent plasma cells to produce IgE antibody (Section D2). IL-5 is also produced by Th2 cells and mast cells and is important to B-cell activation and in induction of B-cell class switch to IgA antibody. It also has a role in eosinophil growth and differentiation. IL-10, which is produced by Th2 cells and macrophages, induces B-cell activation and Th2 responses and inhibits Th1 responses, perhaps by enhancing IL-4 production and/or by suppressing macrophage activity and production of IL-12, a Th1-stimulatory cytokine.

IL-1, IL-6, IL-8, and IL-12, together with TNFα, are a group of cytokines that are pro-inflammatory, acting as important mediators of inflammation. These are produced by macrophages in response to a number of stimuli, one of which is lipopolysaccharide (LPS) in the walls of Gram-negative bacteria. IL-1, IL-6, and TNFα have activities that include (i) increasing body temperature and lymphocyte activation, which decrease pathogen replication and increase specific immune responses; (ii) mobilization of neutrophils for phagocytosis; (iii) induction of release of acute phase proteins (CRP, MBL), thus complementing activation and opsonization.

In addition to activation of vascular endothelium (in preparation for neutrophil chemotaxis), IL-1 also induces systemic production of IL-6. IL-8 increases access for, and chemotaxis of, neutrophils, but also activates binding by integrins that facilitate neutrophil binding to endothelial cells and migration into tissues. Like IL-1, TNFα also activates vascular endothelium and is able to increase vascular permeability. It activates macrophages and induces their production of nitric oxide (NO). Although produced by monocytes and macrophages, TNFα is also produced by some T cells. Finally, IL-12, which is produced by both B cells and macrophages, activates NK cells that then produce IFNγ, a cytokine important for inducing differentiation of Th0 cells to Th1 cells (Section F5).

## Chemokines

This group of more than 50 small, closely related cytokines (MW 8–10 kDa) are primarily involved in chemoattraction of lymphocytes, monocytes, and neutrophils (Table 5). They are made by monocytes/macrophages, but also by other cells including endothelial cells, platelets, neutrophils, T cells, keratinocytes, and fibroblasts. Chemokines can be divided into four different groups (α, β, γ, and δ) based on unique aspects of their amino acid sequence, and in particular the position of conserved cysteine residues. One group has two adjacent cysteines (CC), a second has two cysteines separated by another amino acid (CXC), another has one cysteine, and the last has two cysteines separated by three other amino acids ($CX_3C$). For the most part, CC chemokines such as monocyte chemotactic protein (MCP-1) are chemotactic for monocytes, inducing them to migrate into tissues and become macrophages, whereas CXC chemokines, such as IL-8, are chemotactic for

**Table 5.  Representative chemokines**

| Class | Name | Source | Chemoattractant for activation of |
|---|---|---|---|
| CXC (α) | IL-8 | Mo, MØ, Fb, Kr | Naive T cells, PMNs |
| | NAP-2 | Platelets | PMNs |
| | MIP-1b | Mo, MØ, En, PMNs | CD8 T cells |
| CC (β) | MCP-1 | Mo, MØ, Fb, Kr | Memory T cells, Mo |
| | Rantes | T cells | Memory Th cells |
| C (γ) | Lymphotactin | | Lymphocytes |
| $CX_3C$ (δ) | Fractalkine | | Lymphocytes, monocytes, NK cells |

Mo, Monocytes; MØ, macrophages; En, endothelial cells; Fb, fibroblasts; Kr, keratinocytes; PMNs, neutrophils.

neutrophils, inducing them to leave the blood and migrate into tissues. Some of these chemokines are also chemotactic for T cells. Chemokines are produced in response to an infectious process or to physical damage and not only direct cells to the source of infection/damage, but may also enhance their ability to deal with tissue damage.

Receptors for chemokines are all integral membrane proteins with the characteristic feature that they span the membrane seven times. These molecules are coupled to G (guanine-nucleoside-binding) proteins that act as the signaling moiety of the receptor. Although most of these receptors can bind more than one type of chemokine, they are usually distributed only on particular cell populations, permitting different chemokines to have selective activity.

Some chemokines, for example IL-8 and MCP-1, have been shown to work by first binding to proteoglycan molecules on endothelial cells or on the extracellular matrix. On this solid surface they then bind blood neutrophils or monocytes, slowing their passage and directing them to migrate down a chemokine concentration gradient toward the source of the chemokine (Section B4). Although the role that each plays in immune defense and pathology is still being clarified, it is evident that these molecules are potent agents for activating and directing effector cell populations to the site of infection and/or tissue damage as well as for controlling leukocyte migration in tissues.

## Interferons

Interferons are pro-inflammatory molecules that can mediate protection against virus infection, and are thus particularly important in limiting infection during the period when specific humoral and cellular immunity is developing. They can be divided into two main groups: type I IFNs (IFNα and IFNβ) and type II IFN (IFNγ) also called immune IFN (Table 6).

### Table 6. The interferons

|  | Type I (IFNα and β) | Type II (IFNγ) |
| --- | --- | --- |
| Chromosomal location | 9 | 12 |
| Origin | All nucleated cells, especially fibroblasts, macrophages, and dendritic cells | NK cells, NKT cells, and Th1, γδ and CD8+ T cells |
| Induced by | Viruses, other cytokines, some intracellular bacteria and protozoans | Antigen-stimulated T cells |
| Functions | Antiviral, increases MHC class I expression, inhibits cell proliferation | Antiviral, increases MHC I and II expression, activates macrophages |

IFNα and IFNβ are produced by many different cells in response to viral or bacterial infections, especially by intracellular microbes. At least 12 different, highly homologous species of IFNα are produced, primarily by infected leukocytes but also by epithelial cells and fibroblasts. In contrast, a single species of IFNβ is produced, normally by fibroblasts and epithelial cells. The pro-inflammatory cytokines IL-1 and TNFα are potent inducers of IFNα/β secretion, as are endotoxins derived from the cell wall of Gram-negative bacteria.

The receptor for both IFNα and IFNβ is the same and found on most nucleated cells. Binding of IFNα and IFNβ to this receptor inhibits protein synthesis and thus viral replication as a result of the induction of the synthesis of inhibitory proteins and of preventing mRNA translation and DNA replication. In addition, these interferons inhibit cell proliferation, increase the lytic activity of NK cells, and induce increased expression of MHC class I and other components of the class I processing and presentation pathway leading to induction of antigen-specific CTL responses against virally infected cells. Induction of MHC class I is also important for protection of uninfected cells from killing by NK cells (Section B1). The importance of IFNα/β in innate defense against viral infections is indicated by animal studies in which treatment of virus-infected mice with antibodies to IFNα/β resulted in death.

In contrast to the broad and rather nonspecific antiviral activity of IFNα/β, IFNγ is primarily a cytokine of the adaptive immune system. It is important not only for its antiviral activity but also its major role in regulation of the development of specific immunity and in activation of cells of the immune system. Produced primarily by NK cells and Th1 cells, IFNγ plays a critical role in induction of Th1 immune responses. That is, early in the development of a specific immune response, IFNγ is involved in inducing Th0 cells to differentiate to Th1 cells which make more IFNγ and provide help for development of CTL responses (Section F5) and for antibody class switching to IgG (Section D3). In addition, Th1 cells or CTLs responding to peptides presented in MHC molecules produce IFNγ that acts both locally and systemically to activate monocytes, macrophages, and PMNs, which are then better able to kill intracellular pathogens. In particular, IFNγ increases the expression of Fc receptors for IgG on macrophages and PMNs (Section D8) as well as MHC class II expression on a wide variety of cells. This enhances the phagocytic function of these cells as well as the antigen-presenting capabilities of professional APCs. IFNγ enhances macrophage killing of intracellular bacteria and parasites probably as a result of its stimulation of their production of reactive oxygen and reactive nitrogen intermediates.

## Other cytokines

Of the many other cytokines that are important to immune defense, several are particularly noteworthy (Table 7). A group of CSFs, including granulocyte–monocyte CSF (GM-CSF), granulocyte CSF (G-CSF), and monocyte CSF (M-CSF), drive the development, differentiation, and expansion of cells of the myeloid series (Section A5). GM-CSF induces expansion of myeloid progenitor cells and their commitment to the monocyte/macrophage and granulocyte lineage, after which G-CSF and M-CSF induce specific

**Table 7. Other cytokines**

| Cytokine | Produced by | Activity |
|---|---|---|
| GM-CSF | MØ, T cells | Stimulates growth, differentiation and activation of granulocytes, Mo, MØ |
| G-CSF | Mo, Fb, En | Stimulates PMN development |
| M-CSF | Fb | Stimulates Mo, MØ development |
| TGFβ | Mo, T cells, Co | Inhibits cell growth and inflammation |
| TNFβ (lymphotoxin) | T cells | Cytotoxic to T, B, and other cells |

Mo, Monocytes; MØ, macrophages; En, endothelial cells; Fb, fibroblasts; PMN, neutrophil; Co; chondrocytes.

commitment to the granulocyte or monocyte lineages, respectively, and then their subsequent expansion. These factors, especially G-CSF, are important clinical tools in a number of disease situations as they can be used to expand myeloid effector cell populations critical to defense against pathogens.

TGFβ is produced by a variety of cells, including monocytes, macrophages, T regulatory cells, fibroblasts, and chondrocytes, and plays an important role in suppressing immune responses, as it can inhibit activation of macrophages and growth of B and T cells. TNFβ (lymphotoxin) is a molecule that is cytotoxic to a variety of cell types, including ineffectual chronically infected macrophages.

# B3 Recognition of microbes by the innate immune system

## Key Notes

**Introduction to pattern recognition receptors (PRRs)**

Receptors of the innate immune system interact with, and facilitate removal of, groups of organisms with similar structures. Pattern recognition receptors (PRR) are present on macrophages, PMNs, myeloid and plasmacytoid dendritic cells, and some epithelial cells. They recognize molecular patterns (pathogen-associated molecular patterns, PAMPs) associated with microbes, and act not only as a first line of defense against microbes, but also to prime the adaptive immune system. Cellular PRRs include Toll-like receptors, NOD-like receptors, mannose receptors, scavenger receptors, and CD14.

**Toll-like receptors**

Toll-like receptors (TLRs) are a family of germline-encoded cell-surface or cytoplasmic proteins that recognize components of bacteria, viruses, and fungi. They are present on macrophages, myeloid and plasmacytoid dendritic cells, and some epithelial cells, where they not only signal the presence of a pathogen, but also trigger the expression of co-stimulatory molecules and effector cytokines important for development of adaptive immune responses.

**NOD-like receptors**

NOD-like receptors (NLRs) are cytoplasmic receptors that recognize components of the peptidoglycans of Gram-positive and Gram-negative bacteria, leading to cell signaling and production of pro-inflammatory cytokines such as IL-1$\beta$ and IL-18. These receptors are present in macrophages, myeloid and plasmacytoid dendritic cells, and some epithelial cells. RIG-1-like cytoplasmic receptors detect RNA derived from viruses in the cytoplasm.

**Mannose receptors**

Mannose receptors (MRs) are C-type lectin receptors that recognize a $Ca^{2+}$-dependent, mannosyl/fucosyl pattern. They are expressed on macrophages, dendritic cells, and endothelial cells. They mediate phagocytosis of microbes and processing and presentation of microbial peptides on MHC class II molecules, thus permitting induction of specific antimicrobial T- and B-cell responses. Another C-type lectin receptor, DC-SIGN, also has specificity for mannose-like carbohydrates on viruses, bacteria, and fungi, and like MR is an important PRR on macrophages and DCs.

| CD14 | This molecule is expressed on monocytes/macrophages, binds LPS on Gram-negative bacteria, and facilitates destruction of the microbe and induction of secretion of cytokines involved in triggering adaptive immune responses. |
|---|---|
| **Scavenger receptors (SRs)** | SRs on macrophages recognize carbohydrates or lipids in bacterial and yeast cell walls, as well as damaged, modified, or apoptotic self cells, and mediate their removal. At least seven different SR able to bind microbes have been identified, as have some of their ligands, including lipid A of lipopolysaccharide and lipotechoic acid. |
| **Soluble pattern recognition receptors** | Several soluble proteins are able to bind to PAMPs, including the collectins (e.g., MBL) and ficolins, which bind mainly to carbohydrates leading to complement activation. The galectins can directly bind to the ligands on bacteria and mediate their death. |
| **Related topics** | (B1) Cells of the innate immune system          (H1) The microbial cosmos |

## Introduction to pattern recognition receptors (PRRs)

In order to recognize and remove the multitude of invading microbes, the innate immune system possesses a number of cellular or soluble receptor molecules that not only act as a first line of defense, but also are important to the development of an adaptive immune response. Unlike the antigen-specific receptors expressed by lymphocytes (Section C1), which recognize specific epitopes (Section A4), these PRRs have evolved to recognize the pathogen-associated molecular patterns (PAMPs) of a wide variety of microbes and to facilitate their removal. The cellular PRRs include the Toll-like receptors, NOD-like receptors, mannose receptors, scavenger receptors, and CD14, all expressed by macrophages (Figure 1). Soluble PRRs include MBL (a collectin), ficolins, and galectins. A summary of key features of the PRRs is shown in Table 1. These receptors sense the "danger" signals given by invading microbes, but also are able to recognize some self components and are therefore important in recognition and clearance of dead (apoptotic) and dying cells.

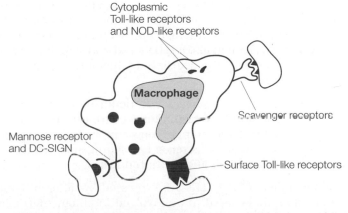

Figure 1. Some of the pattern-recognition receptors expressed by macrophages.

**Table 1. Pattern recognition receptors**

| Name | Specificity | Location |
|------|-------------|----------|
| Mannose receptor | Mannosyl/fucosyl structures | MØ, endothelial cells, DC |
| DC-SIGN | Mannosyl/fucosyl structures | MØ, DC |
| Toll-like receptors | LPS, peptidoglycan, glucans, teichoic acids, arabinomannans ssDNA, DNA, etc. | DC, MØ, epithelial cells |
| NOD-like receptors | Peptidoglycans/MDP | MØ, DC and some epithelial cells |
| RLRs | Viral RNA | MØ, DC |
| CD14 | LPS | MØ |
| Scavenger receptors | Carbohydrates/lipids | MØ, DC, endothelial cells |
| Soluble PRRs | | |
|   Mannose receptor | Carbohydrates | MØ, DC |
|   Ficolins, galectins | | |

DC, dendritic cells; LPS, lipopolysaccharide. RLRs, RIG-1-like receptors; MØ, macrophages

Interestingly some of these same PRRs have also been found associated with lymphocytes. They play a role in enhancing T-cell-dependent antibody response but to date their precise role is unclear. This again highlights the interdependence of two separate immune systems—innate and adaptive.

## Toll-like receptors

Toll-like receptors (TLRs) are a family of closely related proteins that all have an extracellular leucine-rich repeat domain and a cytoplasmic domain that mediates signal transduction of a variety of effector genes. They are surface or cytoplasmic receptors that recognize different components of bacteria, viruses, and fungi (Table 2). Thus, different TLRs are able to recognize major molecular signatures of pathogens, including peptidoglycan, lipoteichoic acids (Gram-positive bacteria), LPS (Gram-negative bacteria), arabinomannans, and glucans. Of particular importance, these germline-encoded molecules of the innate immune system are not only able to signal the presence of a pathogen, but also trigger expression of co-stimulatory molecules and effector cytokines. In so doing, they prepare the cell for its involvement in the development of an adaptive immune response, for example, enhancement of antigen presentation if the cell stimulated is a macrophage or dendritic cell (Section F2). It is believed that signaling via TLRs of phagocytic cells does not directly induce phagocytosis but rather induces phagocytosis through secondary effects of cytokines.

As well as being expressed by macrophages and dendritic cells, some TLRs are also expressed by epithelial cells lining the respiratory, gastrointestinal, and genitourinary tracts (Section C3) where they are important in triggering secretion of antimicrobial peptides (Section B2). Moreover, some TLRs recognize self ligands, which is important in clearing damaged, dying, or dead cells and their products.

Many of the family members of the TLRs are expressed by different lymphocytes and their subpopulations and are believed to be involved in enhancing/maintaining their function, although the exact mechanisms have not yet been elucidated.

**Table 2. Toll-like receptors (TLRs): their cellular distribution and specificity**

| Receptor | Ligands | Microbes recognized | Cellular distribution |
|---|---|---|---|
| TLR1 | Lipopeptides | Bacteria | Monocytes/MØ, DC |
| TLR2 | Lipoteichoic acid peptidoglycan, zymosan | Both Gram-negative and positive bacteria, some fungi | Monocytes/MØ, PMN, DC, mast cells, FDC |
| TLR3* | dsDNA | Viruses | MØ, DC |
| TLR4 | LPS | Gram-negative bacteria, some fungi | MØ, PMN, DC, mast cells, FDC |
| TLR5 | Flagellin | Bacteria | Monocytes/MØ, DC |
| TLR6 | Lipopeptides | Bacteria | Monocytes/MØ, mast cells |
| TLR7/8* | ssDNA | Viruses | Monocytes/MØ, DC |
| TLR9* | dsDNA | Bacteria | Monocytes/MØ, DC |
| TLR10 (CD290) | Unknown | | Plasmacytoid DC |

DC, myeloid and plasmacytoid dendritic cells; FDC, follicular dendritic cells; LPS, lipopolysaccharide; PMN, polymorphonuclear cells; MØ, macrophages; *intracellular TLRs found in endosomes.

## NOD-like receptors

NOD-like receptors (NLRs) are nucleotide-binding domain, leucine-rich repeats containing cytoplasmic proteins. The two archetypal NLRs, NOD-1 and NOD-2, recognize different components of the peptidoglycans of Gram-positive and Gram-negative bacteria. NOD-2 binds to muramyl dipeptide (MDP) and signals the cell to produce various cytokines such as the pro-inflammatory IL-1β and IL-18. NLRs are present in macrophages, dendritic cells, and some epithelial cells, where they are believed to synergize with TLRs to enhance signaling. NOD-2 is found in PMNs. Other cytoplasmic receptors include retinoic-acid-inducible gene-I (RIG-I)-like receptors (RLRs), which belong to the RNA helicase family that specifically detects RNA species derived from viruses in the cytoplasm.

## Mannose receptor

Mannose receptor (MR) is a 180 kDa transmembrane molecule expressed on macrophages, conventional dendritic cells, and subsets of endothelial cells. A C-type lectin receptor, MR has eight carbohydrate recognition domains, at least some of which have different pattern recognition motifs, making this one receptor fairly broad in the number and range of ligands it can recognize. Its $Ca^{2+}$-dependent, mannosyl/fucosyl recognition pattern permits it to interact with a variety of pathogens that enter through mucosal surfaces (Table 1). Because the mannose receptor is expressed on macrophages throughout the body, it is likely to be one of the first of the innate receptors to interact with microbes (Figure 1). Furthermore, this receptor mediates phagocytosis and destruction of microbes even before the adaptive immune response is induced. Another C-type lectin receptor, DC-SIGN, was originally shown to be an adhesion molecule for ICAM3 and hence the name (dendritic-cell-specific intercellular adhesion molecule-3-grabbing non-integrin). This receptor also has specificity for mannose-like carbohydrates (PAMPs) on viruses, bacteria, and fungi, and like the MR is an important PRR on macrophages and DC (Figure 1 and Table 1).

In addition to its role as a front-line receptor mediating destruction of a wide range of organisms, the MR represents an important direct link to the adaptive immune system. Thus, microbes bound by the MR are internalized and degraded in endosomes. Peptides from the microbe are loaded on MHC class II (and MHC class I, Section F2) molecules for display on the surface of these APCs so that T cells of the adaptive immune system can now recognize microbial determinants, thus permitting induction of microbe-specific T- and B-cell responses.

## CD14

CD14 is a phosphoinositolglycan-linked cell surface receptor on macrophages that acts as a co-receptor along with TLR4 to bind to lipopolysaccharide (LPS), a unique bacterial surface structure found only in the cell walls of Gram-negative bacteria, for example *E. coli*, *Neisseria*, and *Salmonella*. The core carbohydrate and lipid A of LPS are virtually the same for these microbes and are the targets for binding by CD14. Binding of LPS on a Gram-negative bacterium to macrophage CD14 and TLR4 facilitates destruction of the microbe as well as induction of secretion of various cytokines involved in triggering a wide array of immune responses.

## Scavenger receptors (SRs)

Scavenger receptors (SRs) are a group of transmembrane cell-surface molecules that mediate binding and internalization (endocytosis) of microbes (both Gram-negative and Gram-positive) as well as certain modified, damaged, or apoptotic self cells. These molecules are expressed on macrophages and dendritic cells as well as on some endothelial cells and have specificity for polyanionic molecules and the cells with which they are associated. At least seven different SRs that may interact with microbes have been identified, including SR-A I and II, MARCO, SR-CL I and II, dSR-C1, and LOX-1. Of note, SR-A has apparent specificity for the lipid A component of LPS and of lipoteichoic acid, which are associated with bacteria. Another SR, LOX-1, not only binds oxidized low-density lipoprotein (LDL) and therefore appears to play a role in atherogenesis, but can also recognize certain microbes (e.g., *S. aureus and E. coli*) and may be important in innate immunity.

## Soluble pattern recognition receptors

There are a number of proteins in soluble form that are able to bind to PAMPs. These include, among others, the collectins and galectins. Collectins belong to the superfamily of collagen-containing C-type lectins. The already described acute phase protein, mannose-binding lectin (MBL), attaches to mannose on glycoproteins and glycolipids of microbes and induces complement activation via the lectin pathway. Ficolins are similar to the collectins and have a fibrinogen domain in addition the collagen domain. These recognize *N*-acetylglucosamine and activate complement via the lectin pathway.

Galectins have recently been shown to bind to glycans and directly kill some bacteria.

Thus, a large number of carbohydrate-binding molecules, using a number of different effector mechanisms, exist and play a role in innate immunity.

# B4 Innate immunity and inflammation

## Key Notes

**Overview of inflammation**

Inflammation is the process by which the body deals with an insult from physical or chemical agents and invasion by microbes. There are two types of inflammation based on the duration of the response and the prominent inflammatory cell type. Acute inflammation is of short duration and the result of an initial response, primarily by PMNs, to infection. Once the microbe is removed through cells of the innate or adaptive systems, inflammation subsides and the critical repair process ensues. Chronic inflammation may last for months or years, and is usually due to the persistence of a microbe, in a viable or inert state. The immune cells involved are lymphocytes, macrophages, and plasma cells.

**Initiation of acute inflammatory responses**

Acute inflammation is caused initially by the release of inflammatory mediators from tissues, microbes themselves, or from other cells, including mast cells and macrophages. The complement cleavage products C3a, C4a, and C5a also induce inflammation. Mast cells are central to the acute inflammatory process through release of histamine, other vasoactive amines, and pro-inflammatory cytokines that result in vascular changes. Tissue macrophages play a role in generation of pro-inflammatory cytokines (including IL-1 and TNFα) via recognition of microbes through their pattern recognition receptors.

**Vascular changes**

Inflammatory mediators cause changes in tight junctions in endothelial cells resulting in the passage of fluid (and thus molecules, including antibacterial proteins, antibodies, etc.) and phagocytic cells (PMNs) from the blood to the site of infection. PMNs leave the blood as a result of their recognition of adhesion molecules displayed on the endothelial cells. The expression of these adhesion molecules is induced by pro-inflammatory cytokines released from macrophages. This process involves the capture and rolling of PMNs, followed by their activation, flattening, and extravasation.

**Termination of the response and repair**

Once the offending insult, for example a microbe, has been removed or controlled, inhibitors of the pro-inflammatory cytokines (soluble receptors and anti-inflammatory cytokines such as IL-4, IL-10, and TGFβ) dampen inflammation and tissue repair mechanisms become

|  | activated. Also, macrophages produce collagen and growth factors important in the repair process. | |
|---|---|---|
| **Related topics** | (B1) Cells of the innate immune system | (D8) Antibody functions |
| | (B2) Molecules of the innate immune system | (K2) IgE-mediated (type I) hypersensitivity: allergy |

## Overview of inflammation

Inflammation is the process by which the body deals with an insult from physical or chemical agents and invasion by microbes. It is recognized by its cardinal signs, including redness, heat, swelling, and pain.

There are two types of inflammation based on the duration of the response and the prominent inflammatory cell type. **Acute inflammation** is generally of short duration, lasting from minutes to a few days, and is the result of an initial response by cells of the immune system (primarily PMNs) to remove an infectious agent. **Chronic inflammation**, which may last months to years, usually results from the persistence of a microbe in a viable or inert state and involves lymphocytes, macrophages, and plasma cells of the immune system.

Apart from physical and chemical agents and microbes, immune mechanisms themselves can lead to inflammatory responses (hypersensitivity reactions), for example allergies and granulomatous lesions (Section K). A summary of the main stimulators of acute inflammation is shown in Figure 1.

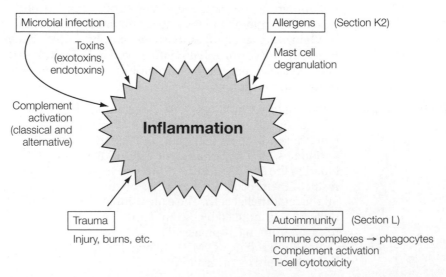

Figure 1. Activators of acute inflammation. The activation phase of the acute inflammatory response may be initiated by trauma, infection, allergy, and autoimmune reactions, although the latter is more often associated with the chronic form of inflammation. While the initiating events may be different, the overall inflammatory response is similar, with the exception of inflammation caused by IgE/mast cell interactions in which case the response may be immediate and more systemic.

Following the successful elimination of a microbe, some tissue damage remains that has to be repaired. This repair process is an important part of the overall inflammatory response.

## Initiation of acute inflammatory responses

Acute inflammation follows the release of inflammatory mediators from microbes, damaged tissues, or immune cells, including mast cells and macrophages (Table 1). Complement cleavage products C3a, C4a, and C5a can trigger release of histamine from mast cells. Moreover, C5a is chemotactic for PMNs, and C3a, C4a, and C5a increase neutrophil and monocyte adherence to endothelial cells.

**Table 1. Source and mechanism of induction of inflammatory mediators resulting from infection**

| Source of initiating factors | Mechanism of induction |
|---|---|
| Exotoxins | **Via damage to tissues:** |
| | Direct effect on vascular endothelium of prostaglandins and leukotrienes |
| | **Via pattern recognition receptors:** |
| Endotoxins (from Gram-negative bacteria) | Direct effects on macrophages (and mast cells) via TLR4 to release pro-inflammatory cytokines e.g., IL-1, IL-6, IL-12, IL-18, TNFα, and IFNγ |
| Lipopeptides, etc. (from Gram-positive bacteria) | Direct effects on macrophages (and mast cells) to release pro-inflammatory cytokines (as above via TLR2) but also including IL-4 |
| C3a derived by alternative or classical (and lectin) pathway | Causes mast cell degranulation |

Mast cells, which are distributed throughout the body (Section B1), are central to the acute inflammatory process in that, on stimulation, they release histamine and other vasoactive amines that result in the vascular changes seen in acute inflammation. Pro-inflammatory substances released by mast cells include the cytokines IL-1 and TNFα and other molecules including leukotrienes, PAF, and nitric oxide which cause blood vessel dilation and edema. These molecules also increase adhesion of neutrophils and monocytes to endothelium (see below). Vasoactive amines such as histamine can also have an effect on smooth muscle contraction, which is important in defense against worms in the intestine (Section H2). Histamine is released immediately as the result of mast cell degranulation whilst pro-inflammatory cytokines are synthesized *de novo* by mast cells. Tissue macrophages also play a role in generation of pro-inflammatory cytokines (including IL-1, TNFα, and IL-6) via recognition, through their pattern recognition receptors (PRRs), of PAMPs associated with microbes (Section B3). It is now known that mast cells themselves express PRRs (Table 1 in Section B3) that trigger cytokine production, and for some PRRs, degranulation and histamine release.

Crucial to both elimination of a microbe and the healing process is the production of acute phase proteins by the liver through the release of the pro-inflammatory cytokines (Section B2) during the early stages of inflammation.

## Vascular changes

The inflammatory mediators released by tissues, mast cells, and macrophages cause dilation of the blood vessels (vasodilation), which increases blood flow and smooth muscle contraction. These inflammatory mediators also cause rapid alterations in the blood vessel endothelium and induce increased expression of cellular adhesion molecules, which assist in the transfer of blood leukocytes. Overall, changes in tight junctions in endothelial cells occur that permit the passage of fluid (containing antibacterial proteins, clotting factors, and antibodies, etc.) and PMNs from the bloodstream to the site of release of these inflammatory mediators (Figure 2), so as to combat the microbe and/or repair the damage. Vasodilation and increased blood flow result in the redness and heat, and the edema (fluid accumulation) results in swelling. Fluid accumulation together with tissue damage gives rise to pain through pain receptors. Overall, the mast cell plays a major role in acute inflammation initiated by injury or infection by microbes (Figure 2).

PMNs leave the bloodstream as a result of their recognition of adhesion molecules displayed on the endothelial cells. The expression of these adhesion molecules is induced by pro-inflammatory cytokines released from macrophages. PMNs form weak attachments through their sialyl-Lewis X molecules (S-Lx) to E-selectin on the endothelial surface. Binding to chemokines, for example IL-8, activates the PMN to express an active form of LFA-1 that then binds to ICAM-1 on the endothelial surface. This leads to arrest and flattening of the PMN, which then extravasates by squeezing between the endothelial cells. The PMN then migrates into the tissues towards the source of the inflammatory mediators (Figure 3).

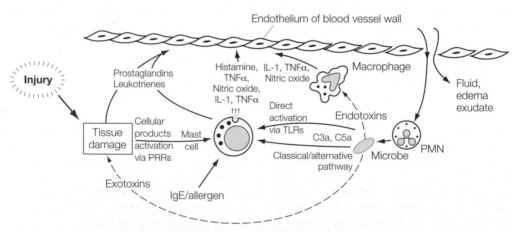

Figure 2. The mast cell in acute inflammation. Microbial products or direct physical damage to blood vessels and tissues leads to release of mediators, for example prostaglandins and leukotrienes, which like mast cell mediators (e.g., histamine) increase vascular permeability and vasodilation. Mast cells release their mediators following microbial activation of complement (classical, alternative, and lectin pathways) via IgE/allergen complexes or through the direct interaction with microbial products or components of dead cells via PRRs. Microbial endotoxins also activate macrophages to release TNFα and IL-1, which have vasodilatory properties. The outcome of this barrage of mediators is the loosening of the endothelial tight junctions, increased adhesion of intravascular neutrophils (and monocytes), and their passage from blood vessels into the surrounding tissues where they can phagocytose the microbes. Serum proteins (fibrinogen, antibodies, etc.) also pass into the tissues and the accumulating fluid (edema) protects the damaged area during repair.

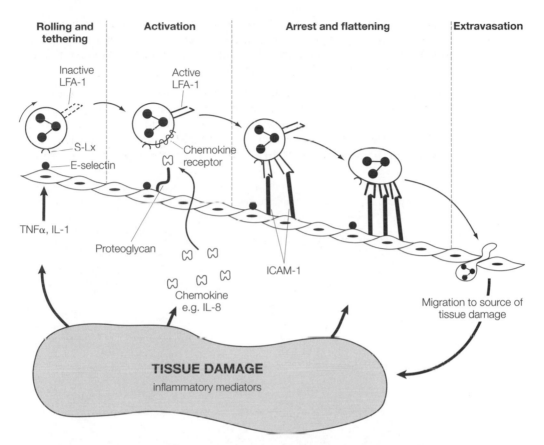

Figure 3. Adhesion to endothelium and extravasation of PMNs. Inflammatory mediators activate endothelial cells to express adhesion molecules. Initially, PMNs begin rolling as they form weak attachments through their sialyl-Lewis X molecules (S-Lx) to E-selectin on the endothelial surface. Binding to chemokines released into the blood vessel and displayed on proteoglycan molecules further slows the PMN down and activates it. This activation results in changing LFA-1 from its inactive to its active form, which can then bind to ICAM-1 on the endothelial surface leading to arrest and flattening of the PMN. The endothelial cells separate and allow the PMN to squeeze through tight junctions between the endothelial cells (extravasation) and into the tissues where inflammatory mediators are being released.

## Termination of the response and repair

Once the offending insult, for example a microbe, has been removed or controlled, inhibitors dampen inflammation and tissue repair mechanisms become activated. Inhibitors of the pro-inflammatory cytokines include their soluble receptors (e.g., receptors for IL 1, TNFα, IL 6, and IL-12), the anti-inflammatory cytokines (IL-4, IL-10, and TGFβ), components of the hemostasis and thrombosis system, and glucocorticoids.

The Th2 cytokine IL-4 downregulates the production of pro-inflammatory cytokines from Th1 cells and TGFβ is a potent inhibitor of many immune functions. Protein C, a component of the hemostasis and thrombosis system, is an anti-inflammatory agent that functions by inhibiting cytokines such as TNFα. Glucocorticoids are well known anti-inflammatory agents and inhibit production of nearly all pro-inflammatory mediators

(Section G6). Other hormones such as α-melanocyte-stimulating hormone reduce fever, IL-2 synthesis, and prostaglandin production, while corticotropin inhibits macrophage activation and IFNγ synthesis. The neuropeptides somatostatin and VIP reduce inflammation by inhibiting T-cell proliferation and migration.

As the inflammatory phase is neutralized by these anti-inflammatory molecules, repair of the damage begins. Various cells, including myofibroblasts and macrophages, both of which make collagen, mend tissues. Macrophage products, including epidermal growth factor, platelet-derived growth factor, fibroblast growth factor, and transforming growth factor, are important in the repair process. In addition, acute phase proteins contribute to the healing process.

# C1 Lymphocytes

## Key Notes

**Specificity and memory**

Lymphocytes provide both the specificity and memory which are characteristic of the adaptive immune response. The two types of lymphocytes involved in the adaptive response are T cells and B cells, both of which have similar morphology. They have specific but different antigen receptors and additional surface molecules necessary for interaction with other cells.

**T lymphocytes**

Large numbers of antigen-specific T cells are produced in the thymus from circulating T-cell precursors derived from stem cells in the bone marrow. Each T cell has receptors specific for only one antigen. These are generated by gene rearrangement from multiple, inherited germline genes. T cells then undergo selection to remove those that are highly self-reactive. In the process, two different kinds of T cells develop. T helper (Th) cells, of which there are two types (Th1 and Th2), express CD4 and provide help for B-cell growth and differentiation. T cytotoxic (Tc) cells express CD8 and recognize and kill virally infected cells. Functionally mature T cells then migrate to secondary lymphoid tissues to mediate protection.

**B lymphocytes and plasma cells**

Hemopoietic stem cell (HSC) differentiation into B cells occurs within the fetal liver and, after birth, the bone marrow. In the bone marrow, B-cell precursors rearrange multiple, inherited, germline genes that encode B-cell antigen receptors (antibodies), thus creating many different B cells, each with a unique specificity for antigen. Many B cells with antigen receptors that react with self are eliminated. In addition, two kinds of B cells (B1 and B2) with different properties develop. IgM is the first antibody expressed on B cells followed by co-expression of IgD. Mature B cells migrate into the secondary lymphoid tissues, where they respond to foreign antigens. When activated by antigen, in most cases with T-cell help, they proliferate in germinal centers and mature into memory cells or into plasma cells that produce and secrete large amounts of antibody.

**Related topics**

(A5) Hemopoiesis –
  development of
  blood cells
(E1) The B-cell receptor
  complex, co-receptors,
  and signaling

(F1) The role of T cells in
  immune responses
(F2) T-cell recognition of
  antigen

## Specificity and memory

Lymphocytes are responsible for the specificity and memory in adaptive immune responses. They are produced in the primary lymphoid organs (Section C2) and function in the secondary lymphoid organs/tissues, where they recognize and respond to foreign antigens. The two main types of lymphocytes, T cells and B cells, mature in the thymus and bone marrow, respectively. In the resting state both types of lymphocytes have a similar morphology with a small amount of cytoplasm (Figure 1). They have specific but different antigen receptors and a variety of other surface molecules necessary for interaction with other cells (Table 1). These include molecules required for their activation and for movement into and out of the tissues of the body. This ability to migrate into the tissues and return via the lymphatic vessels to the bloodstream (recirculation) is a unique feature of lymphocytes.

There are three main functional classes of T lymphocytes: T helper (Th) cells, T cytotoxic (Tc) cells, and T regulatory cells (Tregs). All T lymphocytes have antigen receptors (TCR)

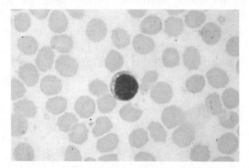

Figure 1. A blood lymphocyte. From Male D, Brostoff J, Roth D & Roitt I (2006) Immunology, 7th ed. With permission from Elsevier.

**Table 1. Characteristics of human B and T cells**

|  | T cells | B cells |
|---|---|---|
| Site of maturation | Thymus | Bone marrow |
| Antigen receptor | TCR | Antibody |
| Requirement of MHC for recognition | Yes | No |
| Characteristic "markers" | All have TCR, CD3<br>Th – CD4<br>Tc – CD8 | Surface Ig, CD19, CD20, CD21, CD79 |
| Main location in lymph nodes | Paracortical area | Follicles |
| Memory cells | Yes | Yes |
| Function | Protect against intracellular microbes<br>Provide help for antibody responses<br>Regulate immune responses | Protect against extracellular microbes |
| Products | Th1 – IFNγ, TNFα<br>Th2 – IL-4, IL-5, IL-6<br>Tc – perforins | Antibodies (B cells mature into plasma cells) |

(Section F2) that determine their specificity and CD3, which is essential for their activation (Section F4). These molecules also serve as "markers" to identify T cells. B lymphocytes make and use antibodies as their specific antigen receptor. They have molecules similar to CD3, that is, CD79a and b, which are important in their activation. B lymphocytes can mature into plasma cells that produce and secrete large amounts of antibody.

## T lymphocytes

### T-cell ontogeny

The thymus is derived from the third and fourth pharyngeal pouches during embryonic life and attracts (with chemoattractive molecules) circulating T-cell precursors derived from hemopoietic stem cells (HSCs) in the bone marrow. In the thymus, these precursors differentiate into functional T lymphocytes under the influence of thymic stromal cells and cytokines. In particular, in the thymic cortex the precursors (now thymocytes) associate with cortical epithelial nurse cells critical to their development. In this site there is major thymocyte proliferation, with a complete turnover of cells approximately every 72 hours. Thymocytes then move into the medulla, where they undergo further differentiation and selection. Most of the thymocytes generated each day in the thymus die by apoptosis, with only 5–10% surviving. Molecules important to T-cell function such as CD4, CD8, and the TCR develop at different stages during the differentiation process (Figure 2).

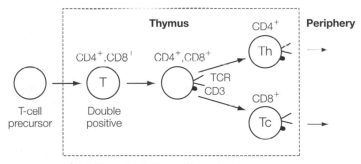

Figure 2. Development of CD4⁺ and CD8⁺ T cells in the thymus.

### Thymus function

The main functions of the thymus as a primary lymphoid organ are to (i) produce sufficient numbers (millions) of different T cells each expressing unique T-cell receptors (generate diversity) such that in every individual there are at least some cells potentially specific for each foreign antigen in our environment; (ii) select T cells for survival in such a way that the chance for an autoimmune response is minimized. It is important to note that T-cell development within the thymus is *independent* of exogenous (foreign) antigens.

### Generation of T-cell diversity in the thymus

Millions of T cells, each with receptors specific for different antigens, are generated by gene rearrangement from multiple (inherited) germline genes (Section F3). Each of the T cells produced in the thymus has only one specificity coded for by its antigen receptor.

### Positive and negative selection

Once produced in the thymus, T cells undergo selection using their newly produced receptors. T cells with receptors that bind weakly to major histocompatibility complex

(MHC) molecules are selected while those with receptors which bind strongly to MHC and self antigens die through apoptosis (central tolerance to self, Section G2) and are removed by phagocytic macrophages.

### Mature T cells and their subsets

T cells which survive the selection process migrate to the peripheral lymphoid tissues where they complete their functional maturation and provide protection against invading microbes. Some T cells reside, at least temporarily, in T-cell-dependent areas of tissues. T cells can be identified using monoclonal antibodies specific for characteristic molecules such as TCR or CD3 (Table 2). These cells function to control intracellular microbes, provide help for B-cell (antibody) responses, and regulate immune responses. Three different kinds of T cells are involved in these functions: T helper (Th) cells, T cytotoxic (Tc) cells, and regulatory T cells (Tregs).

### Table 2. T-cell surface molecules

| Surface molecules | Function |
| --- | --- |
| **The T-cell receptor complex** | |
| TCR | Antigen-specific receptor (most T cells utilize αβ dimers; some use γδ dimers) |
| CD3 (γ, δ, ε, and ζ chains) | Signaling complex associated with the TCR: mediates T-cell activation on binding of TCR to MHC–peptide complexes |
| **Subset markers** | |
| CD4 (on helper and regulatory T cells) | Binds to MHC class II molecules and restricts Th cells to recognizing only peptides presented on MHC class II |
| CD8 (on cytotoxic T cells) | Binds to MHC class I molecules and restricts Tc cells to recognizing only peptides presented on MHC class I |
| **Co-stimulatory molecules** | |
| CD28 | Binds to CD80/CD86 on B cells and APCs and positively regulates T-cell activation |
| CTLA4 | Binds to CD80/CD86 on B cells and APCs and downregulates T-cell activation (present on regulatory cells) |
| CD154 (CD40L): on activated Th cells | Binds to CD40 on B cells and APCs: triggers activation of APC and activation and antibody class switching of B cells |
| **Adhesion molecules** | |
| LFA-1 | Binds to ICAM-1 and facilitates interactions with other cells, including B cells, APCs, and target cells |
| CD2 (LFA-2) | Binds to LFA-3 and facilitates interactions with other cells, including B cells, APCs, and target cells |
| CD45RA (on naive T cells) | Involved in signal transduction |
| CD45R0 (on activated/memory T cells) | Involved in signal transduction |

Th cells provide help for B cells through direct cell surface signaling and by producing cytokines that are critical to B-cell growth and differentiation. In addition to TCR and CD3, Th cells also express cell surface CD4 molecules that bind to MHC class II molecules, an interaction required for their activation by antigen (Section F2). Th cells can be further subdivided into Th1 and Th2 cells based on their ability to help in the development of different immune responses (Section F5), which is in turn related to their cytokine profiles. Tc cells mediate killing of infected cells, primarily those infected with virus. These cells express, in addition to TCR and CD3, a cell surface molecule, CD8, that binds to MHC class I and is important for these cells to interact effectively with virally infected cells. Regulatory T cells also express CD4 and are divided into "natural" Tregs—produced during normal development in the thymus and "induced" Tregs that develop following antigenic stimulation of naive CD4$^+$ T cells. These will be discussed in more detail in Section G. The average percentages of T cells, B cells, and natural killer (NK) cells in human peripheral blood are shown in Table 3. In the majority of T cells, the TCR is composed of an α and β polypeptide chain (Section F3). Some T cells have two different chains for their TCR (γ and δ) and these recognize different antigens from αβ T cells (Figure 3 and Section F2).

**Table 3. T cells, B cells, and NK cells in human peripheral blood**

|  | T cells | | B cells | NK cells |
|---|---|---|---|---|
|  | CD4$^+$ | CD8$^+$ |  |  |
| Percent of lymphocytes | 55 | 25 | 10 | 10 |
| Functional properties | Antigen specific, produce cytokines, memory cells, effector cells | | Antigen specific, produce cytokines, memory cells, plasma cells (antibody factories) | Mediate ADCC, tumor surveillance, no memory, lyse virus-infected cells and tumor cells lacking MHC class I |

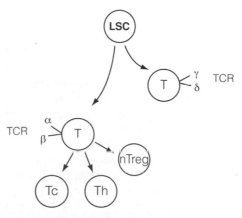

Figure 3. Development of αβ and γδ T cells from lymphocyte stem cells (LSC). Two types of T cells are produced in the thymus, with different TCRs (αβ and γδ). The classical T cells (Th, nTreg, and Tc) utilize αβ for their TCR.

## B lymphocytes and plasma cells

### The bone marrow and B-cell ontogeny

B cells develop from hemopoietic stem cells primarily (perhaps exclusively) in the micro-environment of the fetal liver and, after birth, the bone marrow. The two main functions of the bone marrow as a primary lymphoid organ are to (i) produce large numbers of B cells, each with unique antigen receptors (antibodies) such that, overall, there is suffi-cient B-cell diversity to recognize all of the antigens in our environment (generate diversity); (ii) eliminate B cells with antigen receptors for self molecules. The early stages of B-cell development (like that of T cells) are independent of exogenous antigen. B cells leave the bone marrow and migrate via the bloodstream to the spleen and other tissues, where they mature within the secondary lymphoid organs/tissues and are present in loose aggregates (primary follicles) in lymphoid tissues or in well-defined proliferating foci (germinal centers).

Two kinds of B cells (B1 and B2) have been identified. The B2 cells are produced in the bone marrow (**conventional B cells**) as described above and with the help of Th cells produce IgG, IgA, and IgE antibodies. However, B1 cells arise early in ontogeny, express mainly IgM antibodies encoded by germline antibody genes, mature independently of the bone marrow, generally recognize multimeric sugar/lipid antigens of microbes, and are thymus-independent (Section E2).

### Generation of antigen receptor diversity and negative selection of B cells

Antibodies, like T-cell receptors, are encoded by multiple genes. These genes, which are distinct from the T-cell antigen receptor genes, rearrange during the pro-B-cell stage to create a unique cell surface receptor that defines its specificity for antigen (Section D3). Since rearrangement occurs in millions of different ways in these developing cells, many B cells, each with a different specificity, are generated. This generation of diversity occurs in the absence of foreign antigen and yields large numbers of mature B cells, at least some of which have specificity for each foreign substance or microbe. B cells with specificity for self antigens are induced to die by apoptosis (negative selection) during their immature stage, that is, when they have expressed IgM on their cell surface, but before expression of IgD. As in the thymus, the majority of the B cells in the bone marrow die by apoptosis during development as a result of their production of antigen receptors that cannot be assembled or that are directed against self antigens (Section G2).

### Activated B cells and plasma cells

When activated by antigen and, in most cases, with T-cell help, B cells (Table 4) prolifer-ate and mature into memory cells or plasma cells. Memory cells only produce antibody for expression on their cell surface and remain able to respond to antigen if it is rein-troduced. In contrast, plasma cells do not have cell-surface antibody receptors. Rather, these cells function as factories producing and secreting large amounts of antibody of the same specificity as the antigen receptor on the stimulated parent B cell. The morphology of a plasma cell (Figure 4) is consistent with its primary function—high-rate glycopro-tein (antibody) synthesis. Note the extensive endoplasmic reticulum, mitochondria, and Golgi apparatus. It should also be noted that a plasma cell only produces antibodies of one specificity, one class, and one subclass.

## Table 4. B-cell surface molecules

| Surface molecules | Function |
| --- | --- |
| **The B-cell receptor complex** | |
| Antibody (IgM and IgD on mature B cells) | B-cell receptor (BCR) for antigen |
| CD79a/CD79b (Igα/Igβ) heterodimer | Mediates cellular activation on binding of BCR to antigen |
| **Co-receptors** | All these molecules modulate B-cell activation |
| CD19 | Influences B-cell activation |
| CD20 | $Ca^{2+}$ channel |
| CD21 (complement receptor CR2) | Binds to C3d, C3bi |
| CD32 (FcγRII: Fc receptor for IgG) | Binds to IgG complexed to antigen |
| CD40 | Signals B-cell activation and antibody class switching after engagement of CD40 ligand (CD154) on activated T cells |
| **Molecules required for T-cell activation** | |
| MHC class II molecules | Present peptides to Th cells |
| CD80/CD86 (also called B7.1, B7.2) | Bind to CD28 on T cells to trigger their activation |
| **Adhesion molecules** | |
| ICAM-1 | Binds to LFA-1 and facilitates interaction with T cells |
| LFA-3 | Binds to CD2 and facilitates interaction with T cells |

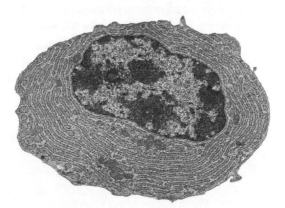

Figure 4. Ultrastructure of a plasma cell. Note the extensive rough endoplasmic reticulum for antibody production. From Male D, Brostoff J, Roth D & Roitt I (2006) Immunology, 7th ed. With permission from Elsevier.

# C2 Lymphoid organs and tissues

---

**Key Notes**

| | |
|---|---|
| **Primary and secondary lymphoid organs** | The thymus and the bone marrow are primary lymphoid organs as T and B cells are produced in these organs/tissues before migrating to the secondary lymphoid tissues, including the spleen, lymph nodes, and mucosa-associated lymphoid tissues (MALT). |
| **Bone marrow** | Bone marrow is the primary source of the pluripotent stem cells (HSCs) that give rise to all hemopoietic cells, including lymphocytes. It is the major organ for B-cell production and maturation and gives rise to the precursor cells of the thymic lymphocytes. |
| **Thymus** | T cells develop in the thymus. Immature T-cell precursors travel from the bone marrow to the thymus (to become thymocytes) where they generate antigen specificity, undergo thymic education, and then migrate to the peripheral lymphoid tissues as mature T cells. |
| **Spleen** | The spleen contains T and B lymphocytes as well as many phagocytes and is a major component of the mononuclear phagocyte system. Its primary function is to protect the body against blood-borne infections and it is particularly important for B-cell responses to polysaccharide antigens. |
| **Lymph nodes** | Lymph nodes are situated along lymphatic vessels and filter the lymph. Like the spleen they contain both T and B lymphocytes as well as accessory cells and are primarily responsible for mounting immune responses against foreign antigens entering the tissues. |
| **Related topics** | (E3) The cellular basis of the antibody response (E4) Antibody responses in different tissues    (G2) Central tolerance (G3) Peripheral tolerance |

---

## Primary and secondary lymphoid organs

The thymus and bone marrow are the primary lymphoid organs in mammals. T and B cells with diverse antigen receptors are produced in these organs. Following selection processes (Sections E3, E4, and F3), they migrate to the secondary lymphoid tissues—the lymph nodes, spleen, and the mucosa-associated lymphoid tissues (MALT) (Figure 1).

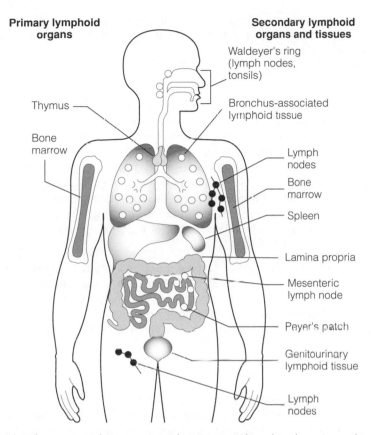

Figure 1. Lymphoid organs and tissues. Lymphocytes produced in the primary lymphoid organs (thymus and bone marrow) migrate to the secondary organs and tissues where they respond to microbial infections. The mucosa-associated lymphoid tissue (MALT) together with other lymphoid cells in subepithelial sites (lamina propria) of the respiratory, gastrointestinal, and genitourinary tracts comprise the majority of lymphoid tissue in the body.

## Bone marrow

During early fetal development blood cells are produced in the mesenchyme of the yolk sac. As the development of the fetus progresses, the liver and spleen take over this role. It is only in the last months of fetal development that the bone marrow becomes the dominant site of hemopoiesis (blood cell formation). Bone marrow is composed of hemopoietic cells of various lineages and maturity, packed between fat cells, thin bands of bony tissue (trabeculae), collagen fibers, fibroblasts, and dendritic cells. All of the hemopoietic cells are derived from multipotent stem cells (HSCs) which give rise not only to all of the lymphoid cells found in the lymphoid tissue, but also to all of the cells found in the blood (Section A5).

Ultrastructural studies show that hemopoietic cells cluster around the vascular sinuses, where they mature, before they eventually are discharged into the blood. Lymphocytes are found surrounding the small radial arteries, whereas most immature myeloid precursors are found deep in the parenchyma. The bone marrow gives rise to all of the lymphoid cells that migrate to the thymus and mature into T cells, as well as to the major population of conventional B cells. B cells mature in the bone marrow and undergo some selection for non-self before making their way to the peripheral lymphoid tissues: there they form primary and secondary follicles and may undergo further selection in germinal centers (Sections E3, E4, and G3).

## Thymus

The thymus is a lymphocyte-rich, bilobed, encapsulated organ located behind the sternum, above and in front of the heart. It is essential for the development and maturation of T cells and of cell-mediated immunity. In fact, the term "T cell" means thymus-derived cell and is used to describe mature T cells. The activity of the thymus is maximal in the fetus and in early childhood and then undergoes atrophy at puberty, although it never totally disappears (Section P2). The thymus is composed of cortical and medullary epithelial cells, stromal cells, interdigitating cells, and macrophages. These "accessory" cells are important in the differentiation of the immigrating T-cell precursors and their "education" (positive and negative selection) prior to their migration into the secondary lymphoid tissues (Section F3).

The thymus has an interactive role with the endocrine system as thymectomy leads to a reduction in pituitary hormone levels as well as atrophy of the gonads. Conversely, neonatal hypophysectomy (removal of the pituitary gland) results in thymic atrophy (Section G6). Thymic epithelial cells produce the hormones thymosin and thymopoietin and in concert with cytokines (such as IL-7) are probably important for the development and maturation of thymocytes into mature T cells.

## Spleen

The spleen (Figure 2) is a large, encapsulated, bean-shaped organ with a spongy interior (splenic pulp) that is situated on the left side of the body below the diaphragm. The large splenic artery pervades the spleen and branches of this artery are surrounded by highly organized lymphoid tissue (white pulp). The white pulp forms "islands" within a meshwork of reticular fibers containing red blood cells, macrophages (MØ), and plasma cells (red pulp). Closely associated with the central arteriole is the "periarteriolar lymphoid sheath" an area containing mainly T cells and interdigitating cells (IDCs). Primary lymphoid follicles, composed mainly of follicular dendritic cells (FDCs) and B cells, are contained within the sheath. During an immune response these follicles develop germinal

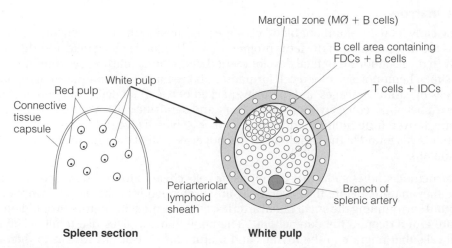

Figure 2. Structure of lymphoid tissue in the spleen. MØ, macrophages.

centers (i.e., become secondary follicles). The periarteriolar lymphoid sheath is separated from the red pulp by a marginal zone containing macrophages and specialized B cells (Figure 2), many of which are specific for polysaccharide antigens. The central arterioles in the periarteriolar sheath subdivide like the branches of a tree. The space between the branches is filled with red pulp, and vascular channels called splenic sinuses. The spleen is a major component of the mononuclear phagocyte system, containing large numbers of phagocytes. Unlike lymph nodes, the spleen only has an efferent lymphatic system. Entry of cells and antigens from the blood is via high endothelial cells of the arteriolar capillaries.

The main immunological function of the spleen is to filter the blood by trapping blood-borne microbes and producing an immune response to them. It also removes damaged red blood cells and immune complexes. Those individuals who have had their spleens removed (splenectomized) have a greater susceptibility to infection with encapsulated bacteria, and are at increased risk of severe malarial infections, which indicates the spleen's major importance in immunity. In addition, the spleen acts as a reservoir of erythrocytes.

## Lymph nodes

Lymph nodes (Figure 3) are small solid structures found at varying points along the lymphatic system, for example groin, armpit, and mesentery. They range in size from 2 to

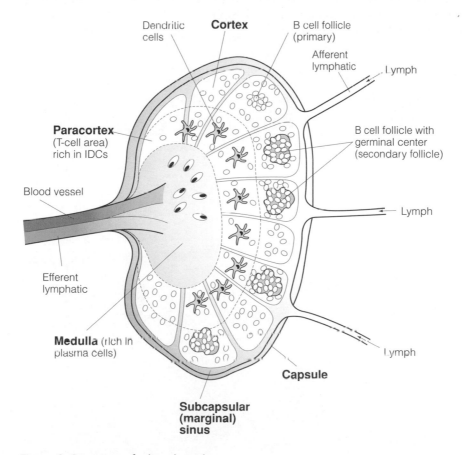

Figure 3. Structure of a lymph node.

10 mm, are spherical in shape and are encapsulated. Beneath the capsule is the subcapsular sinus, the cortex, a paracortical region, and a medulla. The cortex contains many follicles and on antigenic stimulation becomes enlarged with germinal centers. The follicles are composed mainly of B cells and follicular dendritic cells with small numbers of Th cells called T-follicular cells. The paracortical (thymus-dependent) region contains large numbers of T cells interspersed with interdigitating cells.

The primary role of a lymph node is to filter the lymph and then produce an immune response against trapped microbes/antigens. Lymph arriving from the tissues or from a preceding lymph node in the chain passes via the afferent lymphatics into the subcapsular sinus, into the cortex, around the follicles, into the paracortical area, and then into the medulla. A network of "conduits" transport small antigens and chemokines in lymph to different areas within the lymph node. Dendritic cell and lymphocyte processes protrude into these conduits, enabling them to pick up antigen and respond to chemokine signals. Lymph in the medullary sinuses drains into efferent lymphatics and hence through larger lymphatic vessels back into the bloodstream via the thoracic duct.

Lymphocytes enter the lymph nodes from the tissues via the afferent lymphatics and from the bloodstream through specialized post-capillary venules called high endothelial venules (HEVs) that are found in the paracortical region of the node. B cells entering the lymph node through the HEVs migrate to the cortex, where they are found within follicles (B-cell areas). Trafficking of lymphocytes from the blood stream is mediated through specialized adhesion molecules on both the lymphocytes and high endothelial cells of the blood vessels (Section C4).

# C3 Mucosa-associated lymphoid tissues

---

**Key Notes**

| | |
|---|---|
| **MALT** | The majority (>50%) of lymphoid tissue in the human body is located within the lining of the respiratory (NALT, BALT), digestive (GALT), and genitourinary tracts, as these are the main entry sites for microbes into the body. |
| **NALT** | Nasal-associated lymphoid tissue (NALT) includes immune cells underlying the throat and nasal passages and especially the tonsils. The architecture of these lymphoid tissues, although not encapsulated, is similar to that of the lymph nodes and consists of follicles composed mainly of B cells. |
| **BALT** | The lymphoid tissue associated with the bronchus (BALT) is structurally similar to Peyer's patches and other lymphoid tissues of the gut. It consists of lymphoid aggregates and follicles and is found along the main bronchi in the lobes of the lungs. |
| **GALT** | Gut-associated lymphoid tissue (GALT) is composed of lymphoid complexes (also called Peyer's patches in the ileum) that consist of specialized epithelium, antigen-presenting cells, and intraepithelial lymphocytes. These structures occur strategically at specific areas in the digestive tract. |
| **Related topics** | (C4) Lymphocyte traffic and recirculation (G2) Central tolerance (G3) Peripheral tolerance (H1) The microbial cosmos (O2) Immune cells and molecules associated with the reproductive tracts |

---

## MALT

The main sites of entry for microbes into the body are through mucosal surfaces. It is therefore not surprising that more than 50% of the total body lymphoid mass is associated with these surfaces. These are collectively called mucosa-associated lymphoid tissues (MALT) and include NALT, BALT, GALT, and lymphoid tissue associated with the genitourinary system (Section O). There are two main kinds of mucosal lymphoid tissue: (i) aggregates of subepithelial lymphocytes "organized" into follicles often with overlying specialized epithelial cells (microfold or M cells—see below); (ii) "diffuse" lymphoid tissue (separate cell populations) within the lamina propria below the epithelial cell layer of the various tracts.

## NALT

The nasal-associated lymphoid system is composed of the lymphoid tissue at the back of the nose (pharyngeal, tonsil, and other tissue) and that associated with Waldeyer's ring (palatine and lingual tonsils). The strategic location of these lymphoid tissues suggests that they are directly involved in handling airborne microbes. Their composition is similar to that of lymph nodes but they are not encapsulated and are without lymphatics. Antigens and foreign particles are trapped within the deep crypts of their lympho-epithelium from where they are transported to the lymphoid follicles (Figure 1). The follicles are composed mainly of B cells surrounded by T cells. The germinal center within the follicles is the site of antigen-dependent B-cell proliferation. The epithelial cells overlying the lymphoid tissue have some of the features of M cells (see below).

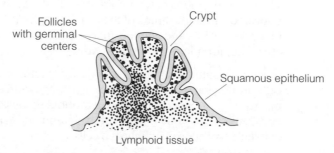

Figure 1. Tonsilar lymphoid tissue. Antigens trapped in the crypts are transported by M-like cells into the subepithelial areas where lymphocytes are stimulated via antigen-presenting cells.

## BALT

Bronchus-associated lymphoid tissue is similar to Peyer's patches in GALT. It is composed mainly of aggregates of lymphocytes organized into follicles that are found in all lobes of the lung and are situated under the epithelium, mainly along the bronchi. The majority of lymphocytes in the follicles are B cells. Antigen sampling is carried out by epithelial cells lining the surface of the mucosa and by way of M cells which transport antigens to underlying APCs and lymphocytes (see below).

## GALT

The primary role of GALT is to protect the body against microbes entering the body via the intestinal tract. It is primarily made up of lymphoid aggregates and lymphoid cells (intraepithelial lymphocytes, IELs) between epithelial cells and within the lamina propria. In order to distinguish between harmful invaders and harmless food, the gut has a "sampling" mechanism that analyzes everything that has been ingested (or in the case of BALT and NALT, inhaled). The analytical, or antigen-sampling, machinery of the gut consists of specialized epithelial cells, M (microfold) cells, and intimately associated APCs (antigen "processing and presenting" cells) (Figure 2). M cells take up foreign molecules and pass them to underlying APCs, which present them in the context of class I and class II MHC molecules to T cells. The helper T cells help to activate B cells, and both T and B cells can migrate to other parts of the gastrointestinal (GI) tract (including salivary glands) and other MALT sites, for example lactating mammary glands and respiratory and genitourinary tracts, and protect these surfaces from invasion by the same microbes (Section E4). Depending on the antigen, the APC and its state, and other factors, tolerance as well as immunity can be induced to the sampled antigen (Sections G4 and I2).

The combination of specialized epithelium and antigen-processing cells plus lymphocytes constitute what are called lymphoid complexes. These are localized structures that occur regularly at specific areas in the digestive tract and are exemplified by Peyer's patches in the terminal ileum. Lymphoid complexes are not distributed uniformly throughout the gut as one might initially expect, but are congregated in several zones (Figure 3).

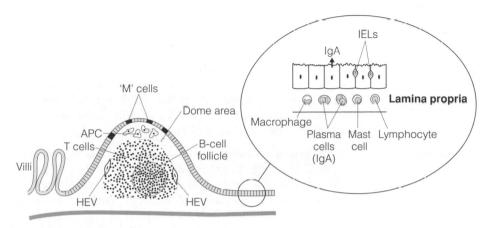

Figure 2. Intestinal lymphoid aggregates: 'M' cells transport luminal antigens into the dome area where they are taken up by antigen-presenting cells (DCs) and processed and presented to T cells entering the site via the high endothelial venules (HEV). These T cells interact with antigen-specific B cells which then migrate via the draining lymph nodes to the subepithelial sites (lamina propria) of the intestinal tract and to locations within the other tracts of the body, that is, the respiratory tract and the genitourinary tract. Insert: Here the B cells develop into IgA-secreting plasma cells and IgA is transported through the epithelium into the lumen of the intestine. CD4+ T cells and, more prominently, CD8+ T cells are present and the latter are frequently seen between the epithelial cells (intraepithelial lymphocytes, IELs). Mast cells, dendritic cells, and macrophages are also present in the lamina propria outside the lymphoid aggregates.

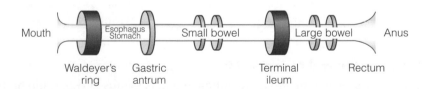

Figure 3. Lymphoid complexes along the gastrointestinal tract. Volume of the rings indicates the relative amount of lymphoid tissue.

# C4 Lymphocyte traffic and recirculation

| Key Notes | |
|---|---|
| **Lymphocyte traffic and recirculation** | T and B cells produced in the thymus and bone marrow, respectively, migrate via the bloodstream to the secondary lymphoid organs/tissues where they carry out their function. They do not stay in one site but continually recirculate through the body in search of antigens. |
| **Trafficking in MALT** | Lymphocytes stimulated in one mucosal organ, for example the GALT, can migrate to the lamina propria of other sites of the mucosal immune system (e.g., lactating mammary glands and salivary glands), and protect these surfaces from invasion with the same microbes. |
| **Mechanisms of lymphocyte traffic** | Lymphocytes and other leucocytes have sets of adhesion molecules ("homing molecules") on their surfaces that interact with their ligands on endothelial and other cell surfaces to allow traffic from the blood stream into lymphoid tissues and inflammatory sites. Migration involves multiple adhesion molecules including selectins, bound chemokines, and integrins. B and T lymphocytes having passed across the endothelium migrate to their specific anatomical sites in response to local release of specific attractive chemokines. Specific lymphocyte homing molecules and endothelial cell addressins allow the targeting of lymphocytes to different lymphoid tissues and extra-lymphoid sites. |
| **Related topics** | (C2) Lymphoid organs and tissues  (C3) Mucosa-associated lymphoid tissues |

## Lymphocyte traffic and recirculation

Lymphocytes produced in the primary lymphoid organs, thymus (T cells), and bone marrow (B cells), migrate via the bloodstream to the secondary lymphoid organs or tissues where they carry out their function. Since these cells have not yet encountered antigen, they are called "naive cells" and do not remain in one secondary lymphoid organ, but continue to recirculate around the body until they recognize their specific antigen (Figure 1). They enter the lymph nodes via the **high endothelial venules** (**HEVs**) and if they are not activated there, they pass via efferent lymphatic vessels into the thoracic duct and hence back into the bloodstream. Both memory and naive cells recirculate through the lymphoid tissues.

T and B cells migrate to different sites within the lymph nodes. T cells reside in the paracortical region, whereas the B-cell domain is the lymphoid follicle (see below). B cells must traverse through the T-cell area to reach the follicle. In the spleen, lymphocytes enter the periarteriolar lymphoid sheath (PALS) by way of the marginal zone (MZ) and leave

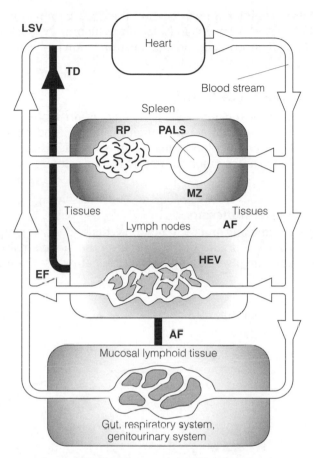

Figure 1. Lymphocyte recirculation. Lymphocytes travel in the blood stream to the spleen where they enter the periarteriolar lymphoid sheath (PALS) via the marginal zone (MZ) and re-enter the blood stream via the red pulp (RP). Lymphocytes enter the lymph nodes via high endothelial venules (HEV) in the paracortical regions and pass via the efferent lymphatics (EF) into the lymphatic system and via the thoracic duct (TD) into the left subclavian vein (LSV). Lymphocytes pass into the mucosal tissues through the HEV and return via the afferent lymphatics (AF) of the draining lymph nodes. Lymphocytes stimulated by microbes in the MALT migrate back to the mucosal tissues where they were initially stimulated. Thus, lymphocytes stimulated in the intestine will migrate back to sites in the lamina propria along the intestine (as well as to other mucosal sites) to protect the body against the specific microbial attack via this route. Arrows indicate the direction of flow.

through the splenic veins (SV) in the red pulp (RP). The lymphoid tissues are dynamic structures, wherein both T and B lymphocytes are continuously trafficking through each other's territories as well as being challenged by antigen on antigen-presenting cells. Lymphocytes are also able to traffic to specific tissues such as the MALT (see below).

## Trafficking in MALT

One of the unique features of MALT is that lymphocytes stimulated in one site can migrate to other sites of the mucosal immune system to protect them against the same antigen or from invasion by the same microbe. Thus, for example, lymphocytes that initially encountered and were stimulated by antigen in the GALT can migrate via the blood to distant sites, including the salivary glands, lactating mammary glands, the respiratory and reproductive tracts, and so forth, and mediate protection in these other MALT tissues (see below).

## Mechanisms of lymphocyte traffic

Lymphocytes and other leukocytes have sets of adhesion molecules ("homing molecules") on their surface that interact with their ligands on endothelial and other cell surfaces to allow traffic from the blood stream into lymphoid tissues and inflammatory sites. This has already been described for neutrophil migration in acute inflammation (Section B4).

### Principles of lymphocyte trafficking

Normal physiological trafficking and recirculation is dependent on lymphocytes migrating into secondary lymphoid organs and tissues from the blood stream and back again via the lymph. Migration involves multiple adhesion pathways, the four main steps of which are shown in Figure 2.

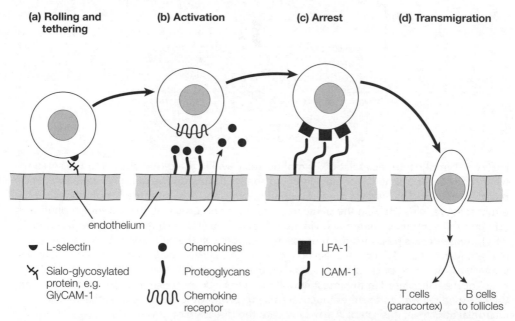

Figure 2. Multiple adhesion pathways of lymphocyte migration: (a) **Rolling and tethering:** the lymphocyte rolls along the endothelial surface and slows down becoming tethered by surface L-selectin (CD62L) binding to sialo-glycosylated ligands on the endothelial surface (addressin—see below). (b) **Activation:** chemokines, either released from the endothelial surface or attached to them through binding carbohydrates on proteoglycans (e.g., heparin sulfate) now bind to lymphocyte chemokine receptors. This results in activation of the lymphocyte causing it to upregulate expression of surface integrins. (c) **Arrest:** integrins (e.g., LFA-1) attach to their receptors on the endothelial surface (e.g., ICAM-1). (d) **Transmigration:** the lymphocyte transmigrates across the endothelium into the lymphoid organ/tissue. This process of diapedesis involves movement of the lymphocyte between adjacent cells of the endothelial cell monolayers into the underlying tissue. Once inside, the T lymphocytes remain and the B lymphocytes migrate to follicles where the FDCs constitutively express CCL13.

1. The lymphocyte rolls along the endothelial surface and is slowed down by its surface L-selectin (CD62L) interacting with their ligands on the endothelial surface (addressin, see below).

2. Chemokines, either released from the endothelial surface or attached to them through binding carbohydrates on proteoglycans (e.g., heparin sulfate), now bind to lymphocyte chemokine receptors.

3. This results in activation of the lymphocyte causing it to upregulate expression of surface integrins, which attach to their receptors on the endothelial surface. The type/amount of lymphocyte integrin expressed plays a role in determining the target specificity of the lymphocyte to different lymphoid tissues.

4. The lymphocyte then transmigrates across the endothelium into the lymphoid organ/tissue. This process of diapedesis involves movement of the lymphocyte between adjacent cells of the endothelial cell monolayers into the underlying tissue.

### Trafficking to lymph nodes

Naive lymphocytes express both L-selectin and the integrin LFA-1. In addition, they express CCR7 (and CXCR4 for B cells), which interacts with the chemokines CCL19 and CCL21 on the endothelium of the HEV, permitting them to traffic into peripheral lymph nodes. When in the node the T lymphocytes stay in the paracortical region, which is rich in the CCR7 binding chemokine, while the B cells (expressing CXCR5) are attracted to the follicular areas through constitutive production of CCL13 by the follicular dendritic cells. Thus, chemokines are important in determining the localization of T and B cells within the lymphoid tissue. If the lymphocytes encounter their specific antigen they stay within the lymph node while naive cells pass through into the efferent lymphatics and hence back into the blood.

### Trafficking to mucosal lymphoid tissues

Lymphocytes that are targeted to the mucosal lymphoid tissues, for example Peyer's patches, express L-selectin and, in addition to LFA-1, another surface integrin—$\alpha_4\beta_7$. This binds to a different addressin, MAdCAM-1, expressed on the surface of HEV endothelial cells in the mucosal tissues (Figure 3). Lymphocytes leave the mucosal areas via the afferent lymphatics into draining lymph nodes and hence back into the lymph and circulation.

### Trafficking to the spleen

Lymphocytes in the bloodstream are attracted across the endothelium of the marginal sinus and migrate to T- and B-cell areas using the same chemokines and receptors used by the different lymphocyte populations in the lymph nodes. Specialized marginal zone B cells remain there (but do recirculate) and are mainly specific for carbohydrate antigens.

### Trafficking of naive, memory, and effector lymphocytes

Most memory cells migrate into lymphoid tissues using the same mechanism as naive lymphocytes. However, many memory cells not only remember the antigen that stimulated them but also where they were stimulated. This is due to "imprinting" in the specific microenvironments resulting in switching on of appropriate surface chemokine receptors through interaction with local dendritic cells. Thus, plasmablasts (B cells) induced in mucosal sites have additional cytokine receptors to allow them to migrate to

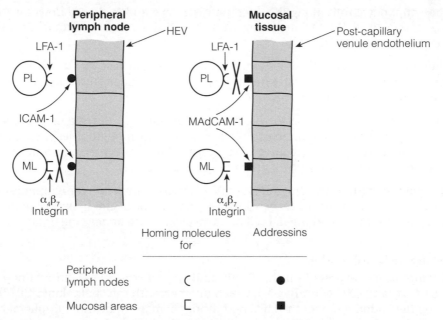

Figure 3. Homing molecules allow trafficking of lymphocytes into specific anatomical locations. Lymphocytes entering peripheral lymph nodes (PL) have specific homing molecules for "addressins" on endothelial cells of the HEV, that is, LFA-1 binding to ICAM-1 (intercellular adhesion molecule-1). These are different from the addressins on endothelial cells in mucosal tissues. Lymphocytes primed in the MALT (ML) have their own homing molecules, which allow them to bind to addressins on endothelial cells of HEV at mucosal sites, for example $\alpha_4\beta_7$ binding to MAdCAM-1 (mucosal addressin cell adhesion molecule-1).

other mucosal sites as described above. In addition, CTLs (CD8+ T cells) have additional chemokine receptors (as well as other receptors) that allow them to migrate into inflammatory sites where they are attracted by inflammatory chemokine/cytokine release to remove virus-infected cells. Antibacterial peptides (e.g., defensins, Section B2) have also been shown to be important chemoattractants for effector cells.

# C5 Adaptive immunity at birth

---

**Key Notes**

| | |
|---|---|
| **Lymphocytes in the newborn** | T and B lymphocytes are present in the blood of newborns in slightly higher numbers than in adults, and many are fully functional. However, their ability to mount an immune response to certain antigens (e.g., polysaccharides) may be deficient, perhaps due to immaturity of some cells, to sequential expression of genes encoding antigen receptors, and/or to maternal antibody. |
| **Antibodies in the newborn** | Maternal IgG crosses the placenta (mediated by specific Fc receptors) and is present at high levels in the newborn. IgG is not synthesized *de novo* by the fetus until birth and IgA not for 1–2 months after birth, whereas IgM is produced late in fetal development. Maternal IgA from colostrum and milk during nursing coats the infant's gastrointestinal tract and supplies passive mucosal immunity. |
| **Related topics** | (D2) Antibody classes      (J2) Primary/congenital (inherited) immunodeficiency |

---

## Lymphocytes in the newborn

Slightly higher than normal numbers of apparently mature T- and B-lymphocyte populations (as well as NK cells) are present in the blood of newborn individuals. Even so, the ability to mount an immune response to certain antigens may be lacking at birth. Thus, children under 2 years do not usually make antibody to the polysaccharides of pneumococcus or *H. influenzae*. In general, the ability to respond to a specific antigen depends on the age at which the individual is exposed to the antigen. There are a variety of explanations for this sequential appearance of specific immunity, including (i) sequential expression of genes encoding receptors for each antigen; (ii) immaturity of some B or helper T-cell populations or of antigen-presenting cells (e.g., macrophages and dendritic cells); (iii) passive maternal antibody that binds antigen and removes it, thereby interfering with the development of active immunity.

Since *Hemophilus* polysaccharide conjugated to tetanus toxoid evokes protective anti-polysaccharide antibodies during the first year of life, the neonatal deficiency is likely to be in the Th-cell population rather than the antigen-specific B cells. Moreover, delayed maturation of the CD4+ Th-cell population may contribute to the generally low levels of IgG leading to immunodeficiency in transient hypogammaglobulinemia (Section J2).

## Antibodies in the newborn

IgM is produced late during fetal development but IgG is not synthesized *de novo* until after birth (Figure 1). IgA begins to appear in the blood at 1–2 months of age. However, **maternal IgG** crosses the placenta into the fetus (mediated by Fc receptor, FcRn) and is

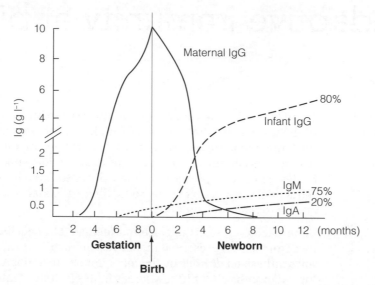

Figure 1. Maternal IgG is actively transported across the placenta and accumulates in the baby's blood until birth. This protective IgG then decreases due to catabolism and disappears completely by about 6–8 months of age. *De novo* synthesis of IgM by the baby occurs first at 6–8 months of gestation and is followed around birth by IgG and later IgA. At 1 year of age, the levels of the baby's IgG, IgM, and IgA are about 80, 75, and 20% of adult levels, respectively.

present at high levels in the newborn. This passive immunity partly compensates for the deficiencies in the ability of the infant to initially synthesize antibody through an immune system in which some components may not be totally mature. Furthermore, maternal IgA obtained by the infant from colostrum and milk during nursing coats the infant's gastrointestinal tract and supplies passive mucosal immunity. As suggested above, this passive immunity may contribute to the infant's unresponsiveness to certain antigens until maternal antibodies are degraded or used up, and are no longer interfering with the development of active immunity.

# D1 Antibody structure

---

**Key Notes**

**Molecular components**

Antibodies, often termed "immunoglobulins," are glycoproteins that bind antigens with high specificity and affinity. In humans there are five chemically and physically distinct classes of antibodies (IgG, IgA, IgM, IgD, and IgE).

**Antibody units**

Antibodies have a basic unit of four polypeptide chains—two identical pairs of light (L) chains and heavy (H) chains—bound together by covalent disulfide bridges as well as by noncovalent interactions. These molecules can be proteolytically cleaved to yield two Fab fragments (the antigen-binding part of the molecules) and an Fc fragment (the part of the molecule responsible for effector functions, e.g., complement activation). Both H and L chains are divided into V (variable) and C (constant) regions—the V regions containing the antigen binding site and the C region determining the fate of the antigen.

**Affinity**

The tightness of binding of an antibody-binding site to an antigenic determinant is called its affinity. The tighter the binding, the less likely the antibody is to dissociate from antigen. Different antibodies to an antigenic determinant vary considerably in their affinity for that determinant. Antibodies produced by a memory response have higher affinity than those in a primary response.

**Antibody valence and avidity**

The valence of an antibody is the number of antigenic determinants with which it can react. Having multiple binding sites for an antigen dramatically increases its binding (avidity) to antigens on particles such as bacteria or viruses. For example, two binding sites on IgG are 100 times more effective at neutralizing virus than two unlinked binding sites.

**Related topics**

(D2) Antibody classes
(D3) Generation of diversity

(E1) The B-cell receptor complex, co-receptors, and signaling

---

## Molecular components

Antibodies are glycoproteins that bind antigens with high specificity and affinity (they hold on tightly). They are molecules, originally identified in the serum, which are also referred to as "immunoglobulins," a term often used interchangeably with antibodies. In humans there are five chemically and physically distinct classes of antibodies (IgG, IgA, IgM, IgD, and IgE).

## Antibody units

All antibodies have the same basic four-polypeptide chain unit: two light (L) chains and two heavy (H) chains (Figure 1). In this basic unit, one L chain is bound, by a disulfide bridge and noncovalent interactions, to one H chain. Similarly, the two H chains are bound together by covalent disulfide bridges as well as by noncovalent hydrophilic and hydrophobic interactions. There are five different kinds of H chains (referred to as $\mu$, $\delta$, $\gamma$, $\epsilon$, and $\alpha$ chains), which determine the class of antibody (IgM, IgD, IgG, IgE, and IgA, respectively). There are also two different kinds of L chains ($\kappa$ and $\lambda$), each with a MW of 23 kDa. Each antibody unit can have only $\kappa$ or $\lambda$ L chains but not both. The properties of the different antibody classes are shown in Table 1.

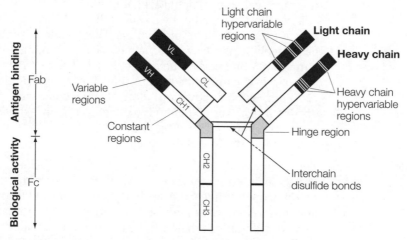

Figure 1. IgG immunoglobulin: basic four-chain structure representative of all immunoglobulins.

## Table 1. Properties of the human immunoglobulins

|  | IgG | IgA | IgM | IgD | IgF |
|---|---|---|---|---|---|
| **Physical properties** | | | | | |
| MW (kDa) | 150 | 170–420 | 900 | 180 | 190 |
| H-chain MW (kDa) | 50–55 | 62 | 65 | 70 | 75 |
| **Physiological properties** | | | | | |
| Normal adult serum (mg ml$^{-1}$) | 8–16 | 1.4–4.0 | 0.4–2.0 | 0.03 | ng |
| Half-life in days | 23 | 6 | 5 | 3 | <3 |
| **Biological properties** | | | | | |
| Complement-fixing capacity | + | – | ++++ | – | – |
| Anaphylactic (type I) hypersensitivity | – | – | – | – | ++++ |
| Placental transport to fetus | + | – | – | – | – |

There are four IgG (IgG1, IgG2, IgG3, IgG4) subclasses, two IgA subclasses (IgA1, IgA2) and two L-chain types ($\kappa$ and $\lambda$); ng, nanograms.

Both H and L chains have intrachain disulfide bridges every 90 amino acid residues, which create polypeptide loops, domains, of 110 amino acids. These domains are referred to as VH, VL, CH1, CH2, and so forth (Figure 1) and have particular functional properties (e.g., VH and VL together form the binding site for antigen). This type of structure is characteristic of many other molecules, which are thus said to belong to the *immunoglobulin gene superfamily*.

The N-terminal half of the H chain and all of the L chain together make up what is called a Fab fragment (Figure 1) that contains the **a**ntigen-**b**inding site. The actual binding site of the antibody is composed of the N-terminal quarter of the H chain combined with the N-terminal half of the L chain. The amino acid sequences of these regions differ from one antibody to another and are thus called variable (V) regions and contain the amino acid residues involved in binding an antigenic determinant. Most of the antibody molecule (the C-terminal three-quarters of the H chain and the C-terminal half of the L chain) are constant (C), that is, have the same amino acid sequence as all other antibodies of the same class and subclass. These C regions do not bind antigen, but rather determine the "biological" properties of the molecule and thus the fate of antigen bound by the antigen-binding site. In particular, the C-terminal half of the H chain, the Fc region (Fragment crystallized), serves other functions, that is, combines with complement, is cytophilic (binds to receptors on certain types of cells, such as macrophages), and so forth. Carbohydrates are also present on antibodies, primarily on the Fc portion of H chains.

## Affinity

Different antibody molecules produced in response to a particular antigenic determinant may vary considerably in their tightness of binding to that determinant (i.e., in their **affinity** for the antigenic determinant). The higher the binding constant the less likely the antibody is to dissociate from the antigen. Clearly, the affinity of an antibody population is critical when the antigen is a toxin or virus and must be neutralized by rapid and firm combination with antibody. Antibodies formed soon after the injection of an antigen are generally of lower affinity for that antigen, whereas antibodies produced later have dramatically greater affinities (association constants 1000 times higher).

## Antibody valence and avidity

The valence of an antibody is the maximum number of antigenic determinants with which it can react. For example, IgG antibodies contain two Fab regions and can bind two molecules of the same antigen or two identical sites on the same particle, and thus have a valence of two. Valence is important for binding affinity, as having two or more binding sites for an antigen can dramatically increase the tightness of binding of the antibody to antigens on a bacteria or virus. This combined effect, avidity, results from synergy of the binding strengths of each binding site. Avidity is the firmness of association between a multideterminant antigen and the antibodies produced against it.

Determining the avidity of an antibody population is difficult, since it involves evaluating some function of the group interactions of a large number of antibodies with a large number of antigenic determinants. Even so, the importance of avidity can be demonstrated both mathematically and biologically. For example, as a result of working together (being on the same molecule) two IgG binding sites are 10–100 fold more effective at neutralizing a virus than two unassociated binding sites, and if the antibody has more binding sites, as in the case of IgM (Section D2), it may be a million times more effective (Figure 2). This can be visualized by considering antibodies with one or two binding sites for a particular antigenic determinant on a microorganism. The antibody with one

| | Fab | IgG | IgM |
|---|---|---|---|
| Binding sites (valence) | 1 | 2 | 10 |
| Relative binding avidity | 1 | 100 | 1 000 000 |

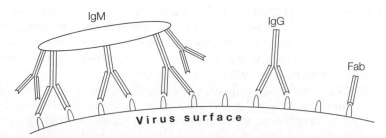

Figure 2. Avidity (tightness of binding) and antibody valence in viral neutralization.

site can bind to, but can also dissociate from, a determinant on the organism. When it comes off, it can diffuse away. However, the antibody with two sites can bind two identical determinants on the organism (each organism has many copies of each protein or carbohydrate). If one binding site dissociates, the other is probably still attached and permits the first site to reform its association with the organism. It therefore follows that the larger the number of binding sites per antibody molecule, the larger the number of bonds formed with an organism, and the less likely it will be to dissociate. Thus, an antibody with a poor intrinsic affinity for an antigenic determinant can, as a result of a large number of combining sites per molecule, be extremely effective in neutralizing a virus or complexing with a microorganism.

# D2 Antibody classes

## Key Notes

**Functional diversity**

Different antibody classes with different biological activities have evolved to deal with antigens (e.g., microbes) with different properties and which enter the body at different sites—through the skin, the gastrointestinal, respiratory, or genitourinary tracts.

**IgG**

IgG immunoglobulins, of which there are four different subclasses (IgG1, IgG2, IgG3, and IgG4), provide the bulk of immunity to most blood-borne infectious agents, and are the only antibody class to cross the placenta to provide humoral immunity to the infant.

**IgA**

IgA is a first line of defense against microbes entering through mucosal surfaces (the respiratory, gastrointestinal, and genitourinary tracts). Secretory (dimeric) IgA is synthesized locally by plasma cells, binds to the poly-Ig receptor on epithelial cells, and is transported through these cells to the lumenal surface where it is released with a portion of the poly-Ig receptor (secretory component, SC). This antibody prevents colonization of mucosal surfaces by pathogens and mediates their phagocytosis.

**IgM**

IgM is a four-polypeptide-chain antigen receptor on B cells and the first antibody produced in an immune response. In the circulation, IgM is composed of five four-chain units with 10 combining sites. It thus has high avidity for antigens and is very efficient per molecule in dealing with pathogens, especially early in the immune response before sufficient quantities of IgG have been produced.

**IgD**

This immunoglobulin functions primarily as an antigen receptor on B cells and is probably involved in regulating B-cell function when it encounters antigen.

**IgE**

Allergic reactions (Section K2) are predominantly associated with IgE. Antigen reintroduced into a previously sensitized individual binds to antigen-specific IgE on "armed" mast cells and triggers release of the pharmacologically active agents (e.g., histamine) involved in immediate hypersensitivity syndromes such as hay fever and asthma.

**Related topics**

| | |
|---|---|
| (C3) Mucosa-associated lymphoid tissues | (E4) Antibody responses in different tissues |
| (C5) Adaptive immunity at birth | (K2) IgE-mediated (type I) hypersensitivity: allergy |

## Functional diversity

Different microbes have different biological properties and can enter the body through different routes (the skin, the gastrointestinal tract, the respiratory tract, or the genitourinary tract). It is likely that the five different antibody classes (IgM, IgD, IgG, IgE, and IgA; Figure 1) and their subclasses have evolved at least partly to facilitate protection against microbes entering at the different sites and with different properties. There is some overlap in their function and in where they are produced, but generally there is a division of labor among the different antibody classes: for example, IgA is the most common antibody in mucosal secretions while IgM is mainly found in the plasma, and both are most effective in those locations.

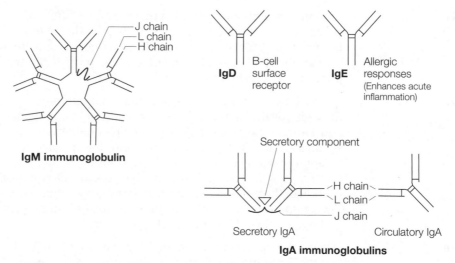

Figure 1. Chain structures of different classes of immunoglobulins.

## IgG

Immunoglobulins of the IgG class have a MW of 150 kDa and are found in both vascular and extravascular spaces as well as in secretions. IgG is the most abundant immunoglobulin in the blood (Table 1 in Section D1), provides the bulk of immunity to most blood-borne infectious agents, and is the only antibody class to cross the placenta to provide passive humoral immunity to the developing fetus and thus to the infant on its birth. The receptor that permits placental transfer from mother to fetus, FcRn, is also responsible for the long half-life of IgG antibodies that allow their protection by temporary sequestration in cells. IgG has two H chains (referred to as γ chains) with either two κ or two λ L chains. Furthermore, there are four different subclasses of IgG (designated IgG1, IgG2, IgG3, and IgG4), which have slightly different sequences in their H-chain constant regions and corresponding differences in their functional activities.

## IgA

This immunoglobulin is present in the serum as a 170 kDa, four-polypeptide (two L and two H) chain protein. More important, it is the major immunoglobulin present in external secretions such as colostrum, milk, and saliva, where it exists as a 420 kDa dimer (Figure 1). In addition to the κ or λ L chains and the IgA heavy chain (designated α), which distinguishes it from IgG or other antibody classes, secreted IgA also contains two other polypeptide chains—secretory component (SC) and J chain (joining chain). SC is part of the poly-Ig receptor involved in the transepithelial transport of exocrine IgA and stabilizes IgA against proteolytic degradation. The two four-chain units composing secretory IgA are held together by the J chain through disulfide bridges. Most IgA is synthesized locally by plasma cells in mammary and salivary glands, and along the respiratory, gastrointestinal, and genitourinary tracts (Section E4). It is then transported through epithelial cells to the lumen. This antibody is a first line of defense against microbial invaders at mucosal surfaces. Of the two subclasses of IgA, IgA2 rather than IgA1 is primarily found in mucosal secretions.

## IgM

IgM is the first antibody produced by, and expressed on the surface of, a B cell. It acts as an antigen receptor for these cells, and is also present as a soluble molecule in the blood. On the B-cell surface this molecule is expressed as a four-chain unit—two μ H chains and two L chains. In the blood, IgM is composed of five four-chain units held together by disulfide bridges at the carboxy-terminal end of the μ chains (Figure 1). J chain is also associated with IgM in the blood and initiates the polymerization of its subunits at the time of its secretion from a plasma cell. Because of its size (900 kDa), IgM is found primarily in the intravascular space (i.e., in the bloodstream). As IgM is the first antibody produced in an immune response, its efficiency in combining with antigen is of particular importance until sufficient quantities of IgG antibody have been synthesized. Although IgM antibodies usually have low-affinity binding sites for antigen, they have 10 combining sites per molecule which can synergize with each other on the same molecule when it binds to a microbe. Thus, the overall tightness of binding of an IgM molecule (avidity) to a microbe is quite high, making antibodies of this class very effective in removal of microbes.

## IgD

IgD is present in low quantities in the circulation (0.3 mg ml$^{-1}$ in adult serum). Its primary function is that of an antigen receptor on B lymphocytes (Figure 1), but it is probably also involved in regulating B-cell function when it encounters antigen. B cells thus can express both IgM and IgD and both are specific for the same antigen. When IgM and IgD expressed on a B cell interact with an antigen for which they are specific, the antigen is internalized, and processed and presented to helper T cells which trigger the B cells to proliferate and differentiate into plasma cells, thus initiating the development of a humoral immune response.

## IgE

IgE is present in the serum at very low levels (nanograms per milliliter), but plays a significant role in enhancing acute inflammation, in protection from infection by worms, and in allergic reactions (Sections B4, H2, and K2). Antibody-mediated allergy is predominantly associated with IgE. After stimulation of the development of IgE-producing plasma cells by an antigen, the IgE produced binds to receptors on mast cells which are specific for the Fc region of IgE. When antigen is *reintroduced* into an individual with such "armed" mast cells, it binds to the antigen-binding site of the IgE molecule on the mast cell, and as a result of this interaction, the mast cell is triggered to release pharmacologically active agents (e.g., histamine). IgE antibodies are thus important components of immediate hypersensitivity syndromes such as hay fever and asthma (Figure 1 in Section K2).

# D3 Generation of diversity

---

**Key Notes**

**Antibody genes**

The DNA encoding immunoglobulins is found in three unlinked gene groups—one group encodes κ L chains, one λ L chains, and one H chains. Each L-chain gene group has multiple different copies of V gene segments and J gene segments and one or more gene segments encoding the constant region. The H-chain gene group has multiple different copies of V, D, and J gene segments and one gene segment for each of the constant regions for the different antibody classes and subclasses.

**Gene rearrangement**

During its development, a single B cell randomly selects from its H-chain gene group one V, one D, and one J gene segment for rearrangement (translocation). It then selects from the κ or λ gene group one V and one J gene segment for translocation. These gene segments then recombine to create a gene (VJ) encoding a binding site for an L chain and a gene (VDJ) encoding a binding site for an H chain.

**Allelic exclusion**

After successful rearrangement of the Ig DNA segments, the cell is committed to the expression of a particular V region for its H chain and a particular V region for its L chain and there is active suppression, allelic exclusion, of other H- and L-chain V-region rearrangements. Each B cell and all of its progeny will therefore express and produce antibodies, all of which have exactly the same specificity.

**Synthesis and assembly of H and L chains**

After successful rearrangement of L- and H-chain DNA, primary L- and H-chain mRNAs are transcribed, and the RNA between the newly constructed V-region gene and the constant region gene spliced out. After translation, the L and H polypeptide chains combine in the endoplasmic reticulum (ER) to form an antibody molecule, which then becomes the antigen-specific receptor for that B cell. In plasma cells, the part of the mRNA encoding the H-chain transmembrane domain, which is important for its membrane expression on B cells, is spliced out and the antibody produced is secreted.

**Differential splicing and class switching**

A mature B cell expresses both IgM and IgD with the same specificity. This results from differential cleavage and splicing of the primary transcript to yield two mRNAs—one for an IgM H chain and the other for an IgD H chain—both of which are translated and expressed on the B-cell surface with L chain. B-cell class switch to IgG, IgA, or IgE requires signals induced by interaction of CD154 on T cells with CD40 on B cells. Cytokines produced by the T-helper cell are critical in determining to which Ig class switching occurs. IL-4 induces switch to IgE; TGFβ (with some involvement of IL-1, 5, and 10) induces switch to IgA; IFNγ induces switch to IgG1. Activation-induced deaminase (AID) is important in this process as it enzymatically creates susceptibility to cleavage and linkage of DNA in the switch regions where the DNA encoding the V-segment gene and the new constant region gene come together. These interactions induce translocation of the VDJ gene segment next to another C-region gene with the loss of intervening DNA. The primary transcript is then spliced to give an mRNA for the new H chain.

**Ways of creating diversity**

Antibody diversity, that is, the generation of antibodies with different specificities, is created at the DNA level as a result of multiple germ line V, D, and J gene segments for heavy chains, and V and J gene segments for light chains; random combination and imprecise joining of these gene segments (i.e., by removal or addition of nucleotides); and by subsequent somatic mutations in the resulting V regions. At the protein level, diversity is created as a result of random selection and pairing of L and H chains.

**B-cell development and selection**

In the pro-B-cell stage, gene segments encoding the variable part of the H chain rearrange and are transcribed with the μ constant region gene. Once IgM H chain appears in the cytoplasm, the pre-B-cell stage, gene segments that encode the variable region of the L chains rearrange. H and L chains combine and are expressed on the surface of the immature B cell. At this stage, B cells with high affinity for self antigens are induced to die by apoptosis (**negative selection**). Surviving B cells mature as indicated by their surface expression of IgD, traffic to secondary lymphoid organs, and are selected to expand by contact with their specific antigen. During the immune response, the overall affinity of antibodies for an antigen increases with time, partly because B cells expressing higher-affinity antibody compete most successfully for antigen and contribute a higher proportion to the antibody pool.

**Affinity maturation**

After class switch, VH and VL genes of activated B cells become more susceptible to somatic point mutations that may generate higher affinity antibodies, allowing these cells to compete most successfully for antigen. These cells clonally expand and differentiate into plasma cells that contribute to the overall antibody pool.

**Related topics**

(C1)  Lymphocytes
(E1)  The B-cell receptor complex, co-receptors, and signaling

(E3)  The cellular basis of the antibody response

## Antibody genes

Three unlinked gene groups encode immunoglobulins—one for κ chains, one for λ chains, and one for H chains, each on a different chromosome (Table 1). Within each of these gene groups on the chromosome there are multiple coding regions (**exons**) that recombine at the level of DNA to yield the DNA coding for a binding site. In a mature B cell or plasma cell, the DNA encoding the V region for the H chain of a specific antibody consists of a continuous uninterrupted nucleotide sequence. In contrast, the DNA in a germ line cell (or non-B cell) for this V region exists in distinct DNA segments, exons, separated from each other by regions of noncoding DNA (Figure 1). The exons encoding the V region of the H chain are V segment (encoding approximately the first 102 amino acids), D segment (encoding 2–4 amino acids), and J segment (encoding the remaining 14 or so amino acids in the V region). For L chains there are only V-segment (encoding the first approximately 95 amino acids) and J-segment (encoding the remaining 13 or so amino acids) exons. In each gene group, there are from 30–65 functional V-segment genes. The D and J regions are between the V and C regions on the chromosome and there are multiple different genes for each, but fewer in number than those encoding the V segment. Thus, DNA segments that ultimately encode the binding site of antibodies have to be moved over distances (translocated) on the chromosome to form a DNA sequence encoding the V region (**gene "rearrangement"**).

The DNA sequences encoding the C region of the L and H chains are 3′ to the V genes, but separated from them by unused J-segment genes and noncoding DNA. Furthermore, each gene group usually has one functional C gene segment for each class and subclass. Thus, the H-chain gene group has nine functional C-region genes, one each encoding μ,

### Table 1. Genes for human immunoglobulins

| Ig polypeptide | Chromosome |
|---|---|
| H chain | 14 |
| κ chain | 2 |
| λ chain | 22 |

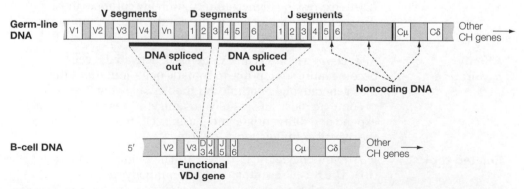

Figure 1. H-chain genes and translocation. In the germ line, and therefore in a cell destined to become a B cell, the H-chain gene loci contains many V-segment genes. In a developing B cell, one of these V segments recombines with one of many D segments, which has already recombined with one of several J segments, to produce a functional VDJ gene. In each B cell, the rearranged gene is transcribed, spliced, and translated into an H-chain protein.

δ, γ1, γ2, γ3, γ4, ε, α1, α2. For the L-chain gene groups, there is one gene segment encoding the C region of κ L chains, but four encoding λ L-chain C regions.

## Gene rearrangement

During its development, a single B cell randomly selects one V, one D, and one J (for H chains), and one V and one J (for L chains) for rearrangement (translocation). Gene segments encoding a portion of the V region are moved adjacent to other gene segments encoding the rest of the V region to create a gene segment encoding the entire V region, with the intervening DNA removed. Gene rearrangement in B cells requires the products of two recombination-activating genes, *RAG1* and *RAG2*, which are only expressed together in developing lymphocytes. These enzymes break and rejoin the DNA during translocation and are thus critical to the generation of diversity.

The H-chain gene group is the first to rearrange, initially moving one of several D-segment genes adjacent to one of several J-segment genes. This creates a DJ combination, which encodes the C-terminal part of the H-chain V region. A V-segment gene then rearranges to become contiguous with the DJ segment, creating a DNA sequence (VDJ) encoding a complete H-chain V region (Figure 1). This VDJ combination is 5′ to the group of H-chain C-region genes, of which the closest one encodes the μ chain. A primary mRNA transcript is then made from VDJ through the μ C-region gene, after which the intervening RNA between VDJ and the μ C-region gene is spliced out to create an mRNA for a complete μ H chain.

After the H chain has successfully completed its rearrangement, one of the V-region gene segments in either the λ or κ gene groups (but not both) is induced to translocate next to a J-segment gene to create a gene (VJ) encoding a complete L-chain V region (Figure 2). For κ chains, the DNA sequences encoding the C region of the L chains are 3′ to the V genes, but separated from them by unused J-segment genes and noncoding DNA (Figure 2a). For λ chains, because the J-segment genes are each associated with a different Cλ gene, translocation of a V gene segment to a J gene segment results in a V region next to a particular Cλ gene (e.g., Cλ2 as shown in Figure 2b). It is important to emphasize that in each B cell, only one of two L-chain gene groups will be used. A primary mRNA transcript is then made from VJ through the L-chain C-region gene, after which the intervening message between VJ and the C-region gene is spliced out to create an mRNA for a complete L chain. It is noteworthy that gene rearrangement on one chromosome for L-chain V regions (or H-chain V regions) does not always result in a functional L chain (or H chain). In this case, rearrangement of L-chain genes (or H-chain genes) is attempted on the other chromosome. As there are two L-chain groups (κ and λ) the cell, theoretically, gets four chances to produce a functional L chain (similarly, the cell gets two chances to produce a functional H chain).

## Allelic exclusion

After successful rearrangement of the Ig DNA segments, the cell is committed to the expression of a particular V region for its H chain and a particular V region for its L chain and *excludes* other H- and L-chain V-region rearrangements. This process is referred to as **allelic exclusion** and is unique to B- and T-cell antigen receptors. If an aberrant rearrangement occurs on the first chromosome the process will continue, that is, the process does not stop if the cell does not get it right the first time. The process stops, however, if the cell gets it right or runs out of chromosomes to rearrange. In fact, following successful VH gene rearrangement on one chromosome there is active suppression of further rearrangement of the other VH gene segments. Similarly, following successful VL gene

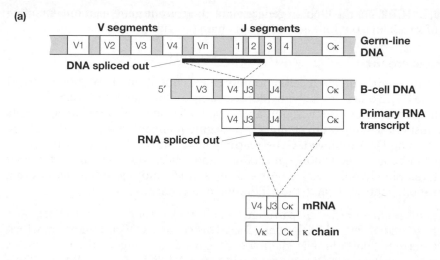

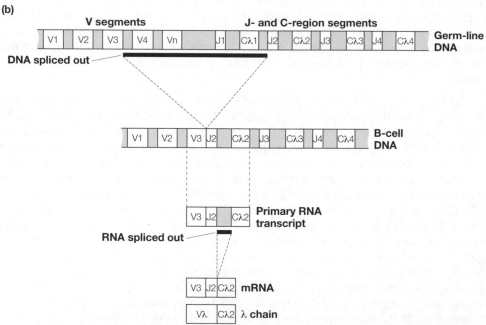

Figure 2. L-chain genes and translocation. During differentiation of a B cell, and after rearrangement of the H-chain genes, one of the two L-chain groups rearrange. In particular, either (a) a germ-line Vκ gene combines with a J-segment gene to form a VJ combination or (b) a germ-line Vλ gene combines with one of the J-segment Cλ gene combinations to form a VJ Cλ combination. The rearranged gene is then transcribed into a primary RNA transcript which then has the intervening noncoding sequences spliced out to form mRNA. This is then translated into light-chain protein.

rearrangement there is active suppression of further rearrangement of other VL gene segments.

Thus, each B cell makes L chains that all contain a V region encoded by the same VJ region sequence and H chains that all contain a V region encoded by the same VDJ sequence. Each B cell will therefore express antibodies on its surface all of which have exactly the same specificity. This cell and all of its progeny are committed to express and produce antibodies with these V regions.

## Synthesis and assembly of H and L chains

After successful rearrangement of both L- and H-chain DNA, L- and H-chain mRNA is produced and translated into L- and H-polypeptide chains that combine in the endoplasmic reticulum (ER) to form an antibody molecule, which is transported to the plasma membrane, along with thousands of identical antibody molecules, as the antigen-specific receptors for that B cell. Since the gene encoding the H chain also contains coding sequences for a transmembrane domain, the H chains produced contain a C-terminal amino acid sequence that anchors the antibody in the cell membrane. In plasma cells, the primary cells that produce antibodies for secretion, the part of the mRNA encoding the H-chain transmembrane domain important for its membrane expression on B cells is spliced out. Thus, the antibody produced by a plasma cell does not become associated with the membrane, but rather is secreted. Furthermore, as plasma cells do not express antibody on their cell surface they cannot respond to antigen.

## Differential splicing and class switching

As indicated above, the first antibody produced by a B cell is of the IgM class. Soon thereafter the B cell produces both an IgM and an IgD antibody, each having the same V regions and thus the same specificity. This is the result of the differential cleavage and splicing of the primary transcript. In particular, a primary transcript is made which includes information from the VDJ region through the Cδ region (Figure 3). This transcript is differentially

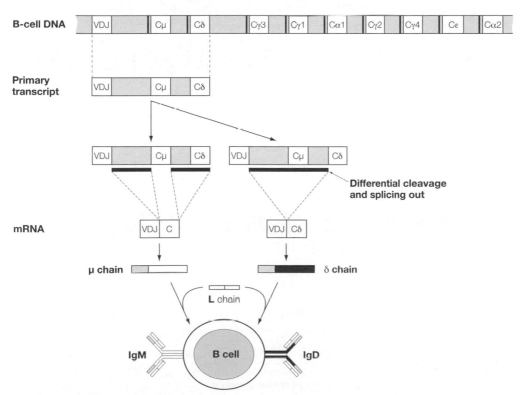

Figure 3. Expression of IgM and IgD on a mature B cell.

spliced to yield two mRNAs—one for an IgM H chain and the other for an IgD H chain. In a mature B cell both are translated and expressed on the B cell surface with L chain.

B cells expressing IgM and IgD on their surface are capable of switching to other H-chain classes (IgG, IgA, or IgE). This isotype (class) switching requires stimulation of the B cell by T-helper cells and in particular requires binding of the CD40 ligand (CD154) on T cells to CD40 on B cells. In addition, the cytokines produced by the T-helper cell determine to which constant region gene class switching occurs. Th2 cells producing IL-4 induce B cells to class switch to IgE; TGFβ (with involvement of IL-1, 5, and 10) induces switch to IgA; IFNγ induces switch to IgG1 (Figure 4). These signals induce translocation of VDJ, and its insertion 5′ to another constant-region gene (Figure 5). Class switch is guided by

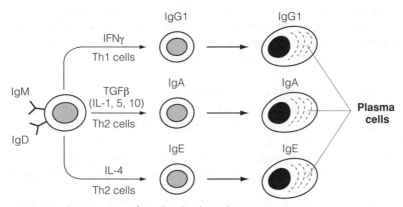

Figure 4. Generation of antibody class diversity.

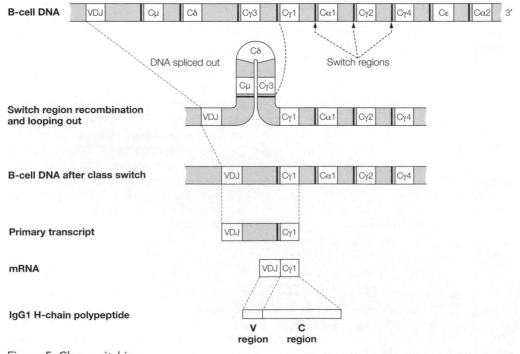

Figure 5. Class switching.

repetitive DNA sequences 5′ to each of the C-region genes and occurs when these **switch regions** recombine. Activation-induced deaminase (AID) is important in this process as it enzymatically creates susceptibility to cleavage and relinking of DNA in the switch regions where the DNA encoding the V-segment gene and the new constant-region gene come together. The intervening DNA is cut out and the resulting DNA on the rearranged chromosome of the B cell that has class switched, and in plasma cells derived from this B cell, no longer contains Cμ, Cδ, or other intervening H-chain C-region genes. A primary transcript is made and the RNA between the VDJ coding region and the new H-chain coding region is spliced out to give an mRNA for the new H chain.

## Ways of creating diversity

Ig diversity (the generation of antibodies with different specificities) is created by several **antigen-independent** mechanisms. In addition, Ig genes in B cells activated by antigen and having received T-cell help undergo mutational events that may increase the affinity of the antibody produced by the B cell. Overall, diversity is generated by:

**Antigen-independent events:**

● At the DNA level as a result of multiple germ line V, D, and J heavy and V and J light chain genes

● At the DNA level as a result of random combination of V, D, and J segments or V and J segments

● At the DNA level as a result of imprecise joining of V, D, and J segments due to nucleotide inclusion or deletion

● At the protein level as a result of random selection and pairing of different combinations of L- and H-chain V regions in different B cells

**Antigen-dependent events:**

● At the DNA level as a result of somatic mutation in the V region, which may create higher-affinity antibody-binding sites

Although rearrangement of the gene segments that will make up the V-region genes occurs in an ordered fashion, they are chosen at random in each developing B cell. As these events occur in a vast number of cells, the result is that millions of B cells, each with a different antigen specificity, are generated. Additional diversity is created during recombination of V and J (L chain) and V, D, and J (H chain) gene segments due to imprecise joining of the different gene segments making up the V region. For example, although translocation of a V gene segment to a J gene segment could occur with all three nucleotides of the last codon of the V segment joining with all three nucleotides of the first codon of the J segment, it is also possible that one or two nucleotides at the 3′ end of the V segment could replace the first one or two nucleotides of the J segment. Such a difference in the position at which recombination occurs can change the amino acid sequence in the antigen-binding area of the resulting V region of the antibody, and thus change its specificity. Furthermore, after antigen stimulation of the B cell, the DNA of its L- and H-chain V regions becomes particularly susceptible to somatic mutation and undergoes affinity maturation (see below).

Diversity is also generated as a result of the fact that any L chain can interact with any H chain to create a unique binding site. Thus, for example, an L chain with a particular VJ combination for its binding site could be produced by many different B cells and interact with the different H chains (i.e., different in their VH region) generated in each of these B cells to create many different specificities.

In sum, almost unlimited diversity is created from a limited number of V-region gene segments. This diversity almost certainly exceeds the amount of diversity needed to bind the immunogens of microbes. Thus, the vast majority of the different B cells generated will never encounter antigen to which they can bind, and will not be stimulated to further development. And yet, such apparent wastefulness is justified by the fact that this mechanism of creation of diversity ensures that there are B cells, and thus antibodies, reactive with virtually all antigens, past, present, or future. When an antigen to which this antibody binds is encountered, the B cell is triggered to divide and to give rise to a clone of cells, each one of which makes, at least initially, the originally displayed antibody molecule (clonal selection, Sections A3 and E3).

## B-cell development and selection

B cells go through several stages in their path toward development of memory or plasma cells. Gene segments encoding the variable parts of the V regions of antibodies rearrange during the **pro-B-cell stage** (Figure 6). Since rearrangement occurs in millions of different ways in these developing cells, many B cells, each with a different specificity, are generated. This generation of diversity occurs in the absence of foreign protein and yields large numbers of mature B cells, of which at least some have specificity for each foreign substance or microbe. The first genes to rearrange encode the variable part of the H chain of the antibody and, together with the genes of the constant part of the molecule (initially genes which code for the µ H chain), are transcribed into H chains. These are expressed on the B-cell surface in association with a surrogate L chain, an event that may be important for further B-cell development. At this **pre-B-cell stage**, the genes that code for the variable region of the L chains rearrange and a light chain is produced. The transcribed H and L chains combine, giving rise to a functional IgM antigen receptor, which is expressed on the surface of this **immature B cell**. It is during this stage that B cells with high affinity for self antigens are induced to die by apoptosis (negative selection). As with T cells in the thymus, the majority of the B cells die during development from production of antigen receptors that cannot be assembled or which are specific for self antigens (Sections G2 and G3).

During an antibody response to an antigen, the overall affinity of the antibodies produced increases with time. That is, antibodies produced in the secondary response have higher affinity for (tighter binding to) the antigen than those produced in the primary response. This is partly due to clonal selection and the presence of significantly more antigen-binding B cells at the time of the secondary response than during the primary response. If the quantity of antigen is limited, that is, insufficient to stimulate all B cells that could bind the antigen, B cells with the highest-affinity antigen receptors will compete most successfully for the antigen and give rise to plasma cells making their higher-affinity antibody. The result is that the affinity of the total pool of antibody is increased and is thus more efficient at effector functions than the antibody produced in the primary response.

## Affinity maturation

After B-cell class switch (stimulated by antigen and T cells) to IgG, IgA, or IgE, the DNA of the L- and H-chain V regions becomes particularly susceptible to somatic mutation. This results in changes in the nucleotides of the DNA and thus corresponding changes in the amino acid sequence of the V regions of the antibody expressed by the B cell. As a result, the B cell may have a different specificity and not bind to, or be stimulated by, the original antigen. However, it often happens that at least some mutations result in

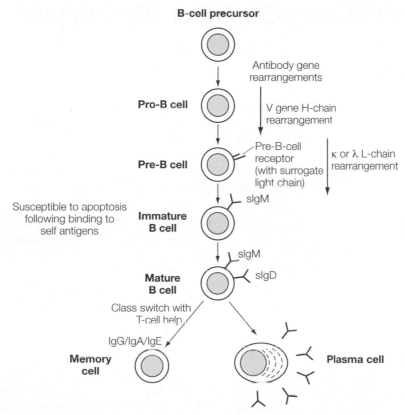

Figure 6. Life history of a B cell. B-cell precursors develop into pro-B cells which begin to rearrange their H-chain V genes. During the pre-B-cell stage the translated H chain peptide assembles with "surrogate" light chain and is expressed on the B-cell surface while κ or λ L-chain genes rearrange and are transcribed and translated into L chains that replace the "surrogate" light chain resulting in expression of surface IgM—the cell's functional antigen receptor. This immature B cell is susceptible to apoptosis/anergy on contact with self antigen. Mature B cells express surface IgD in addition to IgM and migrate to the secondary lymphoid organs and tissues where they respond to foreign antigens by proliferation and development into memory and plasma cells.

amino acid changes that increase the tightness of binding of the antibody on the B cell to its antigen. These B cells will compete more efficiently for antigen than the original B cell, and will differentiate into plasma cells producing a higher-affinity antibody (**affinity maturation**), resulting in an overall increase in the affinity of the antibody population to that antigen. Typical antibodies have binding constants of $10^6$–$10^7$ $M^{-1}$. After successive immunization with limiting antigen they are usually $10^8$–$10^9$ $M^{-1}$ but may be as high as $10^{12}$ $M^{-1}$.

# D4 Allotypes and idiotypes

| Key Notes | |
|---|---|
| **Allotypes** | These are genetic markers on immunoglobulins (Igs) that segregate within the species. If Ig expressing a particular allotype is injected into an individual whose Igs do not express that allotype, an immune response could develop against the allotype. Like blood types, they are inherited in Mendelian fashion but are usually of no functional consequence. |
| **Idiotypes** | These are unique antigenic determinants associated with antigen-binding sites of antibodies and are the result of the different amino acid sequences which determine their specificities. |
| **Related topics** | (G5)  Regulation by T cells and antibody    (M2) Transplantation antigens |

## Allotypes

In addition to class and subclass categories, an immunoglobulin (Ig) can be defined by the presence of genetic markers termed allotypes. These markers are different in different individuals and are thus immunogenic when injected into individuals whose Ig lacks the allotype. Like the ABO blood group antigens, they are determinants which segregate within a species (the Ig of some members of the species have them, others do not). Allotypes are normally the result of small amino acid differences in Ig L- or H-chain constant regions. For example, the Km (*Inv*) marker is an allotype of human κ L chains and is the result of a leucine versus valine difference at position 191. The *Gm* markers are allotypes associated with the IgG H chains. Allotypes are inherited in a strictly Mendelian fashion, and usually have no significance to the function of the antibody molecule, nor are they so immunogenic as to compromise their use in immunotherapy.

## Idiotypes

Antigenic determinants associated with the binding site of an antibody molecule are called idiotypes and are unique to all antibodies produced by the same clone of B cells. That is, although all antibodies have idiotypic determinants, these determinants are different for all antibodies not derived from the same clone of B cells. Thus, the number of different idiotypes in an individual is at least as numerous as the number of specificities. In fact, one's own idiotypes may be recognized by one's own immune system. Thus, idiotypes are immunogenic and can induce immune responses. To that end monoclonal antibodies (Section D5) to idiotypes are being explored as a therapeutic approach to B-cell tumors, all of which would express idiotypes on their cell surface.

# D5 Monoclonal antibodies

---

**Key Notes**

| | |
|---|---|
| **Monoclonal antibodies** | Standardized procedures involving fusion of an immortal cell (a myeloma tumor cell) with a specific predetermined antibody-producing B cell are used to create hybridoma cells producing monospecific and monoclonal antibodies (mAbs). These mAbs are standard research reagents and many have significant clinical utility. |
| **Humanization and chimerization of mAbs** | Most mAbs developed have been mouse, and although useful as research and diagnostic tools, they are not ideal for therapy because of their immunogenicity in humans and their inability to react effectively with effector cells and molecules of the human immune system. This has been dealt with by humanizing these murine Abs or by making fully human mAbs. |
| **Fv libraries** | By randomly fusing heavy (H) and light (L) chain variable (V) region genes from B cells, Fv libraries containing a vast number of binding specificities can be generated and used as a source for creation of specific mAbs. |
| **Related topics** | (C1) Lymphocytes  (N4) Immunodiagnosis  (N6) Immunotherapy of tumors with antibodies |

---

## Monoclonal antibodies

In 1975, Kohler and Milstein developed a procedure (for which they received the Nobel Prize) to create cell lines producing predetermined, monospecific, and monoclonal antibodies (mAbs). This procedure has been standardized and applied on a massive scale to the preparation of antibodies useful to many research and clinical efforts. The basic technology involves fusion of an immortal cell (a myeloma tumor cell) with a specific predetermined antibody-producing B cell from immunized animals or humans (Figure 1). The resulting hybridoma cell is immortal and synthesizes homogeneous, specific mAb, which can be made in large quantities. Thus, mAbs have become standard research reagents and have extensive clinical applications.

## Humanization and chimerization of mAbs

Most mAbs have been developed in mice, and although useful as research and diagnostic tools, they have not been ideal therapeutic reagents at least partly because of their immunogenicity in humans. That is, a murine Ab introduced into a patient will be recognized as foreign by the patient's immune system and a human anti-mouse Ab (HAMA) response will develop that compromises the therapeutic utility of the Ab. In addition, mouse mAbs do not interact as well as human mAbs with effector cells (e.g., Fc receptors on phagocytes) and molecules (e.g., complement) of the human immune system. This has been dealt with in two basic ways.

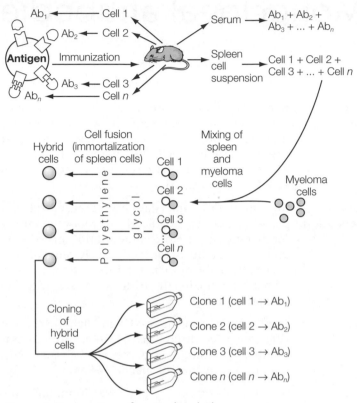

Figure 1. Preparation of monoclonal Abs.

## Humanized antibodies

Murine mAbs can be genetically modified to be more human (Figure 2). In particular the constant regions of the murine IgG heavy (H) and light (L) chains can be replaced at the DNA level with the constant-region genes of human IgG1 H and L chains to create a **chimeric Ab** where only the variable (V) regions are murine. This significantly decreases but does not eliminate the immunogenicity of the Ab. Another approach involves sequencing the mouse Ab VH and VL regions and then inserting the DNA sequences of the antigen binding (hypervariable) regions of these chains into human IgG VH- and VL-chain genes. The resulting "humanized" Ab is 95% human with only the binding regions being murine.

## Fully human mAbs

Various approaches have been used to create fully human mAbs; the most obvious of which, fusing human B cells with myeloma cells, has not been very successful. The most successful approach involves developing human Ab mice by replacing genes for mouse immunoglobulins with genes for human immunoglobulins. Thus, when the mouse is immunized it makes fully human Abs against the antigen, and the B cells making these Abs can be fused with myeloma cells to generate hybridomas making totally human mAbs. Many human mAbs have been made with this technology and some are available for therapy.

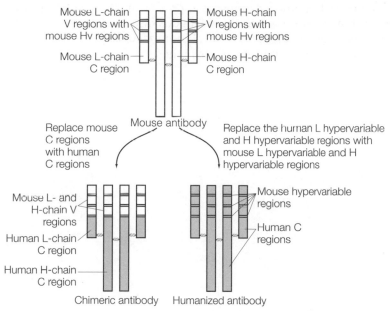

Figure 2. Humanizing and chimerizing mouse monoclonal antibodies. Chimeric mAbs are created by replacing the murine genes for the constant regions of the L and H chains with the corresponding human constant-region genes. Humanized mAbs are created by inserting the gene sequences for each of the antigen-binding hypervariable (Hv) regions of the mouse antibody into the corresponding place in the human antibody genes for the L and H chains.

## Fv libraries

Another way to develop human mAbs initially involves preparing mRNA for the VH and VL regions from a large number of human B cells. From this mRNA, cDNA for each H-chain V region is prepared and joined to the cDNA for each L-chain V region (Figure 3) to create genes that, overall, encode a vast library of different antigen-combining sites (**Fvs**) from which Fvs specific for different antigens can be selected. For example, Fvs can be cloned into bacteriophages (viruses that infect bacteria) and selected for their specificity. Thus, Fvs can be expressed in a replicating bioform and used as a source from which specific

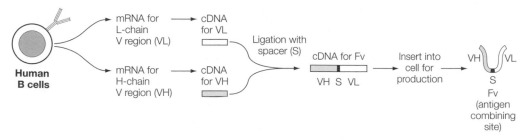

Figure 3. Fv preparation. Human B-cell mRNA for the V regions of L and H chains is prepared using the polymerase chain reaction. From this mRNA, cDNA for each H-chain V region is prepared and joined, with a spacer, to the cDNA for each L-chain V region. This yields a gene encoding the antigen-binding region of the antibody, which is then inserted into a cell for production of a protein, Fv, that is the combining site of an antibody. As this is done using many B cells, a very large number of different combining sites can be created.

mAbs can be created. Another promising approach involves isolating B-cell antibody genes from patients recovering from an infectious disease. These genes can be inserted into cells to produce large quantities of fully human mAbs against a pathogen and can be used for therapy or passive immunity to the pathogen.

# D6 Antigen–antibody complexes (immune complexes)

## Key Notes

**Immune complexes *in vitro***
Combination of antibody (Ab) with a multideterminant antigen (Ag) results in a lattice of alternating molecules of Ag and Ab, which grows until large precipitating aggregates are formed (equivalence). In Ab excess or in Ag excess, there is less lattice formation resulting in more soluble complexes.

**Immune complexes *in vivo***
Introduction of Ag *in vivo* results in an immune response in which there is initially Ag excess. Within days, as Ab is produced, equivalence is reached and the resulting immune complexes are removed by phagocytic cells. After Ag removal, B-cell stimulation stops, and the Ab concentration in the serum decreases as a result of normal catabolism.

**Immune complexes and tissue damage**
Persistence of Ag (microbial or self) may result in continual formation of immune complexes that, with an "overwhelmed" phagocytic system, are deposited in tissues resulting in damage (type III hypersensitivity) mediated mainly by complement and neutrophils.

**Precipitation assays**
Combination of Ab with Ag resulting in lattice formation and precipitation is the basis for qualitative and quantitative assays, including immunoelectrophoresis, for Ag or Ab.

**Agglutination assays**
The interaction of surface Ags on insoluble particles (e.g., cells) with specific Ab to these Ags results in agglutination of the particles. Agglutination can be used to determine blood types, the presence of Ab to bacteria in serum as an indication of previous infection, and in the Coombs test to identify autoantibodies to erythrocytes.

**Related topics**
(B4) Innate immunity and inflammation
(D1) Antibody structure
(D2) Antibody classes
(K3) IgG- and IgM-mediated (type II) hypersensitivity
(K4) Immune-complex-mediated (type III) hypersensitivity
(M2) Transplantation antigens

## Immune complexes *in vitro*

Immunogens have more than one antigenic determinant per molecule (i.e., are multi-determinant). Immunization with antigen therefore results in many antibody populations, each directed toward different determinants on the protein. Since one molecule of Ab (IgG) can react with two molecules of Ag, and one molecule of Ag can react with many molecules of Ab, a lattice or framework consisting of alternating molecules of Ag and Ab can be produced that precipitates. The extent to which a lattice forms depends on the relative amounts of Ag and Ab present (Figure 1). As the amount of Ag added increases, the amount of precipitate and Ab in the precipitate increases, until a maximum is reached, and then decreases with further addition of Ag. When there is both sufficient Ag and sufficient Ab, the combination of Ag and Ab proceeds until large aggregates are formed, which are insoluble and precipitate (**equivalence**). However, in **Ab excess** or in **Ag excess**, less lattice formation occurs and more soluble complexes are formed.

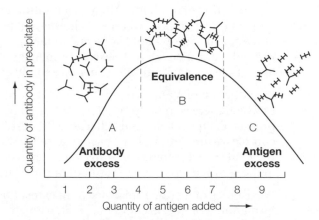

Figure 1. Immune complex formation and precipitation. The same amount of Ab to a protein was added to each of a series of tubes (1–9), followed by the addition of increasing amounts of the Ag to each successive tube. (A) The zone of Ab excess; (B) the zone of equivalence in which all of the Ag and Ab are incorporated into a precipitate; and (C) the zone of Ag excess.

## Immune complexes *in vivo*

These reactions occur *in vivo* during an immune response. Initially, there is Ag excess as no Ab to the Ag is present at the time of first contact with the Ag. Within days, however, plasma cells develop, producing Ab to the Ag which complexes with it (Ag excess). As more Ab is produced, equivalence is reached resulting in large Ag–Ab complexes which are removed by phagocytic cells through interaction with their Fc and complement receptors. Plasma cells continue to produce Ab during their life, increasing the Ab concentration in the serum (Ab excess). However, once Ag has been removed, no further restimulation of B cells occurs and no more plasma cells develop (Section G4). Thus, the Ab concentration in the serum begins to decrease as a result of normal catabolism.

## Immune complexes and tissue damage

If the Ag persists (e.g., with some infectious organisms such as *Streptococcus*) or is self Ag, immune complexes are continually formed and may not be readily removed due to an "overwhelmed" phagocytic system. This can lead to the deposition of immune complexes in tissues and their damage (type III hypersensitivity, Section K4). The complexes activate complement and induce an acute inflammatory response (Section B4). Also, direct interaction of the immune complexes with Fc and complement receptors on the neutrophils causes the release of proteolytic enzymes that damage surrounding tissues.

## Precipitation assays

As previously described, when there is both sufficient Ag and sufficient Ab, the combination of Ag and Ab proceeds until large aggregates are formed which are insoluble in water and precipitate (equivalence). The extent to which a lattice forms depends on the relative amounts of Ag and Ab present. Lattice formation and precipitation are the basis for several qualitative and quantitative assays for Ag or Ab. These assays are done in semisolid gels into which holes are cut for Ag and/or for Ab and diffusion occurs until Ag and Ab are at equivalence and precipitate.

As an example, in **immunoelectrophoresis**, Ags (e.g., serum) are placed in a well cut in a gel (without Ab) and electrophoresed, after which a trough is cut in the gel into which Abs (e.g., horse anti-human serum) are placed. The Abs diffuse laterally to meet diffusing Ag, and lattice formation and precipitation occur permitting determination of the nature of the Ags (Figure 2).

## Agglutination assays

Agglutination involves the interaction of surface Ags on **insoluble particles** (e.g., cells) and specific Ab to these Ags (Figure 3). Ab thus links together (agglutinates) insoluble particles. Much smaller amounts of Ab suffice to produce agglutination than are needed

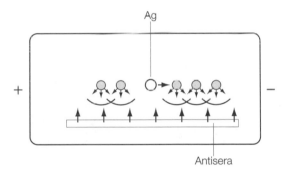

Figure 2. Identification of antigens using gel electrophoresis. Ag (e.g., serum) is placed in a well cut in a gel and subjected to a voltage gradient which causes the various antigens to migrate different distances through the gel dependent on their charge. After electrophoresis, a trough is cut in the gel into which antibodies (e.g., horse anti-human serum) are placed. The antibodies diffuse laterally from the trough until they meet Ag diffusing from its location after electrophoresis. Again, lattice formation and precipitation occur and, based on immunoelectrophoresis of defined standards, the identity of the Ag can be determined.

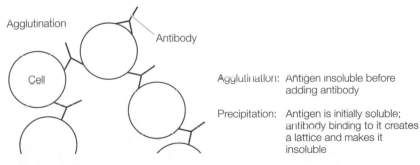

Agglutination:   Antigen insoluble before adding antibody

Precipitation:   Antigen is initially soluble; antibody binding to it creates a lattice and makes it insoluble

Figure 3. Agglutination.

for precipitation. For this reason, agglutination is useful in determining blood group types or if Ab to bacteria is present in blood as an indication of infection with these bacteria. Since IgM has 10 binding sites, whereas IgG has two, IgM is much more efficient at agglutinating particles or cells.

In some instances, such as when an autoantibody has been produced against a given cell type, the cells will have human Ab bonded to them, and thus can be identified in what is known as a **Coombs test** by a second Ab (an Ab to human immunoglobulin) which will cause agglutination of the cells. In an **indirect Coombs test**, the presence of circulating Ab to a cell surface Ag is demonstrated by adding the patient's serum to test cells (e.g., erythrocytes) followed by addition of Ab to human Ab.

# **D7** Immunoassay

---

## Key Notes

**Antibodies and assays**

A variety of assays have been developed that provide specific measurement of Ag or Ab, both of which may be of considerable research and clinical relevance. Ab to a microbe in the serum of a patient demonstrates infection by the organism. Abs also permit localization, quantitation, and characterization of Ags.

**ELISA/RIA/IHC**

The presence and concentration of a specific Ag or of an Ab to a specific Ag can be determined by radioimmunoassay (RIA) or enzyme-linked immunosorbent assay (ELISA). Ag attached to a solid surface captures the Ab, with which it reacts and is quantitated using a labeled second Ab reactive to the first. These assays also permit measurement of the concentration and isotype of Abs specific for a given Ag, such as those reactive with a microbe. Immunohistochemistry (IHC) involves the use of antibodies to examine tissues for abnormal cells (e.g. cancer cells).

**Immunofluorescence and flow cytometry**

Using a fluorescence microscope and Abs labeled with fluorescent molecules, tissue sections can be examined for cells expressing particular Ags (e.g., those which are tumor-associated), including qualitative and quantitative evaluation of several different cell-associated molecules at the same time. Flow cytometers rapidly analyze large numbers of cells in suspension, providing a molecular fingerprint of the cells, while cell sorters separate cell subpopulations for study or therapy.

**Immunoblotting**

Immunoblotting is used to assay for the presence of molecules in a mixture. Western blot analysis involves separating molecules by gel electrophoresis, transferring them to another matrix and detecting the molecule of interest using ELISA or RIA. This assay can be used to confirm the presence of Abs to infectious agents (e.g., HIV) in serum and to analyze products of single cells (e.g., cytokines).

**Affinity purification of Ag and Ab**

Ag coupled to an insoluble matrix (e.g., agarose) specifically binds its Ab, which can then be eluted from the Ag yielding relatively pure Ab in one step. Similarly, Ab coupled to an insoluble matrix permits purification of Ag.

**Related topics**

| (K3) | IgG- and IgM- mediated (type II) hypersensitivity | (L5) | Diagnosis and treatment of autoimmune disease |
| | | (N4) | Immunodiagnosis |

## Antibodies and assays

Methods for measuring antigen–antibody reactions have been well established and include those that have direct biological relevance (Table 1). Other assays have been developed which provide specific qualitative and quantitative measurement of Ag or Ab for both research and diagnostic purposes. Since the immune system recognizes and remembers virtually all Ags that are introduced into an individual, assays which demonstrate the presence of Ab to an organism in the serum of a patient have become a standard way of determining that the patient has had contact with, was infected by, the organism (e.g., the presence of Ab to HIV in the serum of a patient usually means that the patient has been infected with HIV). Alternatively, Abs with defined specificity (e.g., to Ags associated with cancer cells) can be used to determine the presence of disease in a patient. Abs are also extremely important tools in molecular and cellular research as they permit the localization and characterization of Ags.

### Table 1. Effects of combination of antigen and antibody

| | |
|---|---|
| Agglutination | Antigenic particle + specific Ab results in aggregation of particles |
| Precipitation | Soluble Ag + specific Ab results in lattice formation and precipitation |
| C activation | Ag in solution or on particle + specific Ab results in activation of C |
| Cytolysis | Cell + anti-cell Ab + C may result in lysis of the cell |
| Opsonization | Antigenic particle + Ab + C enhances phagocytosis by Mo, MØ, PMNs |
| Neutralization | Toxins, viruses, enzymes, etc. + specific Abs may result in their inactivation |

C, Complement; Mo, monocytes; MØ, macrophages; Ab, antibody; PMNs, polymorphonuclear cells.

## ELISA/RIA/IHC

The presence of Ab to a particular Ag in the serum of a patient can be determined using very sensitive radioimmunoassays (RIA) or enzyme-linked immunosorbent assays (ELISA). Such assays (Figure 1) are of particular value in demonstrating Ab to Ags of infectious agents, for example virus, bacteria, and so forth. The presence of an Ab of a particular isotype (e.g., IgE vs. IgG) can also be determined using a modification of these assays. The radioallergosorbent test (RAST) uses as detecting ligand a radiolabeled Ab to human IgE and permits the measurement of specific IgE Ab to an allergen. ELISA and RIA also provide very specific and sensitive measurement of toxins, drugs, hormones, pesticides, and so forth, not only in serum, but also in water, foods, and other consumer products. Based on these procedures, assays for nearly any Ag or Ab can be readily developed.

Immunohistochemistry (IHC) permits the identification of abnormal cells in tissue samples from patients, especially those who may have cancer. In particular, Abs to specific Ags (e.g., those associated with tumor cells) are added to tissue sections (primary Ab) and after removing unbound Ab a second reagent, usually Ab which is conjugated to a

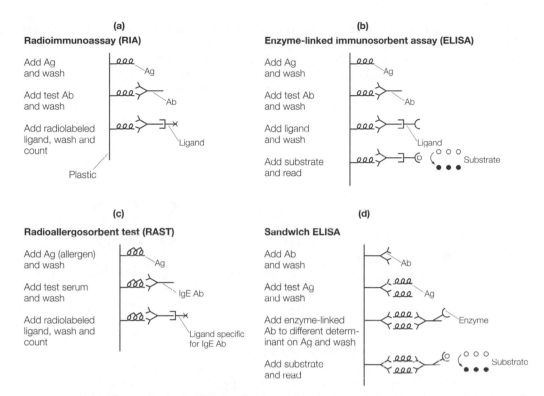

Figure 1. (a) Radioimmunoassay (RIA). Antigen (Ag) is incubated in a microtiter well and small quantities are adsorbed. Free Ag is washed away. Test antibody (Ab) is added, which may bind to the Ag, and unbound Ab washed away. Ab remaining bound to the Ag is detected by a radiolabeled **ligand** (e.g., an Ab specific for the isotype of the test Ab, or staphylococcal protein A which binds to the Fc region of IgG). (b) Enzyme-linked immunosorbent assay (ELISA). This is similar to RIA except that the **ligand** (i.e., the Ab that binds the test Ag) is covalently coupled to an enzyme such as peroxidase. After free ligand is washed away the bound ligand is detected by the addition of substrate which is acted on by the enzyme to yield a colored and detectable end product. (c) Radioallergosorbent test (RAST). This measures Ag-specific IgE in a RIA where the ligand is a labeled anti-IgE Ab. (d) The sandwich assay is done as above except that **Ab** to an antigen is first adsorbed to a microtiter well and unbound Ab washed away. A potential source of Ag is added and what is not bound (captured) by the Ab is washed away. An enzyme-linked Ab to a different determinant on the Ag is then added, followed by washing. The presence of Ag is detected by the change in color of added substrate.

reporter molecule (e.g., an enzyme such as alkaline phosphatase), is added. A substrate that changes color (e.g., 3,3′-diaminobenzidine: DAB) is then added and evaluated by microscopy to identify cells and tissues that are positive. If the reporter molecule is fluorescent, the tissues are examined by a microscope equipped with ultraviolet illumination. In direct IHC the primary Ab is labeled and a secondary Ab is not necessary.

## Immunofluorescence and flow cytometry

The presence and quantitation of an Ag on a cell is usually done using Abs to which a fluorescent marker has been covalently attached. In most cases a mAb is used as it is highly specific for a particular molecule and a particular epitope on that molecule. This type of assay can be done using an Ab to the Ag which is directly fluorescently labeled (**direct** immunofluorescence) or by first incubating the unlabeled Ag-specific Ab with the cells (e.g., a mouse mAb to human T cells) and then, after washing away unbound Ab, adding a second fluorescently labeled Ab that reacts with the first Ab (e.g., a goat Ab to mouse immunoglobulin). This **indirect** immunofluorescent assay (Figure 2) has two advantages, it has higher sensitivity and requires labeling of only one Ab, the second Ab, because, in the example given, it can detect (react with) mouse Ab to other antigens.

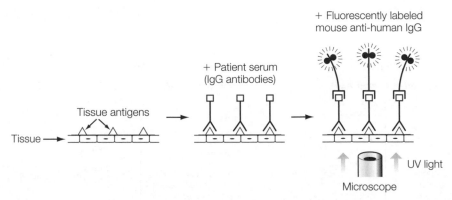

Figure 2. Indirect immunofluorescence assay for autoantibodies. Patient serum is added to tissue sections and the autoantibodies bind to particular autoantigen(s). After washing, fluorescent antibodies to human IgG are added and viewed under a fluorescence microscope. Fluorescence shows where the human antibodies have bound to the tissue autoantigens.

Fluorescent Abs to cell-surface molecules (e.g., those which are tumor-associated) are very useful in examining tissue sections for cells expressing the Ag. This assay is done by incubating the tissue section with the labeled Ab (for direct immunofluorescence) or unlabeled Ab followed by labeled second Ab (indirect immunofluoresence) and then examining the tissue section using a fluorescence microscope. These microscopes irradiate the tissue with a wavelength of light that excites the fluorescent label on the Ab to emit light at a different wavelength. This emitted light can be directly visualized, photographed, and even quantitated. Moreover, it is possible to analyze a tissue sample using several different Abs at the same time, as each Ab could be labeled with a different fluorescent molecule each of which emits light at a wavelength distinct from the others. It is also possible to look for intracellular molecules (e.g., cytokines) by first permeabilizing the cells and then doing the staining and fluorescence microscopy. Using this approach one can develop a molecular fingerprint of the tissue cells.

Although fluorescence microscopy can be, and is, applied to the analysis of **single cell suspensions**, another technologically sophisticated approach, **flow cytometry**, is most often used. This assay uses the same basic staining procedures as described for fluorescence microscopy, followed by automated quantitation of the amount of fluorescence associated with individual cells (Figure 3). In particular, the suspension of stained cells is fed to the flow cytometer, which disperses the cells so they then pass single file through a focused laser beam which excites any fluorescent label associated with the cells. Those stained by the fluorescent Ab emit light that is detected and quantitated by optical sensors and the intensity of fluorescence is plotted in histogram form by a computer. This machine can analyze 1000 cells per second and provide quantitative data on the number

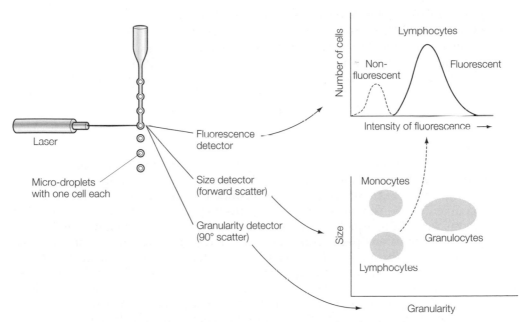

Figure 3. Flow cytometry. After labeling with fluorescent antibody, cells are passed one at a time through a laser beam. Photodetectors measure the amount of fluorescence and plot it as a histogram showing the proportion of nonfluorescent (unstained) and fluorescent (stained) cells. Other detectors simultaneously measure scattered laser light, which is used to generate a "dot blot" in which lymphocytes, monocytes, and granulocytes can be discriminated.

of molecules of a particular kind on each cell. It can also analyze mixtures of cells and provide data on their size and granularity in addition to their expression of specific molecules. Some versions of this machine (**fluorescence-activated cell sorter**) are also able to separate out cells expressing selected amounts of particular Ags into a separate tube for further analysis, culture, or clinical use.

## Immunoblotting

It is possible to combine various separation and detection procedures for identification and analysis of Ags and for evaluating the expression of molecules by single cells. Western blot analysis involves separating Ags by polyacrylamide gel electrophoresis (PAGE) in the presence of sodium dodecyl sulfate (SDS), which results in separation of molecules on the basis of size. These molecules are then transferred to another matrix (e.g., nitrocellulose) to form a pattern on the matrix identical to that on the gel. Enzyme-linked Ab to the molecule of interest is then added, the unbound Ab washed off and substrate added (see ELISA) for visualization. This assay permits specific identification of proteins in a mixture and can be used to confirm the presence of Abs to certain infectious agents (e.g., HIV) in the serum of patients.

Immunoblotting can also be used to assay for the presence of molecules in a mixture as described for the sandwich ELISA. This has now been extended for analysis of products of single cells. For example, to assay for production of a cytokine, Ab to the cytokine is coated onto the nitrocellulose "floor" of a special culture well (see sandwich ELISA), the unbound Ab is washed off, and cells are then plated on top of this Ab. After incubation, an enzyme-linked Ab to a different determinant on the cytokine is added, followed by washing and substrate addition. Wherever a cell produced the cytokine, it will be captured by the first Ab and will then be detected by the second Ab and its conversion of substrate, forming a colored spot on the nitrocellulose (hence the name ELISPOT assay). The nature of the cell producing the cytokine can also be determined by flow cytometry after staining the cells with a fluorescently labeled cell-type-specific Ab (e.g., anti-CD4 for T-helper cells) and an anti-cytokine Ab labeled with a different fluorochrome.

## Affinity purification of Ag and Ab

The specificity of Abs is not only important to the development of many research and diagnostic assays, but can, in some instances, be used to purify, or be purified by, interaction with Ag. This is because Abs do not form covalent bonds when they combine with Ag. Ab coupled to an insoluble matrix (e.g., agarose) specifically binds its Ag, removing it from a mixture of other molecules. After washing to remove all unbound molecules, the Ag can be eluted at low pH and/or at high ionic strength, which breaks the reversible bonds holding it to the Ab. As this can usually be performed without damaging the Ag or Ab, it is possible to obtain relatively pure Ag in one step. Similarly, Ag coupled to an insoluble matrix permits purification of Ab from media or serum. Ab can also be purified based on its binding by proteins (e.g., protein A) isolated from some strains of *Staphylococcus aureus*. Protein A coupled to agarose binds IgG Abs, which can be eluted by decreasing the pH and/or by increasing the ionic strength of the eluting buffer, again without damaging the Ab. Using similar techniques, cell subpopulations with characteristic cell-surface molecules (e.g., immunoglobulin on B cells) can also be isolated (positive selection) or removed (negative selection) from a mixture of cells.

# D8 Antibody functions

## Key Notes

| | |
|---|---|
| **Role of antibody alone** | Antibodies alone can neutralize viruses and toxins if they bind tightly to, and block, epitopes of the toxin or virus critical to their biological activity. Similarly, antibodies can agglutinate microbes and thus prevent them from colonizing mucosal areas. In some cases antibodies can induce programmed cell death (apoptosis). |
| **Role of antibody in complement activation** | IgG or IgM antibodies can, on binding to antigen, activate the **classical** pathway of complement leading to lysis of the cell on which the antigen is located, and/or to attraction of immune cells (chemotaxis) which phagocytose the antigen-expressing cells. |
| **Role of antibody with effector cells** | Phagocytes (PMNs and macrophages) have various receptors, including those for complement component C3b, for the Fc region of IgG (FcγR), and for the Fc region of IgA (FcαR). These receptors enhance binding to, and phagocytosis or ADCC of, antibody and/or complement opsonized microbes. Binding of antigens (e.g., allergens) to IgE already bound to Fc receptors for IgE on mast cells results in degranulation and subsequent enhancement of the acute inflammatory response. |
| **Related topics** | (B1) Cells of the innate immune system <br> (B2) Molecules of the innate immune system <br><br> (B4) Innate immunity and inflammation <br> (H2) Immunity to different organisms |

## Role of antibody alone

Antibodies alone can, in some instances, neutralize, and thus protect against, viruses and toxins. However, their effectiveness depends on the specificity and affinity. That is, antibodies must react with the part of the toxin or virus critical to its biological activity, and must bind tightly enough to prevent interaction of the toxin or virus with the cell-surface receptor through which it gains entry. Similarly, antibodies, primarily of the IgA class, can bind to bacteria and inhibit their attachment to mucosal epithelial cells. They can also cause agglutination of the microbe and thus prevent colonization of mucosal areas (Section D2). In addition, antibodies specific for certain molecules on the surface of cells can induce programmed cell death (apoptosis).

## Role of antibody in complement activation

The ability of antibody to protect against infection is, in many instances, greatly enhanced by, or dependent on, the complement system. As described in Section B2, the complement system is a protective system common to all vertebrates. In man it consists of a set of over 20 soluble glycoproteins, many of which are produced by hepatocytes and monocytes. These molecules are constitutively present in blood and other body fluids and may be present in large amounts, especially C3, the pivotal molecule of the complement system. The component molecules (C) include C1 (C1q, C1r, C1s), C2, C3, C4, C5, C6, C7, C8, and C9, as well as a set of molecules that are primarily associated with the alternative pathway, including factor B and factor D (Section B2). On appropriate triggering, these components interact sequentially with each other. This "cascade" of molecular events involves cleavage of some complement components into active fragments (e.g., C3 is cleaved to C3a and C3b), which contribute to activation of the next component, ultimately leading to lysis of, and/or protection against, a variety of microbes.

When an antibody of the IgG or IgM class (Table 1 in Section D1) attaches to an antigen, the **classical pathway** of complement is activated leading to complement-mediated lysis of the microbe (or other cell) on which the antigen is located. In addition, complement activation can also lead to attraction of immune cells (chemotaxis), and to opsonization and phagocytosis of the cell on which complement is being activated (Section B2). The classical pathway can also be activated by an Ag–Ab lattice.

### Sequence of activation

Formation of a site to which the first component of complement (C1) can bind requires a single bound antibody of IgM, or two IgG molecules bound in close proximity to each other. The C1q component of the C1 complex (C1q, C1r, C1s) then binds to the Fc regions of the cell-bound antibodies (Figure 1). This results in activation of C1, which then catalyzes the cleavage of C4 and C2, pieces of which (C4b and C2b) then bind to the cell surface forming a new cell-bound enzyme, C3 convertase (C4b+C2b). C3 convertase then cleaves C3 into C3a and C3b. C3b binds to the cell surface, forming a C4b–2b–3b complex. The cleavage of C3 into C3a and C3b is the single most important event in the activation of the complement system. This may be achieved by two different cleavage enzymes, C3 convertases—one as a component of the alternative pathway (Section B2), the other a part of the classical pathway. One of the fragments, C3b, is very reactive and can covalently bind to virtually any molecule or cell. If C3b binds to a self cell, regulatory

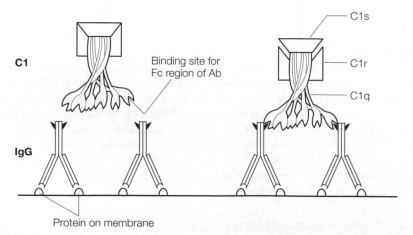

Figure 1. Initiation of complement activation by binding of C1 to antibody (Ab). The CH2 domains of the Fc regions of adjacent IgG molecules, bound to repeating antigenic determinants on a membrane, interact with the C1q subunit of C1. This results in the activation of C1r and C1s subunits, exposing an enzymatic active site.

molecules associated with this cell (see below) inactivate it, protecting the cell from complement-mediated damage.

For the classical pathway, the C4b–2b–3b complex governs the reaction and binding of the next complement components, C5, C6, C7, C8, and C9 to the cell surface (Table 1). More specifically, C5b is crucial to formation of the "membrane attack complex" (MAC), C5b–C6–C7–C8–C9, which mediates lysis of the microbe. The sequence of activation of the C5–9 components is the same as that described for the alternative pathway (Section B2), and leads to functional and structural damage to the membrane as a result of the formation of pores created by insertion of C9 complexes into the membrane.

**Table 1. Sequence of complement activation by the classical pathway leading to cell lysis**

| T (target cell) + A (antibody) | TA complex |
| --- | --- |
| TA + C1q,r,s | TAC1 |
| TAC1 + C4 | TAC1,4b + C4a |
| TAC1,4b + C2 | TAC1,4b,2b + C2a |
| TAC1,4b,2b + C3 | TAC1,4b,2b,3b + C3a |
| TAC1,4b,2b,3b + C5 | TAC1,4b,2b,3b,5b + C5a |
| TAC1,4b,2b,3b,5b + C6 + C7 | TAC1,4b,2b,3b,5b,6,7 |
| TAC1,4b,2b,3b,5b,6,7 + C8 | TAC1,4b,2b,3b,5b,6,7,8 |
| TAC1,4b,2b,3b,5b,6,7,8 + C9 | TAC1,4b,2b,3b,5b,6,7,8,9 Lysis of T |

## The major functions of the complement system

The classical pathway has the same biological activities and major functions as the alternative pathway, including:

- Initiation of (acute) inflammation by direct activation of mast cells
- Attraction of neutrophils to the site of microbial attack (chemotaxis)
- Enhancement of the attachment of the phagocyte to the microbe (opsonization)
- Killing of the microbe activating the membrane attack complex (lysis)

The components of the complement system most important to these main functions are the inflammatory peptides C3a and C5a (anaphylatoxins), derived from C3 and C5, respectively. C3a and C5a bind to receptors on mast cells, causing them to release (degranulate) pharmacological mediators such as histamine, which result in smooth muscle contraction and increased vascular permeability (Sections B2, B4, and K4). C5a is also chemotactic and attracts neutrophils (PMNs) to the site of its generation (e.g., by microbial attack). It also causes PMN adhesion, degranulation, and activation of the respiratory burst.

Also important, C3b (and C4b) act as opsonins marking a target for recognition by receptors on phagocytic cells. These receptors (e.g., complement receptor, CR1 = CD35) are expressed on monocytes/macrophages, PMNs, and erythrocytes. PMNs attracted to a site of complement activation by C5a find and bind to C3b through their cell-surface complement receptors, an interaction that greatly enhances internalization of the microbe by these cells. Thus, complement not only leads to lysis of a microbe, but attracts phagocytes and identifies, using C3b, what these cells should phagocytose. Even organisms resistant to direct lysis by complement may be phagocytosed and killed. Binding of C3b-containing complexes to CR1 on erythrocytes shuttles immune complexes to the mononuclear phagocytes of the liver and spleen, facilitating their removal. Finally, C5b through C9, the MAC, and especially C9 produce "pores" in the target cell membrane. These pores have diameters of about 10 nm and permit leakage of intracellular components and influx of water that results in disintegration (lysis) of the cell.

## Regulation

The complement system is a powerful mediator of inflammation and destruction and could cause extensive damage to host cells if uncontrolled. However, complement components rapidly lose binding capacity after activation, limiting their membrane-damaging ability to the immediate vicinity of the activation site. The complement system is also tightly regulated by inhibitory/regulatory proteins. These regulatory proteins (Table 2) include C1 inhibitor, factor I, C4b-binding protein, factor H, decay-accelerating factor (DAF), membrane cofactor protein (MCP), and CD59 (protectin). They protect host cells from destruction or damage at different stages of the complement cascade. Because regulatory proteins are expressed on the surface of many host cells but not on microbes, they limit damage to the site of activation and usually to the invading microbe that initiated complement activation.

### Table 2. Regulatory proteins of the complement system

| Protein | Function |
| --- | --- |
| C1 inhibitor | Binds to C1r and C1s; prevents further activation of C4 and C2 |
| Factor I | Enzymatically inactivates C4b and C3b |
| C4b-binding protein | Binds to C4b displacing C2b |
| Factor H | Displaces C2b and C3b by binding C4b |
| DAF | Inactivates C3b and C4b |
| MCP | Promotes C3b and C4b inactivation |
| CD59 | Prevents binding of C5b,6,7 complexes to host cells |

### Activation equals inactivation

Because the activated complement components are unstable and also readily inactivated by complement regulatory proteins, the activity of complement is short-lived. Therefore, activation of complement is equivalent to its inactivation. Thus, depressed complement levels in an individual may indicate that complement is being used up faster than it is being produced, suggesting chronic activation of complement perhaps resulting from continuous *in vivo* formation of antigen–antibody complexes.

### Role of antibody with effector cells

A variety of effector cells have receptors for the Fc region of antibodies. Phagocytes (PMNs, macrophages, and eosinophils) utilize their Fc receptors (FcR) for IgG (FcγR) or IgA (FcαR) to enhance phagocytosis of antibody-opsonized microbes.

In addition, these FcR can mediate killing of cells through antibody-dependent cellular cytotoxicity (ADCC). PMNs, monocytes, macrophages, eosinophils, and NK cells can kill antibody-coated target cells directly (Figure 2). That is, in ADCC, lysis of the target cell does not require internalization (although that may also happen) and involves release of toxic molecules (e.g., TNFα, Section B2) at the surface of the target.

Enhanced phagocytosis can also be mediated by phagocyte receptors for the complement component C3b, which is generated by antibody-mediated activation of the complement sequence (classical pathway) or on activation by certain microbes of the alternative path way of complement. Mast cells and basophils have FcR for IgE (FcεR), which on binding of IgE-coated antigens or cells can trigger degranulation and subsequent enhancement of the acute inflammatory response. Overstimulation of mast cells/basophils by this mechanism leads to pathology (Section K2).

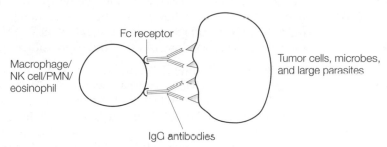

Figure 2. Antibody-dependent cellular cytotoxicity (ADCC) of an antibody-coated target cell. Several effector cell populations have Fc receptors (FcR) for IgG. Antibody-coated microbes attach to macrophages or PMNs through these receptors, and their resulting crosslinking leads to release of toxic substances. This extracellular killing probably occurs prior to phagocytosis of opsonized microbes through FcR or complement receptors. This also occurs when the antibody-coated target is too large to be phagocytosed, for example a worm. Eosinophils are particularly important in killing worms by this mechanism (Section H2). Macrophages, PMNs, and eosinophils can also use IgA FcR for ADCC. NK-cell-mediated death of virus-infected cells and tumor cells can be enhanced through ADCC.

# E1 The B-cell receptor complex, co-receptors, and signaling

## Key Notes

| | |
|---|---|
| **The B-cell receptor (BCR) complex** | The BCR complex consists of the antigen receptor, Ig, in association with two other polypeptides, Igα and Igβ (CD79a and CD79b). Igα and Igβ are signaling molecules for the BCR and are also required for assembly and expression of Ig. |
| **B-cell co-receptors** | Co-receptors, including CD32 and the co-receptor complex (CD19, CD21, and CD81), associate with the BCR complex especially when both the BCR and one or more of the co-receptors are linked through an antigen–complement or antigen–antibody complex. Depending on which molecules are ligated, signaling by the Ig–Igα/Igβ complex is enhanced or inhibited. |
| **Receptor–ligand interactions** | Lymphocytes need to be activated in order to carry out their function. Binding of the lymphocyte to an antigen via its antigen receptor, *signal 1*, is necessary, but not sufficient, to stimulate it and may lead to anergy. Accessory and co-stimulatory molecules on the surface of B cells are required for cell–cell interaction and the signal transduction events leading to activation (*signal 2*). |
| **Signaling by co-receptors** | B-cell signaling is initiated through the Igα/Igβ complex associated with the BCR and results in phosphorylation of tyrosine activation motifs (ITAMs). This is followed by an ordered series of biochemical events involving kinases and phosphatases. These events are modulated by signals from co-receptors. Second messengers lead to activation of transcription factors and thus to activation of lymphocyte function. |
| **Related topics** | (B1) Cells of the innate immune system    (C1) Lymphocytes    (E2) B-cell activation |

## The B-cell receptor (BCR) complex

As described in Section D2, the receptor for antigen on B cells is immunoglobulin. Initially cells make IgM and then IgD, which are both displayed on the surface of a mature B cell. Although these Igs are transmembrane molecules, the cytoplasmic domain of each is only three amino acids long, too short to signal the cell when antigen binds to the antibody. However, this membrane-bound Ig is associated with two other polypeptides on the B cell: Igα (CD79a) and Igβ (CD79b) (Figure 1). These small-molecular-weight (20

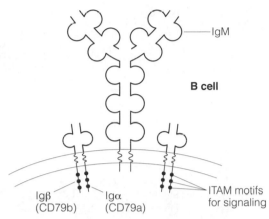

Figure 1. The B-cell receptor (BCR) complex.

kDa) transmembrane molecules are the signaling molecules for the BCR. When IgM and IgD (or other Ig isotypes on the B cell) are cross-linked by binding to antigen, Igα and Igβ transduce signals which begin to prepare the cell for a productive interaction with T-helper cells. Igα and Igβ are also required for assembly and expression of immunoglobulin, and thus of the B cell receptor complex, in the plasma membrane.

## B-cell co-receptors

Molecules associated with the BCR complex are expressed early in development to enable assembly of a functional antigen receptor on the B-cell surface. Other molecules important for B-cell functions, including MHC class II molecules that present antigen to T cells, also develop early in the life of a B cell. Moreover, co-receptors on the surface of B cells can, depending on which molecules are ligated, enhance or inhibit signaling by the Ig–Igα/Igβ complex. In particular, when the BCR co-receptor complex, which includes CD19 (a signaling molecule), CD21 (complement receptor 2, CR2), and CD81, associates with the antigen at the same time that the BCR interacts with antigen, B-cell activation is enhanced. However, if the BCR binds antigen with which soluble Ab has also interacted, and then binds through its Fc region to CD32 (a receptor on B cells for the Fc region of IgG, FcγRIIB), BCR signaling by the Igα/Igβ complex is inhibited (Figure 2).

## Receptor–ligand interactions

Lymphocytes need to be activated in order to carry out their function. At the molecular level this means receiving a message from outside the cell via interaction with a cell-surface receptor. This signal is then passed through the cytoplasm to the nucleus (signal transduction) to induce the gene transcription required for cell proliferation and synthesis and release of effector molecules, for example cytokines and antibodies. Although binding of the lymphocyte to an antigen via its antigen receptor (*signal 1*) is necessary to stimulate the cell, it is not sufficient and usually results in anergy if signal 2 is not also provided. Certainly, the binding of accessory cell surface molecules (e.g., B7-1 (CD80), B7-2 (CD86), CD40, and LFA-1) on B cells with their counter receptors on T cells is important, as these interactions increase the avidity of cell–cell interaction. Of particular significance, co-stimulatory molecules (some of which are also accessory molecules) modulate the signal transduction events leading to activation by providing the critical second signal (*signal 2*). Although B cells can be activated with or without the requirement of T cells,

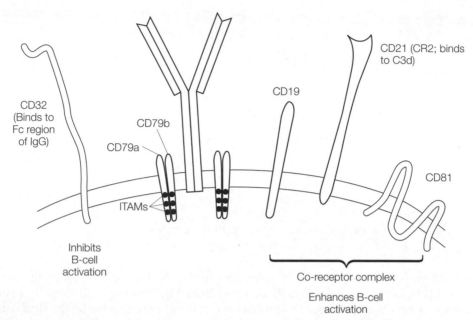

Figure 2. The BCR complex, its co-receptor complex (CD21 and CD81, which binds to complement component C3d), and the CD32 co-receptor (which binds the Fc region of IgG).

multimeric antigens can stimulate B cells directly, while responses to protein antigens require T-cell help.

## Signaling by co-receptors

The antigen receptors on B cells do not have intracytoplasmic tails of sufficient length and amino acid composition to act as signaling molecules. Thus, B-cell signaling is initiated through the CD79a/b associated with the BCR. These molecules have immunoreceptor tyrosine-based activation motifs (ITAMs) that are phosphorylated by kinases to initiate the activation process. Similar to the early events leading to activation of T cells, the molecules of the BCR complex become associated with the enzymes involved in phosphorylation within cholesterol-rich areas of the membrane termed "lipid rafts" (Section F4). An ordered series of biochemical events then occurs via kinases and phosphatases, which is modulated by signals from other co-receptor cell-surface molecules. Second messengers are produced which are eventually responsible for activation of transcription factors inside the nucleus and for production of cell cycle proteins and molecules required for lymphocyte effector functions. Cytokines induce proliferation and further differentiation of activated B cells.

# E2 B-cell activation

## Key Notes

**Two kinds of B cells**

Two B-cell groups can be distinguished based on their requirement for T-cell help in order to proliferate and differentiate. *B1 cells* are T-cell-independent (T-I), produce mainly IgM antibodies for secretion, and generally recognize multimeric sugar/lipid (T-I) antigens of microbes. *B2 cells* (conventional B cells) are T-cell-dependent (T-D) and are responsible for the development of IgG, IgA, and IgE antibody-mediated immunity.

**Thymus-independent (T-I) antigens**

Although B-cell responses to most antigens require T-cell help, activation of B cells by certain antigens does not. These **T-independent** antigens are of two types, both of which generate primarily IgM antibodies of low affinity. Type 1 antigens are bacterial polysaccharide mitogens that activate B cells independently of their antigen receptors. Type 2 antigens are linear, polymeric, and poorly degradable antigens (e.g., pneumococcus polysaccharides) that are picked up, persist on the surface of macrophages (but are not expressed on MHC molecules), and directly stimulate B cells through cross-linking of their surface receptors.

**Thymus-dependent (T-D) antigens**

T-helper (Th) cells induce B cells to produce antibodies, to switch the isotype of antibody being produced, and to undergo affinity maturation. Binding of most antigens to antigen receptors on B cells provides one signal while the cytokines produced by Th cells, and the engagement of complementary surface molecules on the *cognate* B cell, provides the second signal resulting in B-cell activation. Th cells recognize antigenic peptides on the surface of antigen-specific B cells that are associated with class II MHC molecules. On interaction with B cells, Th cells express CD40L, which triggers B-cell activation via its CD40 surface receptor. Activated B cells reciprocally co-stimulate T cells via CD28, to produce IL-2, IL-4, and IL-5. As a result, both the T cells and B cells clonally expand and differentiate.

**Biochemical events leading to B-cell activation**

Igα/Igβ molecules transmit the first signals following B-cell interaction with antigen through ITAMs of their intracytoplasmic tails. Co-receptors of the B-cell receptor complex modulate these initial signals (Section E1). Cross-linking of membrane receptor antibodies on B cells by T-I antigens induces clustering of co-receptors leading to multiple signals enhancing activation of kinase networks in IgM-producing B cells. B cells primed by a T-D antigen receive a second signal when the Th-cell CD40L binds to

CD40 on the B cell. Together with cytokines such as IL-2, IL-4, and so forth, this signaling induces proliferation, class switching, and differentiation of the B cells into plasma or memory cells.

| Related topics | (C1) Lymphocytes | (G6) Neuroendocrine regulation of immune responses |
|---|---|---|
| | (E1) The B-cell receptor complex, co-receptors, and signaling | (J2) Primary/congenital (inherited) immunodeficiency |
| | (F2) T-cell recognition of antigen | |
| | (G5) Regulation by T cells and antibody | |

## Two kinds of B cells

Two B-cell groups can be distinguished based on their requirement for T cell help in order to proliferate and differentiate. *B1 cells* arise early in ontogeny, produce mainly germ-line-encoded IgM antibodies for secretion, and mature independently of the bone marrow. These cells generally recognize multimeric sugar/lipid antigens of microbes and are T-cell-independent (T-I), that is, they do not require T-cell help in order to proliferate and differentiate in response to antigen.

*B2 cells* are the *conventional* B cells primarily responsible for the development of humoral (antibody-mediated) immunity. They are produced in the bone marrow, and are T-cell-dependent (T-D), that is, they require T-cell help in order to proliferate and differentiate in response to antigen. B2 cells eventually give rise to plasma cells that produce IgG, IgA, and IgE antibodies.

## Thymus-independent (T-I) antigens

Although B-cell responses to most antigens require T-cell help, activation of BI cells by certain antigens does not. These B1 cells recognize and respond to T-I antigens and produce primarily IgM antibodies of low affinity, whereas T-D antigens generate much higher-affinity antibodies of the other antibody classes. T-I antigens are of two types.

### Type 1 antigens

Bacterial polysaccharides have the ability to provide signal 1 through their specific antigen receptors and provide signal 2 through a component that is mitogenic (i.e., is able to induce proliferation) and bypasses the early biochemical pathways initiated through the antibody receptor. The polysaccharide interacts with B cells and, at sufficiently high concentrations, drives their activation (Figure 1).

### Type 2 antigens

Some linear antigens that are not easily degraded and have epitopes spaced appropriately on the molecule, for example pneumococcus polysaccharide, can directly stimulate B cells in a T-cell-independent fashion. These antigens persist on the surface of splenic marginal zone and lymph node subcapsular macrophages and directly stimulate B cells through cross-linking of their surface receptors (Figure 1). Although activation is independent of T cells, cytokines produced by T cells can amplify these responses.

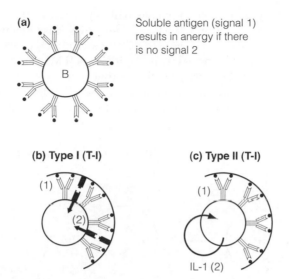

Figure 1. Activation of B cells through T-cell-independent antigens. (a) Soluble antigen interaction with the B-cell antigen receptor (antibody) results in anergy (signal 1 only). Signal 2 for B cells is provided by a mitogenic component (arrows) of the type I antigen (b) and via autocrine activity of IL-1 (arrow) for type II antigen (c).

## Thymus-dependent (T-D) antigens

The production of antibody to most antigens, and in particular those including a protein component, requires the participation of T cells. More specifically, Th-cell help is required for B2 cells to proliferate, differentiate, and produce antibodies. In addition, Th cells induce both class switching and affinity maturation. To accomplish this, Th cells directly engage the **cognate** B cell, trigger its activation through cell-surface receptors, and produce critical cytokines important for its proliferation and differentiation. This T-cell–B-cell *collaboration* is necessary because binding of most (non-multimeric) antigens to antigen receptors on B2 cells provides one signal that, in the absence of a second signal, results in B-cell anergy, that is, turns off the B cell. Cytokines produced by the Th cells, and the engagement of complementary surface molecules, provide the essential second signals to the B cell resulting in its activation (Figure 2).

More specifically, B cells capture antigen via the membrane antigen receptors IgM and IgD. This feature of B cells, to capture, process, and present specific antigen, makes them

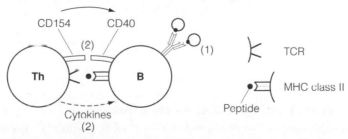

Figure 2. T-cell activation of B cells. T cells provide the second signal to B cells via ligation of CD40 by CD 154 (CD40 ligand) and via cytokines.

unique amongst the antigen-presenting cells which normally take up antigen via scavenger and other receptors (Section B3). The antigen cross-links receptors on the cell surface and is then endocytosed; it is degraded via the exogenous processing pathway, and peptides from the antigen become associated with class II MHC molecules (Figure 3 and Section F2). Th cells whose TCR are specific for that peptide–MHC complex, recognize and bind to B cells via TCR–MHC antigen and adhesion molecule interactions and the resulting signals and cytokines induce B-cell activation, proliferation, and differentiation.

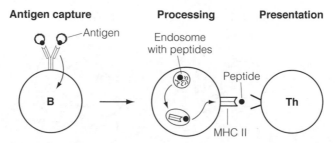

Figure 3. Activation of B cells through T-cell help. Antigens captured by B cells are processed and presented to Th cells, which provide the 2nd signal required for B cell activation.

Once triggered via the TCR, the Th cell expresses CD40 ligand (CD40L), which interacts with the B-cell surface molecule CD40 to induce activation of the B cell (Figure 4). The activated B cell, through B7-1/B7-2 (CD80/CD86) binding to CD28 reciprocally co-stimulates the Th cell. At this time, both the T cell and B cell are stimulated. T cells then produce cytokines including IL-2 (autocrine growth factor for the Th cells) and IL-4 and IL-5 (growth and differentiation factors for the activated B cells). As a result, both the T cell and B cell clonally expand and differentiate. Moreover, ligation of B cell CD40 by CD40L on T cells rescues B cells from death in germinal centers (Section E3).

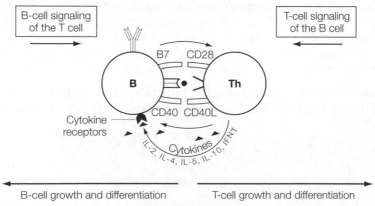

Figure 4. Reciprocal activation of T and B cells.

## Biochemical events leading to B-cell activation

The transmembrane surface antigen receptors on B cells (antibody), like the TCR on T cells, have short intracytoplasmic tails unable to transduce signals themselves. Therefore, on engagement of the B-cell antigen receptor, other molecules associated with these receptors mediate signaling. In particular, the ITAMs associated with CD79a/b (Igα/Igβ) of the BCR complex (Section E1) are phosphorylated during the early stages of activation and initiate the B-cell signaling cascade. Other members of the B-cell co-receptor complex (Figure 5) can enhance the strength of cell–cell interaction and modulate the initial signals mediated through antigen binding. For example, CD21, a complement receptor that binds C3d, provides an additional positive signal in B-cell activation if C3d is associated with antigen. The tetraspanin molecule CD81, part of the B-cell co-receptor complex, is important in lipid raft formation that is essential for cell activation. As with T-cell activation, these processes in B cells are mediated through phosphatases and kinases.

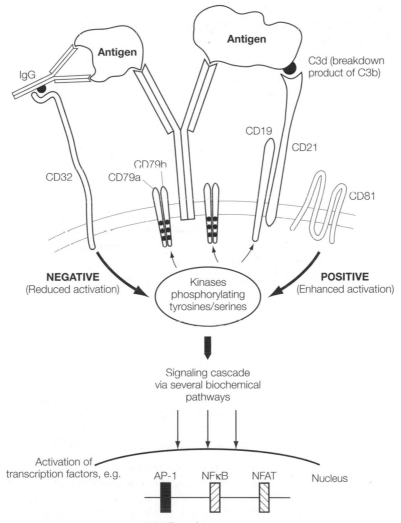

Figure 5. Activation of B cells via the BCR and co-receptors. Binding to antigen results in phosphorylation of the ITAMs of the intracytoplasmic chains of Igα and Igβ and subsequent activation of kinases. This leads to a signaling cascade via several biochemical pathways culminating in activation of transcription factors in the nucleus. Binding of CD21 to antigen bound to complement (C3d) regulates signaling via CD19 and CD81 in a positive way, while interaction of CD32 with IgG antibody bound to antigen and the antigen receptor provides a negative signal. This is mediated through regulation of the signaling pathways.

The importance of one kinase, btk, is indicated by the observation that mutation of the gene encoding it results in the absence of mature B cells (Bruton's agammaglobulinemia, Section J2).

B cells, like T cells, require two signals for their activation. Binding of soluble antigen to the antibody receptor, followed by cross-linking of receptors on the B-cell surface (signal 1), induces B-cell activation to the point where antigen is presented on its surface in MHC class II molecules. Alone, signal 1 results in apoptosis. However, proliferation is induced if a second signal, that for further B-cell activation, is provided by interaction of the B cell with a T-helper cell reactive with antigen presented by the B cell in MHC class II. This interaction results in T-cell expression of CD40L, which binds to CD40 on the B-cell surface to provide signal 2, and to induce class switching. Other cytokines produced by Th cells induce appropriate signals important to differentiation of the B cells into plasma or memory cells (Figure 6).

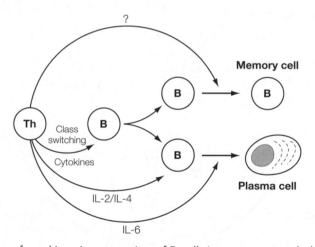

Figure 6. The roles of cytokines in maturation of B cells into memory and plasma cells.

After B-cell activation and following class switching to IgG, B cells are susceptible to regulation by **concomitant** binding of FcγRII (CD32) and the B-cell antigen receptor. In particular, further activation of these cells may be inhibited by their binding of antigen already attached to IgG from a previous immune response to the antigen (Figure 5). That is, binding of the co-receptor CD32 to IgG on the antigen, at the same time that another part of the antigen is bound by the B-cell antigen receptor, can transmit a negative signal to the B cell and prevent its activation ("negative feedback," Section G4).

T-I antigens, which do not induce IgG responses (because their CD40 molecules are not ligated), receive their second signal via the mitogenic component of T-I type I antigens. The second signal via T-I type 2 antigens is through binding of repeating antigenic units by the BCR, which leads to clustering of co-receptors. In these cases, the signals transduced are quite different from those resulting in activation of T-D B cells.

# E3 The cellular basis of the antibody response

## Key Notes

| | |
|---|---|
| **Selection and activation of B cells** | Antigen introduced into an individual binds specifically to B cells with receptors for that antigen. In the presence of T-cell help these B cells clonally expand and some differentiate into plasma cells that make antibody specific for the antigen triggering the response. |
| **Primary and memory responses** | On first exposure to antigen, a primary immune response develops resulting in production of IgM antibodies. This is usually followed by an IgG immune response within 4–5 days. This response is self-limiting and stops when antigen is no longer available to stimulate B cells. When antigen is reintroduced the response is more rapid, as there are more antigen-specific and antigen-responsive memory B cells, and usually results in IgG production. |
| **Multiclonal responses** | Antibodies produced by a single cell are homogeneous, but the response to a given antigen involves many different antibody-producing cells and thus, overall, is very heterogeneous (i.e., multiclonal). This heterogeneity is important to the effectiveness of an antibody response to a microbe. |
| **Cross-reactive responses** | Similar or identical antigenic determinants are sometimes found in association with different molecules or cells. This cross-reactivity is important: (i) in protection against organisms with cross-reactive antigens; and (ii) in autoimmune diseases induced by infectious organisms bearing antigens cross-reactive with normal self antigens (e.g., streptococcal infections which predispose to rheumatic fever). |
| **Related topics** | (C1) Lymphocytes<br>(D2) Antibody classes<br>(E2) B-cell activation<br>(F5) Clonal expansion and development of memory and effector function     (G5) Regulation by T cells and antibody<br>(G6) Neuroendocrine regulation of immune responses |

## Selection and activation of B cells

When antigen is introduced into an individual, B cells with receptors for that antigen bind and internalize it into an endosomal compartment, and process and present it on MHC class II molecules to helper T cells (Section E2). These B cells are triggered to proliferate, giving rise to large numbers of daughter cells. Some of the cells of these expanding clones serve as memory cells, others differentiate and become plasma cells (Section E2) that make and secrete large quantities of specific antibody. For example, on introduction of an antigen (e.g., Ag5) into a person (Figure 1), more than $10^6$ B cells have the opportunity to interact with it. However, very few of these B cells have receptors specific for this antigen. B5 binds Ag5, internalizes it, and processes and presents it on MHC class II molecules on the surface of this B cell. T-helper cells with specific receptors for peptides from Ag5 in MHC class II bind to this complex and stimulate this B cell to clonally expand and differentiate into memory B cells, and plasma cells that produce soluble antibody to Ag5. In addition, direct T-cell interaction with the B cell induces class switching that, depending on the type of T-helper cell (Th1 vs. Th2) and the cytokines it secretes, determines the class of antibody (IgG, IgA, or IgE) produced (Sections D3 and G5).

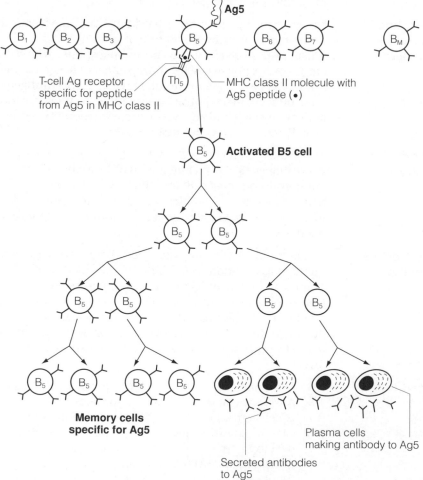

Figure 1. Clonal selection, memory cells, and plasma cells.

## Primary and memory responses

An antigen, when introduced into an individual who has not previously encountered it, will induce the development of a **primary** immune response within 4–5 days (Figure 2). This response initially results in the production of IgM to the antigen, followed by a class switch

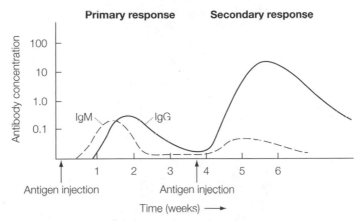

Figure 2. Kinetics of the immune response.

to IgG or another antibody isotype. The duration of antibody production and its isotype depend on the quantity of antigen introduced and its mode of entry. The antibody produced reacts with the remaining antigen, forming complexes and/or precipitates that are eliminated by phagocytes. Plasma cells can have a short life span or be relatively long-lived, dying through the process of apoptosis, but make and secrete antibody throughout their lives. If enough antigen is introduced initially, there could be re-stimulation of antigen-specific B cells, the development of more plasma cells, and thus production of more antibody. Eventually, when the antigen has been removed and none remains to stimulate B cells, the antibody response will reach its peak and the concentration of antibody in the circulation will begin to decrease as a result of the normal rate of catabolism of the antibody.

At the time antigen is reintroduced, more antigen-specific B cells exist in the individual compared with the period before primary introduction of antigen. Moreover, these cells have differentiated to more antigen-responsive memory B cells. Thus, when antigen is reintroduced a secondary (**memory** or anamnestic) antibody response occurs which is characterized by:

- A shorter lag period before significant levels of antibody are found in the serum
- The presence of many more plasma cells
- A higher rate of antibody production, and a higher serum concentration of antibody
- Production mainly of antibodies of the IgG or IgA class
- Higher-affinity antibodies

## Multiclonal responses

Although antibodies produced by a single cell and its daughter cells are identical (homogeneous or monoclonal), the response to a given antigen involves many different clones of cells and thus, overall, is very heterogeneous (multiclonal). Considering the size of an antigenic determinant, the number of determinants on a molecule, and the number of different molecules on a microbe, the total response to a microbe results in a large number of different antibodies (Figure 3). Even antibodies against a single well-defined antigenic determinant are heterogeneous, indicating that the immune system is capable of producing many different antibodies to a single antigenic determinant. This heterogeneity is essential for many of the protective functions of antibodies (Section D8).

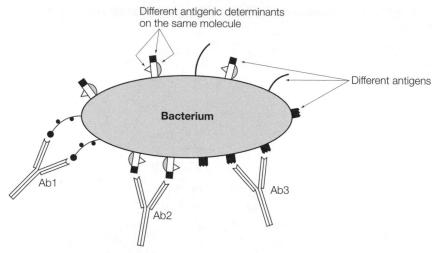

Figure 3. A heterogeneous antibody response against bacteria.

## Cross-reactive responses

Occasionally, a similar or identical antigenic determinant is found in association with widely different molecules or cells. This is termed cross-reactivity. Thus, the presence in most individuals of antibodies directed toward blood group carbohydrates other than their own is a result of the presence on certain microbes of carbohydrate antigens which are very similar, if not identical, to the blood group antigens. Infection with such an organism causes the production of antibodies directed toward the antigenic determinants of the microorganism, including these carbohydrate antigens (Table 1).

The development of immunity to one organism could, in some instances, protect against infection by another organism with cross-reactive antigens. Many vaccines are effective because of similar or identical determinants expressed by (i) both virulent and avirulent strains of the organism; or (ii) toxic molecules and their nontoxic derivatives. Natural or innate antibody to a wide variety of molecules is probably a result of the same phenomenon. In addition, some autoimmune diseases are due to infection by organisms bearing antigens that are cross-reactive with normal self antigens. Group A β-hemolytic streptococcal infections can lead to rheumatic fever as a result of the development of antibodies to certain streptococcal determinants that are similar to molecules in heart tissue. These antibodies may then react with, and damage, not only the microbe but also heart muscle cells (Sections K3 and L3).

## Table 1. Examples of clinically relevant cross-reactivity

| Immunogen | Cross-reactive antigen | Importance |
|---|---|---|
| Tetanus toxoid | Tetanus toxin | Protection vs. bacterial toxin |
| Sabin attenuated strain of poliovirus | Poliovirus | Protection vs. pathogenic virus |
| Various microorganisms | Type A and type B RBC carbohydrates | Transfusions |
| β-hemolytic streptococcus | Heart tissue antigens | Rheumatic fever |

# E4 Antibody responses in different tissues

## Blood

The localization and mechanism of elimination of antigen depend to a large extent on its route of entry. When introduced into the bloodstream, antigens are eventually trapped in the spleen. The antigen is endocytosed by splenic B cells, macrophages, and dendritic cells which process and present pieces of the antigen (antigenic determinants) on MHC class II molecules. Specific Th1-helper cells recognize these MHC–peptide complexes and provide help to B cells and induce class switching to primarily IgG.

## Mucosa

On penetrating the mucosal epithelium, the antigen comes into contact with lymphocytes underlying the mucosal areas, including those in the tonsils and Peyer's patches. As in the spleen, B cells interact with antigen through cell-surface antibodies which function as their antigen-specific receptor. T cells interact with antigen that has been processed and presented by B cells, and a humoral immune response is stimulated. In this case, the T-helper cell population is a Th2 cell that usually induces B-cell class switch to IgA, but sometimes to IgE. Dimeric IgA (mainly of the IgA2 subclass, Section D2) is released from plasma cells, binds to the poly-Ig receptor on epithelial cells, and is transported through the cell to the lumen, where it has its primary protective role (Figure 1).

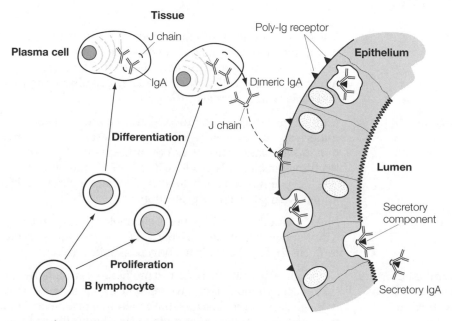

Figure 1. Transport of IgA across the epithelium.

## Lymphatics

Antigen introduced into tissues is channeled through the lymphatics to the lymph nodes, where again B cells, macrophages, or dendritic cells trap and process the antigen and present it to T cells for initiation of specific immune responses. Antigen is also picked up by dendritic cells (Langerhans cells) in the dermis, processed, and carried via the lymphatics to the draining lymph nodes, where it is presented to T-helper cells. B and T cells are concentrated in different parts of the lymph nodes, the B cells in follicles and the T cells in the paracortical areas. On activation T cells migrate toward the lymphoid follicles to meet with B cells that have migrated to the edge of the follicles. The interaction of B and T cells in this area induces further B-cell activation and their migration back into the follicle, where they proliferate and differentiate. The center of each follicle is the germinal center and is made up of rapidly dividing B cells.

## Germinal centers as sites of B-cell maturation

Germinal centers are unique well-defined proliferating foci within the secondary lymphoid tissues where three important processes in B cell maturation occur—the generation of memory cells, antibody class switching, and the maturation of antibody affinity (Figure 2). Primary B-cell follicles in secondary lymphoid tissues, for example lymph nodes and spleen, are made up of aggregates of B cells. When B cells in the primary follicle

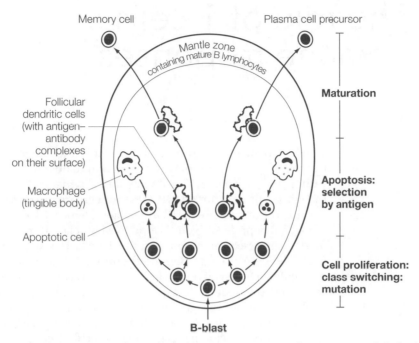

Figure 2. B-cell maturation in the germinal center (GC). A B cell in a primary follicle, having been activated with T-cell help (now a B-cell blast), begins to proliferate and initiate GC formation. During proliferation the cells undergo class switching and somatic mutation of antibody V genes. B cells expressing new antigen receptors with the same or higher affinity for the same antigens will be selected for proliferation and differentiation. Those B cells with new receptors not able to bind the antigen may die by apoptosis and be taken up by tingible body macrophages. Those with antigen receptors with high affinity for the displayed antigen will survive to leave the GC either as memory cells or as plasma cell precursors. Plasma cell precursors either mature locally, but outside the GC, into plasma cells or leave via the bloodstream to mature into plasma cells in bone marrow, lymph nodes, spleen, or mucosa-associated lymphoid tissue depending on the Ig class of the precursor and the secondary lymphoid site in which it was induced.

are stimulated by antigen associated with follicular dendritic cells (FDCs) and receive T-cell help, they proliferate and begin to form the germinal center. Germinal centers are formed from a small number of activated B cells. These B cells begin to lose their surface IgM and IgD, and switch to IgG (usually in the spleen or lymph nodes) or to IgA (usually in mucosal tissues). During this time, there is hyper-mutation of the variable-region genes, and receptors with slightly different amino acid sequences appear on the surface of these B cells. Some of these modified receptors are unable to bind the same antigen that triggered them, and B cells with these receptors will therefore not be able to be re-stimulated by this antigen. However, some modified receptors may bind more strongly to this same antigen, and B cells with these higher-affinity receptors for the antigen would be selected (as they compete best for antigen), survive, and proliferate. Some of these cells mature into memory cells that stay in the mantle of the germinal center or join the re-circulating lymphocyte pool, while others mature into plasma cells that only synthesize and secrete one class of specific antibody (Section D3). Small numbers of 'follicular' T cells are also present and contribute to the production of long-lived antibody responses.

# F1 The role of T cells in immune responses

<table>
<tr><td colspan="2">Key Note</td></tr>
<tr><td>Overview</td><td>T cells have evolved to protect us against intracellular microbes (viruses and some bacteria) and to help B-cell responses. The specific T-cell receptor for antigen (TCR) recognizes protein antigens that have been processed into peptides and bound to MHC molecules. Helper (CD4⁺) T cells recognize peptide antigens bound to MHC class II molecules on dendritic cells, macrophages, and B cells. Cytotoxic (CD8⁺) T cells recognize peptides bound to MHC class I molecules. The T-cell repertoire is generated and selected for survival in the thymus. Recognition of the peptide antigen by the TCR results in signaling leading to transcription of genes encoding cytokines and their receptors, for example IL-2, required for clonal expansion of specific T cells. Effector molecules such as IFNγ that activate macrophages and induce IgG class switch are produced by Th1 cells. IL-4 is produced by Th2 cells and is mainly important for B-cell proliferation.</td></tr>
<tr><td>Related topics</td><td>(H2) Immunity to different organisms</td></tr>
</table>

## Overview

Cell-mediated immunity is mediated by T cells, which distinguishes it from immunity mediated by antibodies (humoral immunity). These terms evolved from the finding that immunity to certain antigens could be transferred to other animals by either cells, if they were of the same inbred strain, or antibodies. T cells have evolved to protect us against intracellular microbes (viruses and some bacteria) and to help B-cell (antibody) responses against extracellular microbes. They do this by monitoring the cells of the body for foreign antigens attached to MHC molecules.

Foreign antigens in host cells are broken into linear peptides (processed), which are displayed by major histocompatibility complex (MHC) molecules expressed on their cell surface. Unlike antibodies, which recognize the three-dimensional shape of antigens (Section A4), the T-cell antigen receptor (TCR) only recognizes linear antigens (peptides) bound to MHC molecules; that is, T cells cannot directly recognize or bind to microbes or their unprocessed molecules. Helper (CD4⁺) T cells recognize peptide antigens in the context of MHC class II molecules that are expressed by dendritic cells, monocytes, macrophages, and B cells. Cytotoxic (CD8⁺) T cells recognize peptides associated with MHC class I molecules. This differential requirement for CD4 and CD8 relates to the fact that CD4 and CD8 molecules themselves attach to the nonpolymorphic (nonvariant) part of the MHC class II and MHC class I molecules, respectively.

Association of antigens with either one or the other of the two classes of MHC molecules is the result of antigen being processed by different cellular pathways—the exogenous pathway for MHC class II or the endogenous pathway for MHC class I. The T-cell repertoire generated is very large but each T cell recognizes only one specific foreign peptide. This occurs during normal thymus development where the T cells are "educated," that is, selected for survival or eliminated if self-reactive. There are two kinds of T-helper cells, each with different functions in the immune response, that are defined by their cytokine profiles. Th1 cells help macrophages to kill intracellular microbes, help the development of cytotoxic T cells to kill virus-infected cells and induce B-cell activation and class switch to IgG. Th2 cells are mainly involved in helping B cells to develop into memory cells and plasma cells that produce antibodies.

T cells need to be activated in order for them to carry out their function. Recognition of the peptide antigen by the TCR is not sufficient to activate the cells, as accessory molecules are also required together with co-receptors involved in signaling events. Signaling leads to transcription of genes coding for cytokines and their receptors; for example, IL-2 is required for clonal expansion of specific T cells. Effector molecules such as IFN$\gamma$, which activates macrophages and induces B cells to class switch to IgG, are produced by Th1 cells. IL-4 is produced by Th2 cells and is important for B-cell proliferation and class swith to IgE. Enzymes and molecules involved in killing by CD8$^+$ cells are also induced during activation.

# F2 T-cell recognition of antigen

## Key Notes

**T-cell receptor (TCR) for antigen**

The TCR for antigen is only found on the T-cell membrane and is composed of two polypeptide chains, α and β. Each of these glycoproteins is made up of constant and variable regions, like those of Igs, and together the α- and β-chain variable regions constitute the antigen-binding site. Some T cells express a TCR consisting of γ and δ chains. These cells have some of the characteristics of α and β T cells, but have specificity for unconventional antigens such as heat shock proteins and phospholipids.

**The T-cell receptor complex**

The T-cell receptor complex consists of the antigen receptor, the αβ or γδ dimer, and CD3, a signaling complex composed of γ, δ, and ε chains (and a separate signaling moiety made up of two ζ chains). CD4 on T cells binds to the nonpolymorphic region of MHC class II on APCs, restricting Th cells to recognizing only peptides presented on MHC class II molecules. CD8 on cytolytic T cells binds the nonpolymorphic region of MHC class I, restricting killing to cells presenting peptide in MHC class I.

**Structure of MHC molecules**

MHC molecules are glycoproteins. MHC class I molecules are composed of an α polypeptide chain that has three domains and is anchored into the cell membrane, and a small peptide $\beta_2$-microglobulin. The two outer domains form a "binding groove" into which an antigenic peptide can bind. MHC class II molecules have an α and a β chain—the outer amino acids of which form the peptide-binding grove.

**Nature of MHC binding peptide**

The polymorphic regions of MHC class I and class II are the peptide-binding domains of these molecules and bind peptides ranging from 8 to 10 and 10 to 20 amino acid residues, respectively. Anchor residues on the peptides bind to residues in the class I and II grooves and vary for different MHC alleles. This forms one of the bases for the genetic control of immune responses.

**Genes encoding MHC molecules**

Both MHC class I and class II molecules are encoded by genes that are highly polymorphic and are closely linked together on chromosome 6 in man. These and other genes are critical to antigen processing and presentation, as suggested from their distribution within the MHC gene locus. There are three loci for MHC class I genes (HLA-A, B, and C) and four for MHC class II genes (HLA-DP, DQ,

| | |
|---|---|
| | and DR, and another nonpolymorphic locus for HLA-DM (involved in antigen processing)). Other class II genes code for tapasin, LMP, and TAP, which are involved in antigen processing. MHC class III genes include those coding for some complement components and TNF and are not directly involved in antigen processing and presentation. |
| **Cellular distribution of MHC molecules** | MHC class II molecules are expressed on APCs (B cells, dendritic cells, and macrophages) for the activation of CD4$^+$ helper T cells. MHC class I molecules are expressed on all nucleated cells, permitting cytolytic T cells to recognize cells infected with intracellular pathogens. Cytokines modulate the expression of MHC class I and II molecules. |
| **Antigen processing and presentation** | In the **MHC class I processing pathway** (**endogenous pathway**), peptides that bind to MHC molecules are mainly derived from viruses that have infected host cells. Viral proteins are processed by proteasomes and peptides transported into the endoplasmic reticulum from the cytosol where they become associated with MHC class I molecules. Vesicles containing the MHC class I–peptide complex move to the surface and are recognized by CD8$^+$ cytotoxic T lymphocytes (CTLs). In the **MHC class II processing pathway** (**exogenous pathway**), endocytosed pathogens are exposed to endopeptidases that break down the proteins into peptides that associate with MHC class II molecules produced in the endoplasmic reticulum. Vesicles containing the MHC–peptide complex move to the surface where they are displayed to CD4$^+$ helper T cells. |
| **Related topics** | (A4) Antigens             (M2) Transplantation<br>(C1) Lymphocytes                 antigens<br>(F3) Shaping the T-cell<br>         repertoire |

## T-cell receptor (TCR) for antigen

There is as much diversity of the TCR as of antibody receptors, but unlike the B-cell antigen receptor (Ig), the TCR is only found on the T-cell membrane and not in the serum or other body fluids. Two different groups of T cells can be defined based on their use of either αβ or γδ for their TCRs. Both groups develop in the thymus.

### Alpha/beta (αβ) T cells

αβ T cells are the "conventional" T cells that undergo positive and negative selection in the thymus (Section F3) and constitute the majority of human peripheral T cells. These cells complete their functional maturation in the secondary lymphoid tissues and provide protection against invading microbes. They are found in specific compartments within secondary lymphoid tissues and recirculate around the body in search of their specific antigens (Section C4). These T cells function to control intracellular microbes and to provide help for B-cell (antibody) responses. Two different kinds of αβ T cells are involved in these functions: T-helper (Th) cells and T-cytotoxic (Tc) cells.

The α and β polypeptide chains of the TCR have molecular weights of 50 and 39 kDa, respectively. Like antibody, each chain is made up of constant regions and variable regions that constitute a T-cell antigen-binding site (Figure 1). However, as previously indicated, the TCR, unlike antibodies, does not recognize native antigen, but can only bind processed antigen presented on MHC molecules. The genes coding for TCR polypeptide chains are members of the Ig superfamily.

Another functional type of αβ CD4⁺ T cell produced in the thymus is the regulatory T cell (Treg, Section G3).

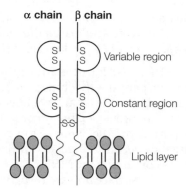

Figure 1. T-cell antigen receptor αβ dimer.

## Gamma/delta (γδ) T cells

γδ T cells have a similar granular morphology to NK cells (Section C1) and represent a small proportion (around 2–10%) of all peripheral blood T cells. They express a TCR consisting of γ and δ chains with similar but different V, D, and C gene segments from those that code the αβ TCR. Unlike the αβ TCR, which is coded for by many different random combinations of variable genes with D and C gene segments (Section F3), these cells use a limited number of V genes but still appear to have a broader specificity for recognition of unconventional antigens such as heat shock proteins, phospholipids, and phosphoproteins. Unlike αβ T cells, they do not generally recognize antigens in association with classical MHC class I and II molecules. There is evidence that γδ T cells are cytotoxic and regulatory and subsets of them appear to have specific tissue locations. In this regard, one subset (using one particular set of Vδ genes) is located mainly at epithelial surfaces and may control microbes there through cytotoxic activity and cytokine production. They may also be involved in responses against self antigens. Their exact functions are, however, still unclear and thus are a focus of intense research. Since γδ T cells are produced in the thymus, have a T-cell receptor and express T-cell-associated molecules, they are classified as T cells. However, because they mainly show lack of MHC restriction for antigen recognition and their specificity is not restricted to peptides, they do not appear to be conventional T cells and probably, like NKT cells (Section B1), interface between innate and adaptive immunity.

## The T-cell receptor complex

The T-cell receptor complex consists of the antigen receptor, the αβ or γδ dimer, associated with several other polypeptides important in T-cell signaling and recognition. In particular, the TCR is associated with CD3, which itself is composed of several polypeptides γ, δ, ε, and ζ. These polypeptides are essential for the assembly of the TCR and ε and ζ chains are also part of the signaling complex (Figure 2). Two other molecules, CD4 and CD8 (considered as co-receptors), also play a role in T-cell recognition of antigen. CD4 binds to the nonpolymorphic region of MHC class II and restricts Th cells to recognizing only peptides presented on MHC class II molecules (Figure 3). Similarly, CD8 on cytolytic T cells binds the nonpolymorphic region of MHC class I, restricting these killer T cells to recognize only cells presenting peptide in MHC class I molecules.

## Structure of MHC molecules

Although molecules coded for by the MHC were originally identified based on their role in transplant rejection, they actually evolved to present foreign antigens to T cells. MHC class I molecules are glycoproteins composed of two chains: an α polypeptide chain that has three domains and is anchored into the cell membrane, and a small peptide $\beta_2$-microglobulin that is important for allowing the α chain to be expressed on the cell surface. The two outer domains form a "binding groove" into which an antigenic peptide can bind (Figure 4). MHC class II molecules are glycoproteins composed of two separate chains: an α and a β chain. The outer amino acids on both chains form a 'groove', which

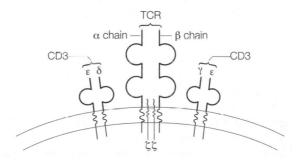

Figure 2. The TCR complex consists of the antigen receptor, the αβ or γδ dimer, associated with several other polypeptides involved in T-cell signaling. The signaling complex is composed of CD3γ, δ, and ε polypeptide chains and a separate homodimer of ζ chains.

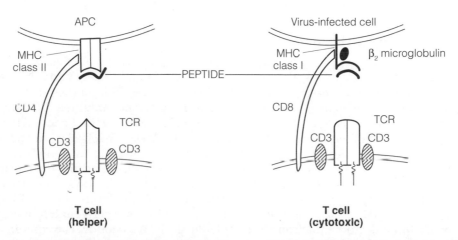

Figure 3. CD8 and CD4 recognition of MHC class I and class II molecules, respectively. CD4 on T-helper cells binds to the nonpolymorphic region of MHC class II; CD8 on cytolytic T cells binds the nonpolymorphic region of MHC class I.

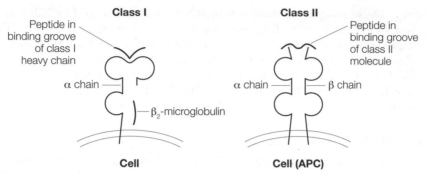

Figure 4. MHC class I and class II binding peptide for presentation to T cells.

binds to antigenic peptides. MHC molecules in man are called HLA molecules (human leukocyte antigen) since they were first discovered on the surface of leukocytes.

## Nature of MHC binding peptide

Polymorphic residues within the peptide-binding groove/site of each of the MHC molecules make contact with the antigenic peptide. The peptides that are bound by MHC class I and class II molecules are short, ranging from 8 to 10 amino acid residues (for MHC class I) and 10 to 20 amino acid residues (for MHC class II). The sites on the peptides that fasten the peptide to the MHC molecule are the **anchor residues** (Figure 2 in Section A4). The peptide residues in the MHC-binding groove are the same for each allelic form of the MHC molecule. Therefore, each allelic form of MHC molecule is only able to bind peptides bearing specific anchor residues. Thus, depending on the MHC molecules that are inherited, a person might not be able to bind specific peptides from, for example, a virus. If the person's MHC molecules cannot bind the peptides generated from a specific virus, then they will be unable to mount a CD8 response to the cells infected with virus. This forms at least one basis for the genetic control of immune responses. That is, the MHC molecules inherited by an individual ultimately determine to which peptides that individual can elicit T-cell-mediated immune responses, and, at the population level, the polymorphism increases the chances of survival of at least some individuals.

## Genes encoding MHC molecules

Both MHC class I and class II molecules are encoded by genes that are highly polymorphic and are closely linked together on chromosome 6 in man. These and other genes are critical to antigen processing and presentation, as suggested from their distribution within the MHC gene locus (Figure 5). Note that in man there are three different loci for MHC class I genes (HLA-A, B, and C) and four loci for MHC class II genes (HLA-DP, DQ, and DR, and another nonpolymorphic locus for HLA-DM (involved in antigen processing)). Other class II genes code for other molecules involved in antigen processing, including those for tapasin, LMP, and TAP (see below). MHC class III genes include those coding for some complement components and TNF and are not directly involved in antigen processing and presentation. The many alternative (allelic) forms of genes for each subregion of the MHC locus gives rise to many different forms of class I and II molecules each with a binding groove that can bind a different antigenic peptide. The combination of the encoded alleles within the MHC locus is referred to as the haplotype (for haploid, as opposed to diploid). Because genes within the MHC are closely linked, haplotypes are usually inherited intact.

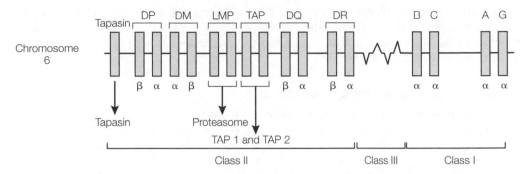

Figure 5. Genes in the MHC locus involved in antigen processing and presentation; in man this locus is on chromosome 6. MHC class I and II genes code for the surface molecules that present antigens to T cells while class III genes code for some complement components and TNF and are not directly involved in antigen processing and presentation. Each MHC class II molecule that presents antigen is made up of two chains coded for by $\alpha$ and $\beta$ genes. HLA-DM is not directly involved in presenting antigen but is involved in processing (Figure 7). Tapasin codes for one of the molecules involved in the assembly of MHC class I molecules. LMP genes code for the two subunits that make up the proteasome, while TAP genes code for two proteins—TAP1 and TAP2 (transporter associated with antigen processing)—involved in antigen processing via the endogenous pathway. MHC class I genes code for single $\alpha$-chain molecules that associate with a small non-membrane-bound protein $\beta_2$-microglobulin, the gene for which is located on chromosome 15.

## Cellular distribution of MHC molecules

MHC class I and MHC class II molecules have a distinct distribution on cells (Table 1) that directly reflects the different effector functions that the cells play. Furthermore, under some conditions (e.g., cytokine activation) the expression of MHC class I and/or II molecules may be induced or enhanced. For example, when B cells take up antigen through their BCR they upregulate their MHC class II molecules to prepare for presentation of peptides to Th cells. Other cells that express MHC class II molecules (dendritic cells, macrophages) are efficient APCs for the activation of Th cells. In contrast, MHC class I molecules are expressed on virtually all cells in humans except for red blood cells. The expression of MHC class I molecules on all nucleated cells permits the immune system to

## Table 1. Expression of MHC class I and II molecules

| Tissue | MHC class I | MHC class II |
|---|---|---|
| T cells | +++ | – |
| B cells | +++ | +++ |
| Monocytes | +++ | + |
| Macrophages | +++ | ++ |
| Dendritic cells | +++ | +++ |
| Neutrophils | ++ | – |
| Liver cells | + | –* |
| Kidney cells | ++ | –* |
| Brain cells | + | –* |
| Red blood cells | – | – |

*Cells of the mononuclear phagocyte system in these organs express MHC class II molecules.

survey these cells for infection by intracellular pathogens and allows their destruction via class I-restricted cytotoxic T lymphocytes (CTLs). It is interesting to note that the absence of class I MHC molecules on RBC may allow the unchecked growth of *Plasmodium*, the agent responsible for malaria.

## Antigen processing and presentation

### Class I processing pathway (endogenous pathway)

To a large extent, fragments of peptides that bind to class I MHC molecules are derived from viruses that have infected nucleated host cells (Figure 6). Some of the viral proteins

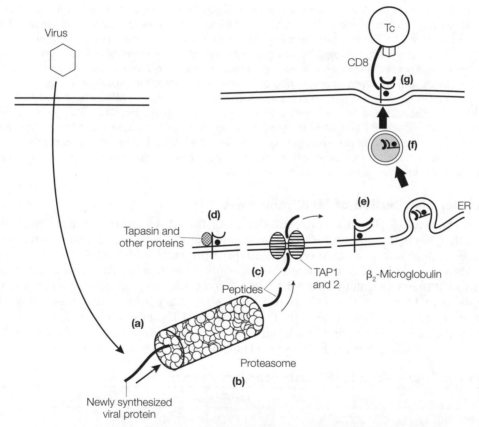

Figure 6. Antigen processing for presentation to CD8+ T cells (**endogenous pathway**). (a) Viruses and other microbes enter nucleated cells and escape into the cytosol where they reproduce. Viral peptides are modified and enter proteolytic factories in one end of a proteasome, of which a cell will have several (called immunoproteasomes when talking about immune peptides). (b) Through endopeptidase activity the viral proteins are broken up into peptides, which exit the other end of the proteasome and are around 8–10 amino acids in length. (c) The peptides are transported into the endoplasmic reticulum through a channel formed by TAP1 and TAP2. (d) Meanwhile, MHC class I molecules have been assembled together with other proteins (such as tapasin to aid assembly) within the endoplasmic reticulum (ER). (e) The transported linear peptides then attach to the binding site on the α chain of the MHC class I molecule. Vesicles pinch off from the ER containing the MHC–peptide complex and (f) traffic to the cell surface, where they fuse with the plasma membrane to (g) display the peptide to specific T cells.

in the cytosol are degraded into peptides of 8–10 amino acids in specialized process-ing factories called proteasomes—also called immunoproteasomes. These peptides are transported into the endoplasmic reticulum through a channel formed by two specific transporter proteins (transporters associated with antigen processing: TAP1 and TAP2). Meanwhile MHC class I molecules have been assembled together with other proteins (such as tapasin to aid assembly) within the endoplasmic reticulum. The transported linear peptides then attach to the binding site on the α chain of the MHC class I mol-ecule. Vesicles pinch off from the endoplasmic reticulum containing the MHC–peptide complex and traffic to the cell surface of the antigen presenting cell, where they fuse with the plasma membrane to display the peptide to specific CD8$^+$ T cells. In general,

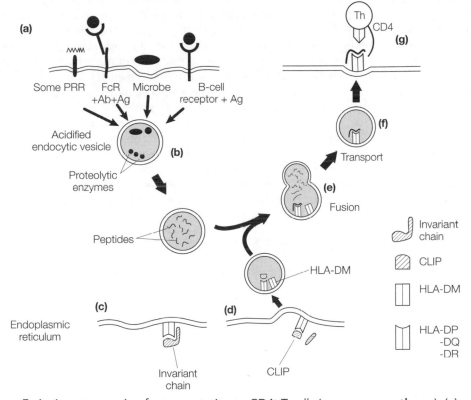

Figure 7. Antigen processing for presentation to CD4$^+$ T cells (**exogenous pathway**). (a) Pathogens can be endocytosed from the environment into endocytic vesicles through a number of surface receptors on the antigen-presenting cells, including some pattern recognition molecules (PRR), through Fc receptors (FcR) binding IgG antibody to which antigen is attached, directly by endocytosis, or by surface antibody receptor in the case of B cells. (b) The acidified endocytic vesicles contain endopeptidases that break down peptides into linear lengths of 10–20 amino acids. (c) Meanwhile, MHC class II molecules, produced within the endoplasmic reticulum, have invariant chain bound into their peptide-binding sites. (d) This chain is enzymatically degraded to leave a smaller protein, CLIP, in the binding site. The vesicle containing MHC class II molecules with CLIP then buds off from the endoplasmic reticulum. An HLA-DM molecule attached to the MHC molecule induces removal of CLIP. (e) On fusion with a vesicle containing antigenic peptides, HLA-DM facilitates their loading onto the binding sites now free of CLIP molecules. (f) The fused vesicles traffic to the cell surface, where they (g) fuse with the plasma membrane, displaying the MHC–peptide complex to CD4$^+$ T cells.

peptides generated in the cytoplasm, that is, the cytosol (as would be the case for cytosolic microbes), become associated with MHC class I that would be recognized by CTLs, distinguished by their expression of CD8.

### Class II processing pathway (exogenous pathway)

While viruses and some bacteria replicate in the cytosol, several types of pathogens, including mycobacteria and *Leishmania*, replicate in cellular vesicles of macrophages. In addition, pathogens can be endocytosed from the environment into endocytic vesicles through a number of surface receptors on the antigen-presenting cells (Figure 7). These acidified vesicles contain endopeptidases such as cathepsin B that break down peptides into linear lengths of 10–20 amino acids. MHC class II molecules are produced within the endoplasmic reticulum and, to protect the peptide-binding site, another molecule, invariant chain, is attached. This chain is enzymatically degraded to leave a smaller protein—CLIP—in the binding site of the MHC molecule. A vesicle containing MHC class II molecules with CLIP then buds off from the endoplasmic reticulum with a nonpolymorphic MHC class II protein HLA-DM attached. This induces removal of CLIP, and on fusion of the vesicle with an endocytic vesicle containing antigenic peptides, HLA-DM facilitates their loading onto the binding sites now free of CLIP molecules. The fused vesicles traffic to the cell surface containing the MHC–peptide complex and are displayed on the surface to Th cells expressing CD4. CD4+ T cells also assist in the destruction of parasites in vesicular compartments, for example mycobacteria, by activating the cells that harbor these pathogens to kill them. For extracellular parasites, CD4+ T cells can activate macrophages to endocytose and destroy the pathogens, as well as instruct B cells to produce antibody to opsonize the pathogens. There are two subsets of CD4+ cells that are potentially involved in these responses (Section F5).

# F3 Shaping the T-cell repertoire

---

**Key Notes**

| | |
|---|---|
| **Generation of T-cell diversity** | Multiple genes code for each of the two polypeptide chains α and β (or γ and δ) of the TCR. Each chain, like those of antibodies, is made up of a V (variable) and a C (constant) region. Three different gene segments—V, D, and J—encode the V region of β and δ chains, whereas two different gene segments—V and J—encode the V region of α and γ chains. As with antibody genes, the T-cell V gene segments rearrange in each developing T cell in the thymus, resulting in a breadth of T-cell diversity similar to that for B cells. Allelic exclusion assures that each T cell will have a single specificity. |
| **Selection of the T-cell repertoire** | In the thymus, those T cells that express a TCR that binds weakly to self MHC are positively selected. Of this group, those that express a TCR that binds strongly to self MHC are eliminated (negative selection). In addition, T cells that recognize self MHC plus self peptides are also removed (negative selection), leaving those T cells that can recognize modified self MHC molecules—self MHC molecules plus foreign peptide—to survive, mature, and become functional T cells in the peripheral lymphoid tissues. |
| **Related topics** | (C2) Lymphoid organs and tissues        (D3) Generation of diversity |

---

## Generation of T-cell diversity

Each of the very large numbers of T cells produced in the thymus has only one specificity, defined by its antigen receptor. Millions of T cells, each with receptors specific for different antigens, are generated by gene rearrangement from multiple (inherited) germ-line genes. Multiple genes code for each of the two polypeptide chains α and β (or γ and δ) of the TCR. Each chain, like those of antibodies, is made up of a V (variable) and a C (constant) region. Three different gene segments—V, D, and J—encode the V region of β and δ chains, whereas two different gene segments—V and J—encode the V region of α and γ chains (Figure 1). The many germ-line genes within each segment, that is, V, D, or J (for β and δ chains) and V and J (for α and γ chains), are separated by noncoding DNA.

During development of αβ T cells the V, D, and J gene segments are rearranged to form a complete V-region gene for the β chain (Figure 2) and V and J gene segments are rearranged to form a complete V-region gene for the α chain. Variability in junction formation and random insertion of nucleotides contributes further to the diversity of variable region gene products (Section D3). As in the case of immunoglobulin rearrangements, the expression of a complete α chain and a complete β chain by the T cell excludes further

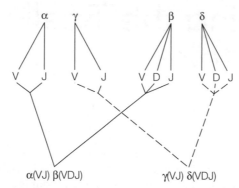

Figure 1. Gene segments involved in formation of the V region of the different polypeptides of the TCR.

rearrangement (allelic exclusion). The cell thus becomes committed to the expression of a single V–C α-chain combination and a single V–C β-chain combination. Together these two chains form an antigen-binding site that determines the specificity of the T cell. Since the rearrangements occur randomly in millions of T cells, considerable diversity of specificity is generated *prior* to antigen recognition. Cells that fail to rearrange functional TCR genes die in the thymus. As with the pre-B-cell receptor (Section D3), a truncated α chain (light chain in BCR) is first produced to display a pre-T-cell receptor. It is thought that this receptor is involved in development of double-positive T cells in the thymus (Figure 3). In some developing T cells, γ and δ gene rearrangements occur, resulting in the T cell expressing γδ TCRs. Here, unlike that seen for αβ T cells, there appears to be a restricted number of Vγ and Vδ genes used for the γδ T-cell repertoire (Section F2).

## Selection of the T-cell repertoire

Upon entry into the thymus, T-cell precursors from the bone marrow begin TCR rearrangements and the receptor is expressed on thymocytes that bear both CD4 and CD8 markers (double-positive thymocytes). T cells that express a TCR that can bind *weakly* to

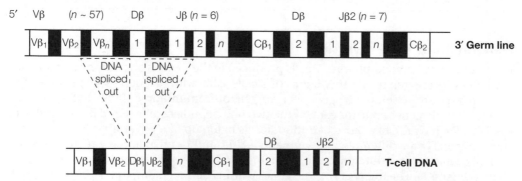

Figure 2. V-region gene for the human TCRβ chain. Similar to antibody genes in B cells, T cells rearrange their TCR genes during development. An example of a complete V-region gene for the β chain is shown. The VDJ exon is transcribed and spliced to join the Cβ gene segment. The resulting mRNA is translated into a β chain of the TCR. There are approximately 57 genes in the Vβ segment and two Dβ genes. There are two sets of Jβ segments with six and seven genes respectively and two Cβ segments. The Cβ genes do not appear to differ functionally from each other, unlike the various immunoglobulin C-region genes (Sections D2 and D3).

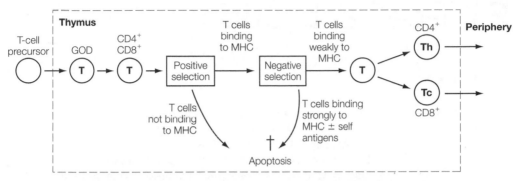

Figure 3. Thymic education. T-cell precursors derived from lymphoid stem cells enter the thymus, where they develop T-cell antigen receptors through multiple gene rearrangements: generation of diversity (GOD). The cells first acquire surface expression of both CD4 and CD8 molecules (double-positive) and then undergo positive selection—when they bind with sufficient avidity to MHC molecules. Those T cells that do not bind to MHC molecules die by apoptosis and are taken up by macrophages. The positively selected T cells that bind strongly to MHC molecules alone or with attached self peptides are negatively selected and die through apoptosis. Those T cells with weak binding to MHC molecules survive and lose either their CD4 or CD8 molecules to mature into single-positive CD4⁺ T-helper and CD8⁺ T-cytotoxic cells that leave the thymus and migrate to the secondary lymphoid organs/tissues.

self MHC are spared from death and are *positively selected* to survive (Figure 3). Therefore, the T-cell repertoire is first selected for cells that can bind self MHC. Of this group, those that express a TCR that binds *very strongly* to self MHC and thus cause problems if they enter the periphery, are induced to die (i.e., are *negatively* selected). This positive and negative selection results in survival and maturation of T cells that can recognize peptides in the context of self MHC (modified self MHC), but cannot react productively with self antigens (Section G2). The failure to rearrange a functional TCR to be negatively selected or to not be positively selected is responsible for the death of the majority (95%) of T cells in the thymus through apoptosis. It is believed that "natural" Tregs (which normally use a conventional αβ TCR) are positively selected in the thymus and then recognize self antigens in association with MHC class II antigens at a *moderately high affinity*, and are not eliminated completely by negative selection (Section G3).

# F4 T-cell activation

## Key Notes

| | |
|---|---|
| **Accessory molecules** | Initial recognition of processed antigen by T cells is via the T-cell antigen receptor. Accessory molecules further link the APC and the T cell, leading to a stronger cell interaction. For example, CD4 binds to the constant-region domain of class II MHC molecules while CD8 binds to class I MHC molecules. Other ligand–receptor pairs such as LFA-1 and ICAM-1 are also important. |
| **Co-stimulatory molecules: two signals required for T-cell activation** | Full activation of antigen-specific T cells requires **two signals**—one signal coming via the TCR and the other signal through engagement of co-stimulatory molecules. T cells receiving one signal via their TCR are turned off (become anergic), while those also receiving the second signal, that is, via T-cell CD28 binding to CD80/86 on the APC, induce T-cell lymphokine production and T-cell proliferation. |
| **Th activation through superantigens** | Some protein products of bacteria and viruses can initiate T-cell activation by directly linking the TCR on T cells to the MHC class II–peptide complex on APCs without the need for antigen processing. These superantigens include *staphylococcal enterotoxins* (SE) that cause common food poisoning and the toxic shock syndrome toxin (TSST). |
| **Early signaling events through the TCR co-receptors** | Contact between TCR, accessory, and co-receptor molecules with antigen-presenting molecules and ligands on the APC is called the "immunological synapse." This specialized signaling domain conveys a signal to the nucleus resulting in specific gene transcription. *Signal transduction* is brought about by phosphorylation and dephosphorylation of particular amino acids, thus activating them in a sequential fashion leading eventually to activation of specific transcription factors in the nucleus and production of functional proteins. CD45 (a phosphatase) on the APC initiates this process by activation of a CD4-associated kinase (lck), which together with Fyn then phosphorylates ITAMs on the $\zeta$ chain of the signaling complex. Binding of ZAP70 to the phosphorylated ITAMs initiates two biochemical pathways. |
| **Related topics** | (C1) Lymphocytes      (G2) Central tolerance<br>(E2) B-cell activation  (G3) Peripheral tolerance |

## Accessory molecules

Pathogens infecting peripheral sites are typically trapped in the lymph nodes directly downstream of the site of infection. Blood-borne pathogens are trapped in the spleen. These secondary lymphoid organs contain APCs (dendritic cells and macrophages) that efficiently trap antigen for processing and presentation. Naive and memory T cells recirculate through these sites looking for appropriately processed antigen. Initial recognition of processed antigen by T cells is via the T-cell antigen receptor.

Accessory molecules provide additional linkages between the APC and the T cell to strengthen their cellular association (Figure 1). CD4 binds to the constant-region domain of class II MHC molecules thereby strengthening the association of the TCR with peptide–class II MHC molecules. Likewise, CD8 binds to class I MHC molecules to strengthen the association of the TCR with class I MHC molecules. CD4 and CD8 are also important in regulating early activation events through signaling. In addition to engagement of these ligand–receptor pairs (Figure 1), additional adhesion molecules, integrins, become engaged. These include intercellular adhesion molecules (ICAMs) and lymphocyte function-associated antigens (LFAs).

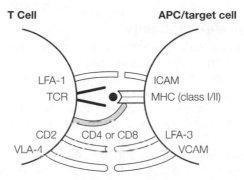

Figure 1. Pairs of molecules which strengthen the association of T cells with antigen presenting and target cells.

## Co-stimulatory molecules: two signals required for T-cell activation

Ligation of the TCR on its own does not stimulate T-cell clonal expansion or lymphokine production. The full activation of antigen-specific T cells requires **two signals**. Signal **one** is provided by the engagement of the T-cell antigen receptor and signal **two** is provided by engagement of a co-stimulatory molecule. The best-characterized co-stimulatory molecule is CD86, which is on many APCs and binds to CD28 on the T cell. Signals emanating from the TCR and CD28 synergize to induce T-cell lymphokine production and T-cell proliferation. If the T cell receives signal 1 (TCR binding) and not signal 2 (co-stimulation) the T cell is turned off (i.e., anergized, Figure 2, Section G3).

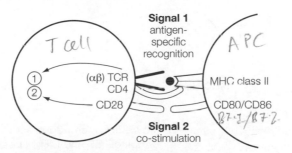

Figure 2. The role of co-stimulation in T-cell activation.

Precursors of CD8$^+$ cytotoxic T cells also need to be activated to develop into mature CD8$^+$ effector T cells containing granzymes and perforin (Section F5). This requires attachment of their TCR to MHC class I–peptide complexes on APCs (signal 1). In addition, a second co-stimulatory signal involving binding of CD86 to CD28 on the CTL is required. Cytokines produced by Th cells and APCs, and ligation of CD40 on the APC, are important for enhancing expression of the co-stimulatory molecules. Note that, although other Th-cell-conditioned APCs may be able to provide the necessary signals for Tc (CTL) activation, dendritic cells are the only cells which have a significant crossover of processed antigen between exogenous and endogenous pathways (Section F2). Moreover, they are, in general, the most efficient of the APCs.

Once activated, mature CD8$^+$ cytotoxic T cells do not, in most instances, need to be further activated to release granzyme and perforin when they come into contact with peptide MHC class I complex expressed by a virus-infected cell.

## Th activation through superantigens

Some protein products of bacteria and viruses are superantigens that bind simultaneously to lateral surfaces of MHC class II molecules (not in the peptide-binding groove) and the V region of the β subunit of the TCR. Superantigens are not processed into peptides as conventional antigens, but are able to bind to a specific family of TCR. In a sense they "glue" T cells to the APC (Figure 3) and cause stimulation of the T cell. However, these T cells are not specific for the pathogen that produced the superantigen, since all members of a particular family of TCR are activated. The consequence of binding to a large percentage of the T cells is the massive production of cytokines leading, in some cases, to lymphokine-induced vascular leakage and shock. Among the bacterial superantigens are the *staphylococcal enterotoxins* (SE) that cause common food poisoning and the toxic shock syndrome toxin (TSST).

## Early signaling events through the TCR co-receptors

The early signaling events are complex and therefore only a simplified outline will be presented. The area of contact between the TCR accessory and co-receptor molecules on the T-cell surface with molecules and ligands on the APC is called the "immunological synapse." This is the specialized membrane region that conveys a signal from the T-cell surface via the cytosol into the nucleus to give rise to specific gene transcription. This *signal transduction* is mediated by TCR molecules, co-receptors, and enzymes that lie in cholesterol-rich areas of the membrane called "lipid rafts" (Figure 4).

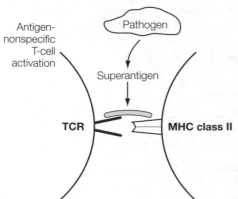

Figure 3. Superantigen activation of T cells by bridging TCR and MHC class II in the absence of specific peptide.

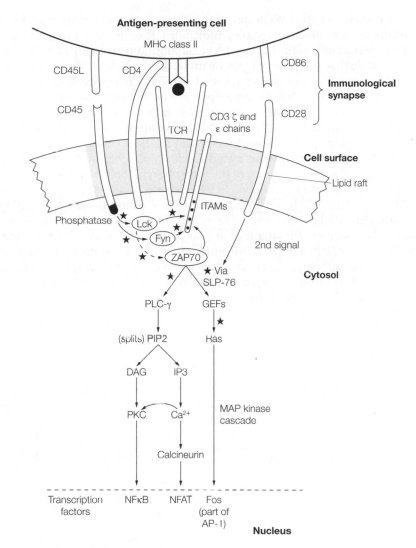

Figure 4. Early biochemical events leading to T-cell activation. The ligation of the TCR results in the initiation of signaling pathways through the action of CD45 (a phosphatase), the kinases Lck and Fyn, and other kinases, including ZAP70. These enzymes "activate" their target molecules by removing (phosphatases) or adding phosphates (kinases). Ligation of CD45 induces activation of Lck and Fyn, which phosphorylate ITAMs on the $\zeta$ chain and CD3. Once phosphorylated, ZAP70 binds to the ITAMs and is activated by Lck to initiate two main biochemical pathways. One is via the phosphatidylinositol pathway where phospholipase C-$\gamma$ (PLC-$\gamma$) cleaves phosphatidylinositol bisphosphate (PIP2) to produce diacylglycerol (DAG) and inositol trisphosphate (IP3). DAG activates protein kinase C (PKC), which activates NF$\kappa$B, which translocates to the nucleus. IP3 increases intracellular $Ca^{2+}$, activating the phosphatase calcineurin, which, in turn, activates NFAT (nuclear factor of activated T cells) and causes it to translocate to the nucleus. The second pathway involves the MAP kinase cascade initiated by RAS through GEFs activated via binding of SL76. This cascade leads to activation of Fos, a component of the AP-1 transcription factor. The second signal delivered via T-cell CD28 interaction with CD86 on the antigen-presenting cell is required to activate the cell. It is thought that CD28 ligation (and indeed ligation of CTLA4, which binds mainly to CD80) modulates these biochemical pathways.

Neither of the two chains of the TCR have intracytoplasmic tails of sufficient length or amino acid composition to act as signaling molecules. Therefore, T-cell signaling is initiated through the longer tails of the $\zeta$ chains, and $\varepsilon$ chains (and probably $\gamma$ and $\delta$ of the associated CD3 molecule that have sets of tyrosine molecules called ITAMs (immunoreceptor tyrosine activation motifs). On ligation of the TCR, CD45 (an endogenous phosphatase) activates (by removal of phosphates) two enzymes, Lck and Fyn, that phosphorylate the ITAMs of the $\zeta$ and CD3 chains. ZAP70 then "docks" with the phosphorylated ITAMs and itself becomes activated leading to further phosphorylation events. Activation of phospholipase C-$\gamma$ via the phosphatidylinositol pathway leads to activation of the transcription factors NFAT (nuclear factor of activated T cells) and NF$\kappa$B and their translocation into the nucleus. Another consequence of the phosphorylation mediated by ZAP70 is the activation of the MAP kinase cascade via SLP-76, guanine nucleoside exchange factors (GEFs), and Ras, which finally leads to activation of Fos—a component of the AP-1 transcription factor. This whole process is very rapid and the multiple phosphorylation and dephosphorylation events in the membrane take place within seconds of ligation of the TCR. The initial signal occurring within the lipid rafts is amplified via the molecules of the different biochemical pathways ("second messengers"), leading finally to transcription of effector molecules, for example, cytokines (IL-2, IL-4, and IFN$\gamma$) and cell cycle proteins (cyclins), required for clonal expansion.

# F5 Clonal expansion and development of memory and effector function

## Clonal expansion and development of memory

Like B cells, T cells are selected by their specific antigen to take part in the immune response (clonal selection). They proliferate, producing a large number of cells (clonal expansion) and develop into effector cells, that is, Th cells or Tc cells, or into memory cells (Figure 1). The first memory cells to be produced (central memory, Tcm) have all the homing and chemokine receptors, like naive cells, to allow them to recirculate to the T-cell areas of the lymphoid tissues. These cells act as a clonal reservoir, are capable of proliferating on subsequent antigen activation, and give rise to effector memory

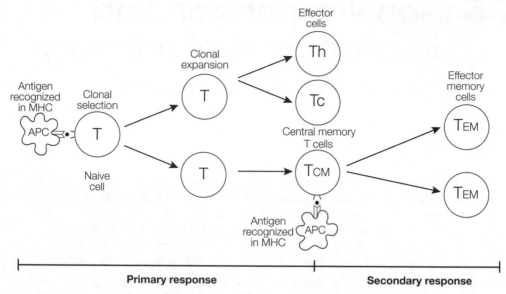

Figure 1. Clonal expansion and development of memory T cells. Specific naive T cells are selected by antigen to proliferate (clonal selection and clonal expansion) and develop effector functions, that is, T-cell help or cytotoxicity (primary response). Central memory T cells (TCM) are also produced that are capable of proliferation and recirculation through lymphoid tissues. On subsequent contacts with specific antigen (e.g., secondary response) they develop into effector memory cells (TEM) that rapidly acquire full function (within a few hours) and are capable of homing to sites of inflammation, where they mediate help or killing.

cells (TEM). They rapidly develop effector function, for example cytotoxic activity, and are capable of migrating into sites of inflammation in nonlymphoid areas. On initial activation by antigen, Th cells produce IL-2, an autocrine growth factor for T-cell pro-liferation and also express IL-2 receptors (Figure 2). This results in clonal expansion. Other surface molecules induced by activation of T cells include CD40L (CD154), which through interaction with CD40 on dendritic cells leads to the production of cytokines (e.g., IL-1, IL-12) required for induction of Th subsets (see below).

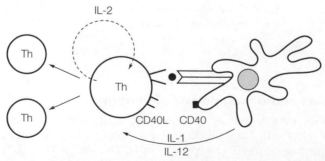

Figure 2. Initial priming of helper T cells through dendritic cells. Th cells recognizing antigenic peptides associated with MHC class II on a dendritic cell produce IL-2 and upregulate IL-2 receptors. IL-2 induces their proliferation, leading to clonal expansion. This priming also results in the development of memory cells.

## Helper T cells

The two main kinds of CD4⁺ T-helper subsets (Th1 and Th2) have different functions. Each develops from uncommitted Th0 cells following initial contact with a microbe. Th1 cells are predominately involved in mediating inflammatory immune responses (through the activation of macrophages), while Th2 cells are primarily involved in the induction of humoral immunity (via the activation of B cells). In this regard, Th0 to Th1 cell development is encouraged when a Th0 cell recognizes a microbial peptide presented by a dendritic cell that is producing IL-12 induced through IFNγ produced in autocrine fashion and from NK cells (Figure 3). In contrast, Th0 cells are encouraged to become Th2 cells under the influence of IL-4, released by dendritic cells exposed to IL4, B cells, and other cells (e.g., mast cells). Thus, following activation by specific peptide antigen, Th1 cells produce cytokines such as IL-2, IFNγ, and TNFα that mainly help cell-mediated responses such as macrophage activation, or help for Tc cells and also IgG class switch and Th2 suppression. Cytokines produced by Th2 cells (IL-4, IL-5, IL-6, and IL-13) are involved mainly in B-cell differentiation and maturation. IL-10 is also produced by Th2 cells and regulates the activity of Th1 cells (Section G5). The transcription factor **Tbet** is activated in Th1 cells, and **GATA3** in Th2 cells. These transcription factors are responsible for the spectrum of cytokines produced by each of these subsets. It is important to

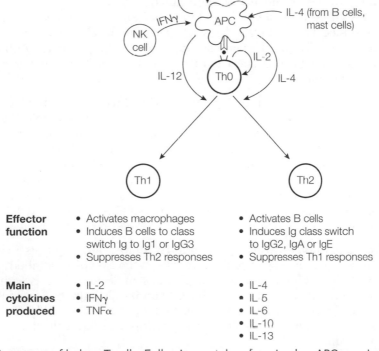

| | | |
|---|---|---|
| **Effector function** | • Activates macrophages<br>• Induces B cells to class switch Ig to Ig1 or IgG3<br>• Suppresses Th2 responses | • Activates B cells<br>• Induces Ig class switch to IgG2, IgA or IgE<br>• Suppresses Th1 responses |
| **Main cytokines produced** | • IL-2<br>• IFNγ<br>• TNFα | • IL-4<br>• IL 5<br>• IL-6<br>• IL-10<br>• IL-13 |

Figure 3. Two types of helper T cells. Following uptake of a microbe, APCs are induced to produce IL-12 through IFNγ that either they produce themselves or is derived from another source (e.g., NK cells). They present microbial-derived peptides in MHC class II molecules to specific Th0 cells. In the presence of IL-12 and IFNγ, Th0 cells differentiate into Th1 cells, whereas in the presence of IL-4 derived from other sources (e.g., autocrine, B cells, or mast cells) the Th0 cells differentiate into Th2 cells.

remember that in most cases, an immune response elicits both Th1 and Th2 activities, although there are some instances where one or the other is more effective in mediating protection (Section H2).

## Th2 cells

Participation of T cells is required for B-cell responses to most antigens (Section E2). T cells most effective in inducing the production of antibody from B cells, especially of the IgA and IgE isotypes, are the helper CD4 Th2 cells. Th2 cells induce B cells to produce antibodies, switch the antibody isotype being produced, and induce antibody affinity maturation. This involves not only cytokines but direct engagement of surface molecules on the T and B cells (cognate interactions) which trigger their activation.

More specifically, Th2 cells recognize antigenic peptides in MHC class II molecules on the surface of antigen-specific B cells and, through interaction with other surface molecules, are activated (Sections F2 and F4). The interaction of CD40L on T cells with CD40 on B cells induces B-cell proliferation and class switching to IgE- or IgA-producing cells (Figure 4 in Section D3). Cytokines produced by Th2 cells, including IL-4, IL-5, and IL-6, act as growth and differentiation factors for B cells. Specialized Th cells in the B-cell follicle (T-follicular cells) are believed to be important in memory B-cell production and class switching.

## Th1 cells

**Role of Th1 cells in macrophage recruitment and activation.** The response to a variety of intracellular parasites is dependent upon functionally intact Th1 cells. For example, the immune responses to *Leishmania* and mycobacteria are severely diminished if the host cannot produce IFNγ and TNFα. This is because, in the absence of these cytokines, infected macrophages cannot become activated to kill the pathogen. Although other cytokines can augment macrophage activities, both IFNγ and TNFα are critical for effective macrophage activation.

Th1 cells when activated also produce chemokines (which assist in the recruitment of monocytes) and colony-stimulating factor (GM-CSF) that induces their differentiation into macrophages at the site of infection. In addition, IL-3 increases the production and release of monocytes from the bone marrow. Also, TNFα from Th1 cells, as well as monocyte/macrophages themselves, alters the surface properties of endothelial cells, promoting access of monocytes to the site of infection (Section B4). The coordinated production of these mediators allows the infiltration of T cells and monocytes into the site of inflammation, where their interaction leads to macrophage differentiation, activation, and the elimination of the pathogen (Figure 4).

**Role of Th1 cells in isotype switching and affinity maturation.** Th1 cells, in addition to T-follicular cells, induce B cells to proliferate, switch the isotype of antibody produced, and undergo antibody affinity maturation (Figure 4 in Section D3). As with Th2 cells, the interaction of CD40L on Th1 cells with CD40 on B cells induces B-cell proliferation and class switching. Cytokines are also important, but in this case IFNγ and TNFα are involved, resulting in signals and help for the development of B cells that produce primarily IgG antibodies.

**Role of Th1 cells in induction of CD8⁺ T-cell cytotoxicity.** Antigen-presenting cells (APCs) initially process and present microbial peptides via the exogenous pathway in association with MHC class II molecules (Section F2). Th1 cells recognize and are activated by interaction with these cells, and in turn influence APC function. They do this by direct cell–cell interaction and signaling, and through release of cytokines

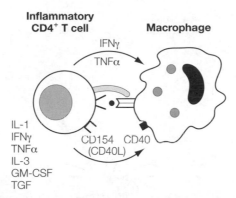

Figure 4. Macrophage activation by CD4+ inflammatory T cells. Cytokines released by Th1 cells, as well as signaling through direct contact of cell surface receptors, increase (a) fusion of lysosomes and phagosomes; (b) production of nitric oxide and oxygen radicals for killing pathogens; and (c) expression of MHC class II molecules and TNF receptors by macrophages. Note that CD154/CD40 interactions are also important in activation of the macrophage.

(e.g., IFNγ). The APCs (e.g., dendritic cells) become "**conditioned**" to present peptides to CTLs via their MHC class I molecules in addition to presentation to Th1 cells through MHC class II (Figure 5). Through some crossover of exogenous antigen into the endogenous pathway some peptides can become associated with MHC class I molecules. The help provided by the Th1 cells induces the development of precursor Tc cells into mature cytotoxic T cells.

### Th17 cells

This recently described Th subset is thought to be related to the Th1 subset, but its induction from Th0 cells is dependent on TGFβ and IL-21 and not IL-12 and IFNγ. The induction of Th 17 cells by TGFβ suggests that they are related to the Treg subset (Section G3). They produce both IL-17 and IL-22 and appear to be of major importance in protection against microbes entering the gastrointestinal tract.

## Cytotoxic T cells (CTLs)

### Recognition of antigen and activation

Peptides derived from viral proteins are processed via the endogenous route and are presented on the cell surface by MHC class I molecules, marking this cell as infected and as a target for CTL killing. CTLs express cell-surface CD8 which binds to the nonpolymorphic region of MHC class I (expressed on all nucleated cells), restricting these killer T cells to recognizing only cells presenting peptide in MHC class I molecules (Section F2). This interaction also serves to stabilize the interaction of the T-cell receptor with specific peptides bound to the polymorphic part of the MHC class I molecule (Figure 6). Other surface co-stimulatory and adhesion molecules such as LFA-1 are important for close interaction of the CTL with the infected cell and for activating its cytotoxic machinery (see below). This activation step also induces the expression of FasL on the CTL, which can interact with Fas expressed on the surface of the virus-infected cell.

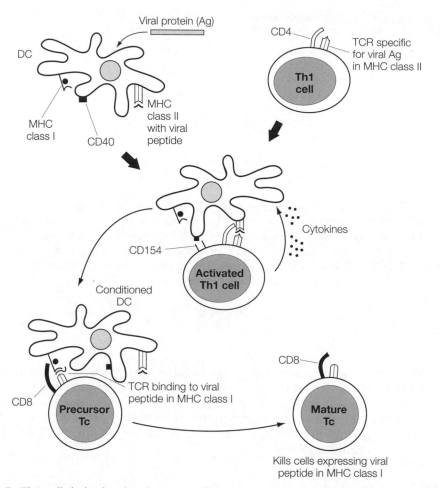

Figure 5.  Th1 cells help the development of cytotoxic T cells. Dendritic cells (DC) take up viral antigens and present viral peptides in MHC class II molecules to antigen-specific Th1 cells. Interaction of specific Th1 cells with peptide on the DC induces the expression of CD154 (CD40L) that binds CD40 on the DC. Other co-stimulatory molecules such as CD86 and CD28 are also involved in the activation process (not shown). Activation results in the release of cytokines, which, together with the cell interactions, "conditions" the DC to present viral peptides in MHC class I to specific Tc-precursor cells, inducing their maturation into Tc cells.

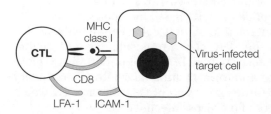

Figure 6.  CTLs recognize peptides associated with MHC class I molecules. CD8 binds to nonpolymorphic MHC class I, stabilizing this interaction and enhancing killing. The interaction of LFA-1 and ICAM-1 is also important in killing of the target.

## Mechanisms of cytotoxicity

Mature CTLs, generated with the help of Th1 cells, contain the cytotoxic machinery required to kill virus-infected cells. In particular, these CTLs are able to induce programmed cell death (apoptosis) of the virus-infected cells through two distinct pathways: (i) release of lytic granules containing perforin and granzymes which enter the target cell; (ii) interaction of FasL on the CTL with Fas on the target cell.

1. *Perforin-induced apoptosis.* CTLs contain large cytolytic granules and are difficult to distinguish morphologically from NK cells (also called large granular lymphocytes; Section B1). These intracytoplasmic granules contain proteases, granzyme A and granzyme B, and perforin, a molecule similar to C9 of the complement pathway (Section D8). On interaction of the CTL with a virus-infected cell, the granules move toward the portion of the membrane close to the point of contact with the target cell. On fusion with the membrane, the granules release perforins, which polymerize in the membrane of the infected cell creating pores that allow entry of the proteases (Figure 7). These enzymes cleave cellular proteins, the products of which initiate induction of programmed cell death (apoptosis). CTLs then resynthesize their granular contents in preparation for specific killing of another infected cell.

2. *Fas-mediated apoptosis.* Nucleated cells of the body infected with some viruses upregulate expression of Fas (CD95). CTLs activated to release their granules by their first encounter with antigen presented by MHC class I molecules are induced to upregulate FasL, which then also allows them to kill specific virus-infected cells by an additional mechanism through interaction with surface CD95 (Figure 8). Note that persistent activation of both Th and Tc cells leads to upregulation of Fas leading to fratricide through activation-induced cell death (AICD). This is believed to be an immune mechanism designed to reduce the pathological immune responses against allergens and autoantigens (Sections K and L).

The importance of apoptosis as a killing mechanism used by the immune system is that targeted cells can be removed rapidly by phagocytes without initiating inflammatory responses. Another mechanism of cell death—necrosis—results from tissue trauma or certain kinds of infection and leads to acute inflammation through the production of inflammatory cellular products.

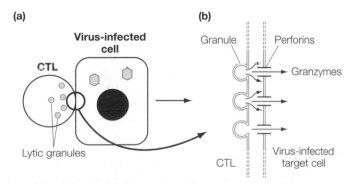

Figure 7. Apoptosis induced by release of lytic granules. (a) Lytic granules containing perforin and granzymes accumulate at the point of contact of the CTL with the virus-infected cell. (b) The granule contents are released and the perforins polymerize in the infected cell membrane, allowing entry of granzymes into the target cell which induce apoptosis.

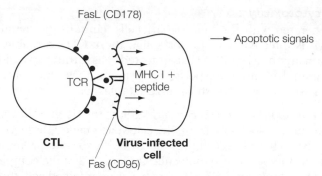

Figure 8. Apoptosis induced by Fas/FasL interactions. CTLs have preformed FasL in their granules, which is rapidly expressed on their surface when they attach via their TCR to the target cell. Ligation of Fas on the virus-infected cell by FasL on the CTL is an additional mechanism for induction of apoptosis.

# G1 Overview

---

**Key Notes**

**Overview**

The immune system has to be tightly regulated, "turned on" in response to "danger" signals, fine-tuned, and "turned off" again when the threat has been removed. These danger signals come mainly from invading "non-self" microbes but can also come from within in the form of self antigens (autoimmunity) when the natural mechanisms for prevention (immunological tolerance) break down.

**Regulatory mechanisms within the innate immune system**

In the innate immune system, phagocytes only recognize self cells if they are damaged or dying; natural killer cells are normally inhibited from killing self cells through inhibitory receptors; complement cannot be activated on the surfaces of normal body cells due to inhibitory molecules.

**Regulation by the adaptive immune system**

Antigen initiates and drives the immune response, the magnitude of which is under the control of many genes especially those in the MHC gene locus. Regulation is influenced by antigen size, closeness to self, persistence and removal, help from T cells, modulation by the antibody produced in both a positive and negative way, and suppression of the response through Tregs. Tolerance to self molecules is initiated at the level of development (central tolerance) and in the periphery mainly through the lack of co-stimulatory signals, through activation-induced cell death and Tregs. The neuroendocrine system also plays an important role in modulating immune responses.

**Related topics**

(B1) Cells of the innate immune system
(B2) Molecules of the innate immune system

(B3) Recognition of microbes by the innate immune system

---

## Overview

It is essential that the immune system be tightly regulated. It has to be "turned on" in response to a "danger" signal, then fine-tuned to give an optimum and appropriate response (in the case of the adaptive immune response), and then "turned off" again when the danger has gone. The danger signals come mostly from invading microbes but also from within the body in the form of self antigens when the natural mechanisms for prevention (immunological tolerance) break down (Section L). In addition, some non-microbial substances such as allergens can pose as danger signals to some individuals and danger/alarm signals are generated within the body from "aberrant" cell death or cell stress resulting from a range of causes such as anoxia. All of the immune cells alerted in response to these danger signals have to be controlled and this involves a variety of regulatory mechanisms.

## Regulatory mechanisms within the innate immune system

There are two ways in which self reactivity is prevented: lack of recognition (ignorance) of self cells (unless they change their surface structure); and the presence of inhibitory structures/receptors on the nonimmune cells.

### Phagocytes

Phagocytes of the innate system, including macrophages and neutrophils, do not normally "recognize" or phagocytose healthy self cells. However, aging (erythrocytes) and dying or dead cells express new surface molecules that are recognized by a number of different receptors on phagocytes, resulting in the removal of these altered self cells. In particular, those cells of the innate system recognize microbes through their pattern recognition receptors (PRRs) (Section B3). Some of these same PRRs are also used to recognize dead or altered self cells for their removal. Examples of these are TLR4 and TLR2 that recognize stress/heat shock proteins. Other PRRs recognize sugars, for example mannose (Section B3).

Target molecules on the surface of mammalian cells that might be recognized by these receptors are either absent or concealed by other structures, for example sialic acids. When an erythrocyte ages, it loses sialic acid, exposing $N$-acetylglucosamine, which the phagocyte now recognizes as non-self and phagocytoses (Figure 1). When nucleated cells die, a large number of new surface molecules are exposed which are recognized by phagocytes. One of these molecules, phosphatidylserine (PS)—a membrane phospholipid—is normally restricted to the inner surface of the cell membrane. When the cell begins to die through apoptosis PS "flips" onto the surface and is recognized by the phagocytes.

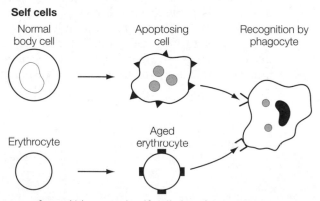

Figure 1. Recognition of aged/damaged self cells by phagocytes

### Natural killer cells

Natural killer cells play an important role in killing virus-infected cells. Killing of noninfected nucleated cells of the body is prevented through a balance in signaling involving killer activation receptors (KARs) and killer inhibitory receptors (KIRs) that recognize molecules on self cells. The inhibitory receptors recognize MHC molecules on normal cells and prevent their killing by NK cells (Sections B1, F2, and N3). However, when certain viruses infect cells they downregulate expression of molecules (MHC class I) recognized by KIRs, giving rise to an overriding activation through KARs leading to death of the infected cells (Figure 2).

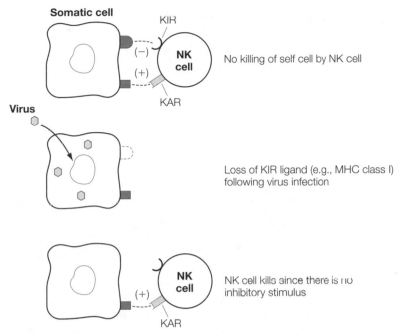

Figure 2. Inhibition of NK cell activity. NK cells recognize self antigens and travel around the body in search of aberrant self cells. When they come into contact with a healthy cell they receive two signals—a positive signal to kill via their KARs (killer activation receptors) and a negative signal via their KIRs (killer inhibitory receptors). These two signals cancel each other out and the NK cell goes on its way. Some viruses inhibit expression of the self molecules recognized by the KIR (e.g., MHC class I, HLA-A, B, C in man), which means that the negative signal is absent and the NK cell carries out its lethal duty.

## The complement system

C3 is activated through the alternative pathway by stabilization of appropriate enzymes on the surface of some microorganisms (Section B2). This cannot occur on self cells since they all have inhibitory molecules (Section D8) on their surface membranes (Figure 3).

## Regulation by the adaptive immune system

A large number of genes are involved in the development of immune responses. Their importance is emphasized when they are absent through mutations and deletions, causing immunodeficiency diseases (Section J). The MHC genes are important in that the different allelic forms (polymorphism) code for MHC molecules with different binding sites that bind the same peptides with different strengths and therefore influence the specificities of the T-cell responses (Section F2).

The nature of the antigen that drives the response is important, because its size, state of aggregation, composition (e.g., protein vs. some sugars that can be persistent), and so forth, significantly influence the type of response and its strength (Section A4). Removal of the antigen and therefore the stimulus results in the response subsiding. Helper T cells require at least two signals to provide help: one from the antigenic peptide, the other via co-stimulatory signals from antigen-presenting cells. They also modulate the functions of other cells, including dendritic cells, NK cells, macrophages, and cytotoxic T cells. Although this modulation is often mediated through cytokines, it may also involve

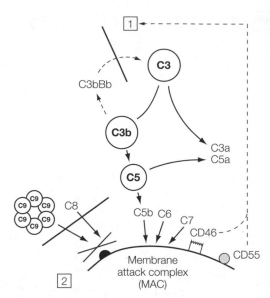

Figure 3. Inhibition of complement activation on self cell surfaces by regulatory proteins. (1) Inhibition of C3 convertase by membrane co-factor proteins (CD46) and decay accelerating factor (CD55). (2) Blocking by CD59 of attachment of C8 and C9 of the membrane attack complex (MAC) to membrane, inhibits active lysis.

direct cell–cell interactions. The influence of Th cells can significantly affect the type of response, depending at least partly on the kinds of cytokines produced and the particular cell participating in the response. Antibody itself can, in some instances, either enhance or inhibit (Sections E2 and G4) further antibody production. Tregs inhibit the immune response to a number of antigens particularly self antigens. Tolerance to self is initiated at the level of development of T and B cells (Sections A5, C1, F3, and G2) in the primary lymphoid organs (central tolerance). Those lymphocytes escaping elimination at this stage are prevented from responding to self through lack of co-stimulatory signals (e.g., those provided by CD80 or CD86 on APCs that are necessary for T-cell activation), resulting in anergy or activation-induced cell death mediated by T cells (peripheral tolerance).

It is also important to note that the immune system does not function in isolation, but is influenced by other body systems. In particular, the neuroendocrine system plays an important role in modulating immune responses.

# G2 Central tolerance

| Key Note | |
|---|---|
| **Central tolerance** | Central tolerance is the process whereby immature T and B cells acquire tolerance to self antigens during maturation within the primary lymphoid organs/tissues (thymus and bone marrow, respectively). It involves the elimination of cells with receptors for self antigens (negative selection). |
| **Related topics** | (D3) Generation of diversity<br>(E2) B-cell activation<br>(F3) Shaping the T-cell repertoire<br><br>(F4) T-cell activation<br>(L3) Autoimmune diseases – mechanisms of development |

## Central tolerance

The fundamental basis for central tolerance is that interaction of antigen with immature clones of lymphocytes already expressing antigen receptors results in their elimination. This theory, for which Burnet and Medawar received the Nobel Prize in 1960, is now recognized to involve a mechanism that causes elimination of self-reactive lymphocytes (**clonal deletion**) on contact with self antigens. Immature precursor T cells derived from bone marrow stem cells migrate to the thymus to mature into immunocompetent T cells. T cells with specificity for self appear during normal development in the thymus as the result of the expression of combinations of V-segment genes (Sections D3, E3, and F3). These self-reactive T cells must be eliminated to prevent autoimmunity.

T cells expressing receptors with weak binding to MHC class I and II antigens are permitted to survive; they are **positively selected** (Figure 3 in Section F3). T cells which bind with high affinity to MHC class I and II, alone or carrying self peptides (Sections F2 and F3), are induced to die through the process of apoptosis. For some time it was unclear how self-reactive T-cell clones with specificity for (auto)antigens expressed in tissues outside the thymus were eliminated by central tolerance. It is now known that an autoimmune regulator (AIRE)—a transcription factor in thymic epithelial cells—is able to switch on a number of genes coding for peripheral tissue antigens normally found in tissues and organs in the body, especially endocrine tissues. Thus, CD4/CD8 double-positive T cells with specificities for these self proteins can be deleted within the thymus (Figure 1).

This **negative selection** leads to elimination of some but not all self-reactive T cells. Cortical epithelial cells are the main players in the positive selection process whereas macrophages and interdigitating dendritic cells (and probably medullary epithelial cells) play a role in negative selection. This "education" process within the thymus leads to suicide of more than 90% of the T cells. Thus, only a small percentage of the T cells generated survive to emigrate to the peripheral tissues. These T cells are mainly specific for foreign non-self peptide antigens but some with affinity for self escape and migrate to the peripheral lymphoid tissues.

A similar process of negative selection occurs during B-cell development in the bone marrow. As in the thymus, receptor diversity for antigen is created from rearrangement of

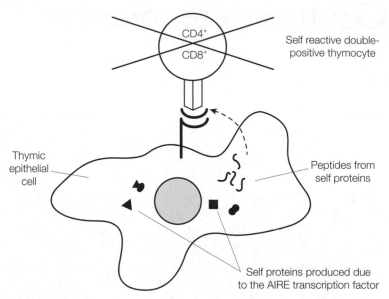

Figure 1. Elimination of self-reactive thymocytes through the transcription factor AIRE. Thymic epithelial cells synthesize proteins normally made by organs and tissues of the body especially endocrine tissues. These peripheral tissue proteins are processed and presented to double-positive thymocytes that have a high affinity for their peptides and eliminated by apoptosis. Other cells such as medullary macrophages are also thought to generate these self proteins through AIRE, or endocytose these proteins and present them through the exogenous processing pathway.

V-segment genes, resulting in some B cells having membrane antibodies with self reactivity. **B-cell tolerance** occurs as a result of clonal deletion, through apoptosis, of immature B cells reactive to self antigens (Figure 2). Also, immature B cells with specificities for soluble self antigens can be rendered unresponsive (tolerant) or anergic and are not eliminated. B cells with IgM receptors that do not react with self antigens or react with low affinity mature and migrate to the periphery, where further maturation occurs.

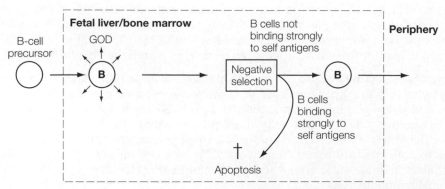

Figure 2. B-cell central tolerance. B-cell precursors develop diverse antigen receptors (GOD) that undergo negative selection with the surviving cells migrating to peripheral (secondary) lymphoid organs/tissues.

# G3 Peripheral tolerance

| Key Note | |
|---|---|
| **Peripheral tolerance** | Since not all self-reactive lymphocytes are eliminated by central tolerance mechanisms (due primarily to the absence of most self antigens in the primary lymphoid organs), self-reactive lymphocytes are anergized or deleted in the peripheral tissues. Peripheral T cells are made unresponsive (anergic) through the absence of the second signal (essential for T-cell activation) given by co-stimulatory molecules (i.e., CD80, CD86) on APCs. Peripheral B cells may become anergic and unable to develop into plasma cells as a result of the absence of co-stimulatory signals from T cells. Moreover, under appropriate conditions, activated T cells expressing Fas ligand (FasL) may kill Fas-expressing B cells (and, perhaps, other T cells) through **activation-induced cell death** (AICD). |
| **Related topics** | (E2) B-cell activation       (L3) Autoimmune diseases – <br> (F3) Shaping the T-cell          mechanisms of <br>       repertoire                 development <br> (F4) T-cell activation |

## Peripheral tolerance

Many self-reactive lymphocytes are not eliminated in the primary lymphoid organs for two reasons. Firstly, many self antigens are neither present in the primary lymphoid organs nor supplied to them via the bloodstream (e.g., "sequestered antigens" such as lens proteins in the eye), and many self antigens expressed as the result of differentiation of cells and tissues in the major organs of the body are not regulated by AIRE. Thus self-reactive T cells that escape negative selection in the thymus have to be deleted or rendered anergic in the periphery where they come into contact with the majority of self antigens. Secondly, self-antigen receptor specificities may be generated in B cells as the consequence of somatic mutation of antibody genes in cells within the germinal centers of secondary lymphoid organs/tissues (Sections C2, D3, and E4). Unlike B-cell antigen receptors, it is believed that TCRs do not normally mutate. The silencing of self-reactive lymphocytes in the periphery might be due to deletion in some cases. However, it is generally believed that the "silencing" is mainly due to **anergy** induced during the initial contact of the T cells with self antigen in the periphery.

### One-signal hypothesis: anergy

Naïve T cells require two main signals to respond to an antigen. One comes via the TCR, the other comes from co-stimulatory molecules. The glycoproteins CD80 and CD86 are essential co-stimulatory molecules, found almost exclusively on professional antigen-presenting cells (APCs). Direct interaction of these molecules on APCs (especially CD86) with CD28 on T cells, together with the resulting cytokines produced by the APC, is required for T-cell activation (Figure 1). It is thought that upregulation of expression of

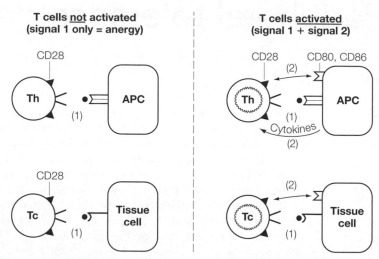

Figure 1. T-cell anergy. Th and Tc cells (including those that are self-reactive) cannot be activated by one signal. Binding of B7 (CD80, CD86) on the APC/tissue cell to CD28 provides a second signal to the T cells leading to their activation.

CD80/86 is a requirement for this interaction, and that engagement of PRRs on the APC through binding to foreign microbes induces this. Thus, in the absence of professional presentation of self antigens and engagement of co-stimulatory molecules (signal 2), the binding of self antigens presented in MHC molecules to the TCR on naive T cells results in anergy. Moreover, if naive T cells do become activated, they express an additional receptor called CTLA-4 which has a greater binding affinity for the CD80 molecules than CD28. Binding of CTLA-4 to CD80 results in a negative signal to the T cells, resulting in inhibition of T-cell activity (Section F4).

Self-reactive B cells require T-cell help in order to respond to T-dependent antigens. Since most self-reactive T cells have been deleted during thymic maturation, self-reactive B cells on contact with self antigens do not receive the required co-stimulatory signals (signal 2) from T-helper cells and consequently become anergic (Figure 2). Engagement of the B-cell co-stimulatory molecules CD40 and CD86 by CD154 and CD28 on T cells, as well as certain cytokines (IL-2, IL-4, IL-5, IL-6), are required for activation (Section E2). Aberrant help by non-self-reactive Th cells providing a second signal for autoreactive B cells may lead to autoimmune responses (Section L).

### Activation-induced cell death (AICD)

Fas/FasL (CD95/CD95L) interaction is directly responsible for AICD. This is important in maintaining immunological as well as physiological homeostasis by eliminating unnecessary cells through apoptosis. Activated T lymphocytes can express both the receptor protein Fas and its ligand (FasL), whereas B cells mainly express Fas. Peripheral tolerance may be facilitated by interaction between activated T cells and B cells (and, perhaps under certain conditions, other T cells) resulting in apoptosis (Figure 3). In addition, T cells activated to kill self cells may themselves be killed by interaction with FasL expressed by certain somatic cells, for example those in the eye and testis (Section L3), thus preventing killing of these self cells. Note that this may also be a strategy used by tumor cells to prevent their demise by cytotoxic T cells.

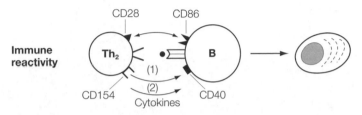

Figure 2. B-cell anergy. B cells require triggering through their CD40 molecules to progress through activation and maturation. Interaction of CD28 on T cells with CD80/86 on B cells is necessary to induce expression of CD154 (CD40 ligand). This binds to CD40 on the B cell, acting as the second signal for activation.

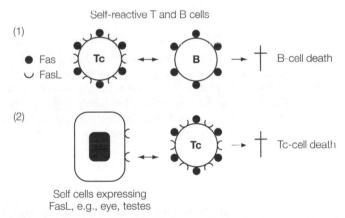

Figure 3. Activation-induced cell death (AICD) in peripheral tolerance. Tc cells may kill self-B cells expressing Fas (1) and because Tc can also express Fas on activation, may themselves be killed by tissue cells expressing FasL:CD178 (2).

## Regulatory T cells (Tregs)

Many immune responses, particularly those directed against self antigens, are regulated through a subset of CD4+ T cells that express a high level of surface CD25 (the α chain of the IL-2 receptor). Activated T cells and B cells also express CD25, but Tregs do not express several markers of activated lymphocytes and characteristically express the forkhead box P3 (FoxP3) transcription factor that controls their development and function. The two main types of Tregs are "natural" or thymus-derived, and "induced" that are generated in the periphery in response to antigens under the influence of specific cytokines, for example TGFβ (Figure 4). Tregs are mostly MHC class II restricted and develop alongside

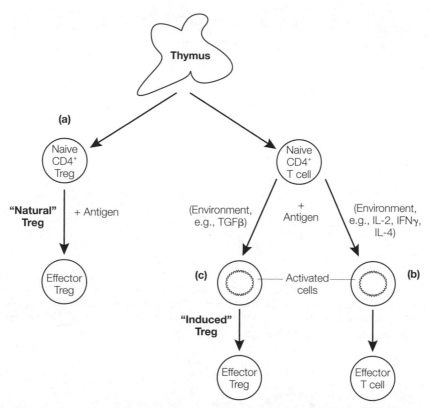

Figure 4. Origin of regulatory T cells (Tregs). Tregs and "conventional" CD4⁺ Th cell are educated in the thymus and migrate to the periphery. (a) The "natural" Tregs, following recognition of (self) antigen, become effector Tregs that suppress immune responses. (b) Naive conventional CD4⁺ Th cells leave the thymus and following cognate interaction with antigen become effector cells. (c) Other conventional CD4⁺ Th cells on recognition of antigen in the presence of particular cytokine environments, for example TGFβ, become "induced" Tregs and effector suppressor cells.

other CD4⁺ T cells in the thymus. They are believed to have a relatively high affinity for MHC during positive and negative selection but many escape death and migrate to the periphery. Most natural Tregs have specificities for self tissue antigens but do not proliferate in response to them, unlike the responses of conventional mature CD4⁺ T cells to non-self antigens; that is, Tregs are anergic. However, they do suppress the response of self-reactive T cells and therefore contribute to peripheral tolerance. The mechanisms by which Tregs inhibit immune responses is unclear, but several possibilities have been suggested (Figure 5).

Tregs constitute 1–4% of adult blood CD4⁺ T cells, decreasing to 0.5–1.5% in healthy aged individuals (Section P). That Tregs play a role in maintenance of self tolerance is demonstrated by the observation that mutations in the gene controlling FoxP3 result in the development of a fatal autoimmune disease, IPEX (immune dysregulation, polyendocrinopathy, enteropathy, X-linked syndrome, Section L). There are also many examples of experimental animal models where removal of these cells results in development of a range of autoimmune diseases.

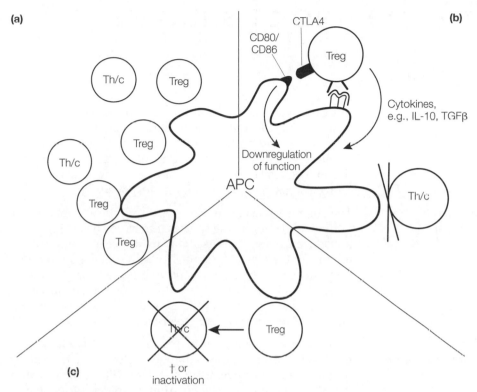

Figure 5. Possible mechanisms for the suppressive function of Tregs. (a) Competition of Tregs with T-helper or T-cytotoxic cells (Th/c) for interaction with APCs could reduce the development of normal effector responses. (b) CTLA-4, expressed by many Tregs, interacts with CD80 (and CD86) on APCs and downregulates their function. Cytokines released by Tregs such as IL-10 and TGFβ can also influence APC function. (c) Tregs might induce apoptosis or inactivation of effector Th/c.

Because of their suppressive effects, much research is directed towards the use of Tregs in treatment of autoimmune diseases (Section L) and allergic diseases (Section K), and enhancing the acceptance of allografts (Section M).

There are several other lymphocyte populations that display regulatory activity, including γδ T cells (Section F) and NKT cells (Section B).

# G4 The role of antigen

## Key Notes

**Overview**

Antigen is required for initiation of immune responses. A number of factors influence development of the immune response, including nature of the antigen, maturity of the immune system (age of host), route of immunization with the antigen, and dose and persistence. Removal of the antigen stops the specific immune response.

**Nature of the antigen**

The chemical makeup and complexity of the antigen, as well as how similar it is to the species into which it is being introduced, determines its ability to induce immunity or tolerance. The closer the similarity with self, the poorer the immune response that develops. Aggregated antigens are taken up more easily by APCs, stimulating a large number of specific T cells. Soluble antigens produce weaker responses and even unresponsiveness (tolerance).

**Maturity of the immune system**

Contact with antigen by immature B cells and T cells produces weaker responses than in adults and is more likely to lead to unresponsiveness (i.e., tolerance). Naive immature lymphocytes express fewer co-stimulatory receptors than mature T cells and developing immune systems have fewer memory T cells able to give stronger immune responses to antigens.

**Route of antigen administration**

Certain antigens given subcutaneously or intramuscularly may be more immunogenic than when given intravenously or intraperitoneally. Antigens fed by the oral route can often induce oral "tolerance" mediated through clonal anergy and deletion, and Tregs.

**Dose, persistence, and removal of antigen**

Small amounts of antigen given over a period of time produce better immune responses than very large amounts that can produce unresponsiveness ("high zone" tolerance). Persistent antigen maintains production of specific immune responses. Removal of antigen by antibody removes the stimulus for immune cells and therefore stops the response.

**Related topics**

| | |
|---|---|
| (A4) Antigens | (G2) Central tolerance |
| (B2) Molecules of the innate immune system | (G3) Peripheral tolerance |
| | (J3) Secondary (acquired) immunodeficiency |
| (C3) Mucosa-associated lymphoid tissues | (L5) Diagnosis and treatment of autoimmune disease |

## Overview

Antigen initiates the immune response via presentation of its peptides by antigen-presenting cells (dendritic cells and macrophages) to antigen-specific Th cells. Th cells then help B cells produce antibody, or CTLs develop. A number of factors influence the magnitude and quality of an immune response. These include nature of the antigen, maturity of the immune system (age of host), route of immunization with the antigen, and dose and persistence. Removal of the antigen stops the specific immune response.

## Nature of the antigen

The more complex the foreign antigen is in composition and structure, and the more dissimilar to the host, the greater the likelihood that a strong immune response will be generated. For example if a microbial antigen is similar in structure to a self antigen then the response to that antigen will be low or absent due to the normal tolerance mechanisms for self antigens (Sections G2 and G3). Microbes have similar antigens to self as escape mechanisms to reduce the host immune response (Section H3). Aggregated antigens or antigens with multiple different epitopes are usually good "immunogens" (i.e., able to induce immunity), whereas soluble antigens are poor immunogens that may induce unresponsiveness (tolerance), especially in immature cells (see below and Section G2). Aggregated antigens are also more likely to be taken up and processed by APCs (especially macrophages), and, as they contain many epitopes, are able to stimulate more T cell responses and the release of appropriate cytokines for their activation. Making an antigen more "immunogenic" is important in developing effective vaccines (Section I).

## Maturity of the immune system

The magnitude of the response produced when immature B cells and T cells respond to an antigen is weaker than in mature immune systems. Contact with antigen by these cells is more likely to lead to unresponsiveness (i.e., tolerance). Responsiveness is also lower in immune-compromised individuals; for example, immunodeficient individuals or animals that are recovering from irradiation have a relatively higher proportion of immature to mature lymphocytes (Section J3). Naive immature lymphocytes express fewer co-stimulatory receptors than mature T cells and developing immune systems have fewer memory T cells that are able to give stronger immune responses to antigens.

## Route of antigen administration

The route of administration of antigen may influence the nature of the immune response; that is, certain antigens given subcutaneously or intramuscularly may be more immunogenic than when given intravenously or intraperitoneally.

Antigens introduced into an individual by the oral route (feeding) can often induce oral "tolerance." At least three mechanisms are involved, including active suppression, clonal anergy, and deletion. Active suppression is probably induced by regulatory T cells (Tregs) through inhibitory cytokines such as TGFβ and IL-10 (Section G3). This is an important mechanism for preventing the body from responding to food antigens that are continually bombarding the intestine and entering the circulation.

## Dose, persistence, and removal of antigen

A very large amount of antigen often produces a poor response or unresponsiveness ("high zone" tolerance). Small amounts of antigen given over time are thought to be good stimulators of immune responses. It is thought that one mode of action of adjuvants is through the slow release of low doses of antigen. Persistent antigen, as found with some viruses and bacteria, maintains production of specific immune responses.

Removal of antigen by antibody (ultimately through phagocytic cells) is the most effective means of regulating an immune response because, in the absence of antigen, the re-stimulation of antigen-specific T and B cells stops. Thus, maternal IgG in the newborn may bind antigen and remove it, thereby interfering with development of active immunity to this antigen. Furthermore, therapy with passive antibodies may interfere with the development of active immunity. For example, antibodies to Rhesus D (RhD) given to RhD-negative mothers carrying RhD-positive fetuses prevents the development of anti-RhD antibodies in future pregnancies. It is thought that one mechanism by which induction of RhD antibodies are inhibited is through antigen removal.

# G5 Regulation by T cells and antibody

## Key Notes

**The role of T cells and cytokines**

The type of immune response is determined by the nature of the antigen, the T cell, and its cytokine products. Th1 cells produce pro-inflammatory cytokines important for killing of intracellular microbes and the generation of T-cytotoxic cells. The anti-inflammatory cytokines IL-4, IL-10, and IL-13, produced by Th2 cells, are important for B-cell proliferation and differentiation and immunoglobulin class switch to IgA or IgE, antibody isotypes important for immune defense of mucosal surfaces. Th1 and Th2 cytokines are self-regulating and inhibit each others' functions. Cytokines promote cell growth, attract specific immune cells (chemokines) or contribute to cell activation. Other cytokines suppress cell proliferation (e.g., TGFβ and IFNα) or inhibit activation of macrophages (e.g., TGFβ).

**The role of antibodies**

Antibodies of the IgM class may regulate the immune response through complement activation. Thus, interaction of antigen–IgM–complement complexes with complement receptors (CD21) on antigen-specific B cells may enhance the response. On the other hand, IgG–antigen complexes may specifically inhibit further responses by antigen-specific B cells as a result of a negative signal, transduced by FcγRII (CD32) on binding of the Fc region of the IgG component of the complex to the B cell.

**Related topics**

(A4) Antigens
(B2) Molecules of the innate immune system
(C5) Adaptive immunity at birth
(D2) Antibody classes
(D4) Allotypes and idiotypes

(E1) The B-cell receptor complex, co-receptors, and signaling
(E3) The cellular basis of the antibody response
(I2) Immunization
(K3) IgG- and IgM-mediated (type II) hypersensitivity

## The role of T cells and cytokines

T-helper cells are an absolute requirement for immune responses to protein antigens in general, and for helping B cells to make the different classes of antibodies. The type of response that develops is, in some instances, determined by the nature of the antigen and its mode of entry, as well as the effect of regulatory CD4⁺ T-helper subsets, Th1 and Th2, and their cytokine products (Section F5). The pro-inflammatory cytokines IL-2, TNFα and IFNγ produced by Th1 cells are important for killing of intracellular microbes and the generation of T-cytotoxic cells, as well as the development of IgG antibodies capable of mediating phagocytosis of extracellular microbes. In contrast, the anti-inflammatory Th2 cytokines IL-4, IL-10, and IL-13 are important for B-cell proliferation and differentiation and immunoglobulin class switch to IgA and IgE as well as the IgG2 response to the polysaccharide antigens associated with encapsulated bacteria such as *Pneumococcus*. Th2 cytokines are also important in helping to eradicate parasitic infections as they lead to the production of IgE and the recruitment of eosinophils, which have powerful anti-parasitic functions (Section H2).

Th1 and Th2 cytokines are self-regulating and also inhibit each other's functions (Figure 1 and Sections B2 and F5). For example, IL-4 and IL-10 downregulate Th1 responses whereas IFNγ has an antagonistic effect on Th2 cells. Downregulatory mechanisms are necessary to prevent collateral damage as well as being energy conserving. Patients with atopy, that is, with a genetic predisposition to having high levels of IgE, are believed to poorly regulate their Th2 cells (Section K2).

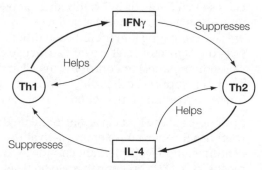

Figure 1. Reciprocal regulation of Th1 and Th2 cells. Th1 cells release IFNγ, which suppresses proliferation of Th2 cells and their IL-4 production. Th2 cells release IL-4 (and IL-10), which suppresses IFNγ production by Th1 cells and their proliferation.

Most cytokines promote growth of particular cell lineages, attract specific immune cells (chemokines), or contribute to cell activation. Other cytokines can be suppressive. TGFβ inhibits activation of macrophages and the proliferation of B and T cells. IFNα also has cell-growth-inhibitory properties. The action of these suppressive cytokines is a primary way that T cells and macrophages regulate immune responses. In addition, the stimulatory and inhibitory actions of cytokines produced by Th1 and Th2 cells on each other also play a major role in determining the type and extent of an immune response (Section B2).

## The role of antibodies

### Enhancement through IgM

Antibodies of the IgM class appear to be important in enhancing humoral immunity. In particular, antigen–IgM–complement complexes that bind to the B-cell antigen receptor stimulate the cell more efficiently than antigen alone (Figure 2). This is probably the result of simultaneous interaction of the C3b component of complement with the CD21 molecule of the antigen receptor complex, which then transduces a positive signal to the B cell. Antigen–IgM–complement complexes also bind to complement receptors on phagocytic cells (Section B3) and are therefore important in the removal of microbial antigens.

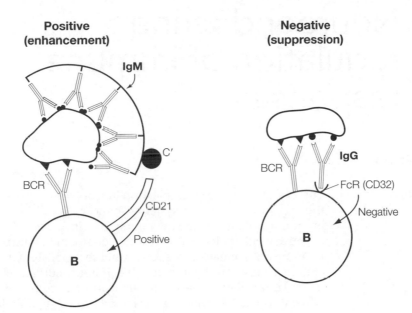

Figure 2. Regulation of B-cell activity by antibody. IgM bound to antigen recognized by the B-cell receptor (BCR) fixes complement which then interacts with CD21, giving a positive signal to the B cell. However, IgG bound to antigen attached to the BCR binds to FcγRII (CD32) and delivers a negative signal to the B cell.

## Negative feedback through IgG

The interaction of IgG–antigen complexes with antigen-specific B cells through the simultaneous binding of both the B-cell antigen receptor and the FcγRII molecule of the B-cell receptor complex can deliver a negative signal to the B cell (Figure 2). Thus, IgG, which is produced later in the antibody response, could interact with antigen (if present) forming a complex that, on binding to antigen-specific B cells, may provide feedback inhibition mediated via FcγRII, decreasing the amount of antigen-specific antibody being produced.

A negative role of IgG that passes across the placenta might explain the unresponsiveness of the newborn to certain antigens (Section C5). Normally, maternal IgG passes across the placenta during fetal life to ensure that the infant at birth has all of the IgG-antibody-mediated humoral immunity of the mother. In addition, maternal IgA obtained by the infant from colostrum and milk during nursing coats the infant's gastrointestinal tract and supplies passive mucosal immunity (Sections C5, D8, and E4). Thus, until these passively supplied antibodies are degraded or used up, they may bind antigen and remove it, thereby interfering with development of active immunity.

The role of IgG in negative control of antibody production is achieved clinically by passive immunization with antibodies to prevent hemolytic disease of the newborn (Section K3). Injection of antibodies to RhD into RhD– mothers before or immediately after birth of an RhD+ infant removes RhD+ erythrocytes that may have passed into the maternal circulation. This prevents the development of RhD antibodies occurring as a result of future pregnancies (Section K3). This results from the simple removal of antigen (RhD+ erythrocytes) such that the mother never develops a memory response to RhD antigen.

# G6 Neuroendocrine regulation of immune responses

## Key Notes

**Overview**

There is a two-way communication link between the nervous and immune systems. In most cases, immunoregulation by the nervous system involves negative feedback loops that result in dampening down of immune (inflammatory) responses and reestablishment of homeostasis. The hypothalamus–pituitary–adrenal axis (HPA) stimulates the release of glucocorticoids from the adrenal cortex. Neuropeptides released from sympathetic and parasympathetic nerve endings are also suppressive. Neuropeptides released from sensory peripheral nerves involved in pain, touch, and temperature perception are generally pro-inflammatory. The mechanism by which modulation is achieved appears to be directly through receptors for neurotransmitters on innate and adaptive immune cells.

**The HPA axis**

The HPA axis mediates release of endogenous glucocorticoids (from the adrenal cortex) that suppress inflammation by inhibiting the release of the pro-inflammatory cytokines IL-1, IL-2, IL-6, IFNγ, and TNFα by dendritic cells and macrophages, decreasing antigen presentation and inhibiting mast cell function. Glucocorticoids also have an effect on cell trafficking of immune cells by inhibiting the production of some chemokines and altering the expression of their homing molecules.

**The peripheral nervous system**

Both the sympathetic and parasympathetic nerve fibers of the autonomic nervous system innervate the systemic organs of the body. Local pro-inflammatory cytokines stimulate the release of norepinephrine (noradrenaline) and neuropeptide Y from the sympathetic fibers. These neurotransmitters can also stimulate the production of chemokines that can potentially increase inflammation. A similar inhibition of cytokine production is seen when acetylcholine is released from fibers of the parasympathetic system. Sensory peripheral nerves also release a number of neuropeptides, including CRH, CGRP, and substance P, which are generally pro-inflammatory and therefore help in removal of pathogens. In contrast, VIP released from

| | |
|---|---|
| | peripheral nerves inhibits production of pro-inflammatory peptides and suppresses expression of TLRs by dendritic cells and macrophages. |
| **Lymphoid organ innervation** | It is well documented that central and peripheral lymphoid organs are innervated, although the role that neurotransmitter release has on development, function, and cell trafficking is unclear. Thymic atrophy during aging is mediated through glucocorticoids and there is an age-associated increase in acetylcholinesterase-positive structures in the human thymus, with older mice having increased noradrenergic sympathetic nerves and a 15-fold increase in the concentration of norepinephrine—suggesting a role for the nervous system in thymic function, especially in old age. |
| **Related topics** | (B)   Cells and molecules of the innate immune system |

## Overview

It has been known for some time that there is a two-way communication link between the immune and nervous systems. Accumulating evidence has led to the conclusion that the nervous system regulates immune responses. In most cases this regulation involves negative feedback loops that result in dampening down of immune (inflammatory) responses and reestablishment of homeostasis. In response to the production of pro-inflammatory cytokines through infection or other stimulants of inflammation, the hypothalamus–pituitary–adrenal axis (HPA) stimulates the release of glucocorticoids from the adrenal cortex that together with epinephrine (adrenaline) produced by the adrenal medulla, result in immune suppression. In addition, norepinephrine (noradrenaline) and neuropeptide Y released from sympathetic nerve endings, and acetylcholine from parasympathetic fibers, are also suppressive. In contrast, sensory peripheral nerves involved in pain, touch, and temperature perception release a number of neuropeptides, including corticotropin-releasing hormone (CRH), calcitonin gene-related peptide (CGRP), and substance P, that are generally pro-inflammatory. The various neurotransmitters released from nerve endings, and those acting systemically, are thought to regulate immune responses through receptors that the cells of the immune system have for these transmitters. In most cases regulation is through modulation of the levels of pro-inflammatory cytokines produced and their effects on both the innate and adaptive immune systems.

## The HPA axis

The HPA axis consists of a set of brain regions (the hypothalamus) and endocrine organs (the pituitary and adrenal cortex) (Figure 1). Infection and other danger stimuli induce the local release of pro-inflammatory cytokines. These cytokines, detected systemically, result in the release of CRH by the hypothalamus, which stimulates the production of adrenocorticotropic hormone by the anterior pituitary. This is released into the blood stream and in turn stimulates the production of endogenous glucocorticoids from the adrenal cortex. Glucocorticoids have wide-ranging regulatory effects on the immune system, including modifying T-cell responses, reducing the number of circulating lymphocytes, monocytes, and eosinophils, suppressing inflammation by inhibiting the release of the pro-inflammatory cytokines IL-1, IL-2, IL-6, IFNγ, and TNFα by dendritic

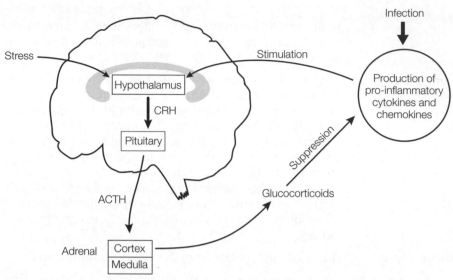

Figure 1. The interconnectivity between the immune and neuroendocrine systems. Infection and stress can affect, either directly or indirectly, both the immune and the nervous systems. Pro-inflammatory cytokines and chemokines produced locally in response to infection stimulate the release of glucocorticoids, which downregulate their production. The cytokines cause the release of corticotropin-releasing hormone (CRH) by the hypothalamus, which stimulates the pituitary to release adrenocorticotropic hormone (ACTH). ACTH causes the adrenal gland to release glucocorticoids from the adrenal cortex which in turn suppress the production of pro-inflammatory cytokines and apoptosis of cells producing the pro-inflammatory cytokines.

cells and macrophages, decreasing antigen presentation, and inhibiting mast cell function (Figure 2). In addition, glucocorticoids induce apoptosis of dendritic cells, macrophages, and T cells. The overall effect is to dampen immune responses. Glucocorticoids also affect trafficking of immune cells by inhibiting the production of some chemokines and altering the expression of their homing molecules (Section C4). Excess production of glucocorticoids through "chronic stress" has been shown to be associated with increased susceptibility to viral infections, prolonged wound healing, and reduced antibody production after vaccination.

## The peripheral nervous system

### The autonomic nervous system

The autonomic nervous system mostly controls body activities not usually under voluntary control. Both the sympathetic and parasympathetic nerve fibers innervate the systemic organs of the body, with the sympathetic fibers mainly innervating lymphoid organs. In the tissues, local inflammation results in production of IL-1 and TNF, resulting in release of norepinephrine (noradrenaline) and neuropeptide Y from the sympathetic fibers. Like systemic glucocorticoids, these neurotransmitters inhibit the production of pro-inflammatory cytokines from DC and macrophages. Whereas the sympathetic nerve fibers release their neurotransmitters locally, epinephrine (adrenaline) and neuropeptide Y are released systemically from the adrenal medulla and generally have an inhibitory effect. Norepinephrine and epinephrine can stimulate the production of some chemokines and therefore can increase inflammation. A similar feedback loop is seen with the

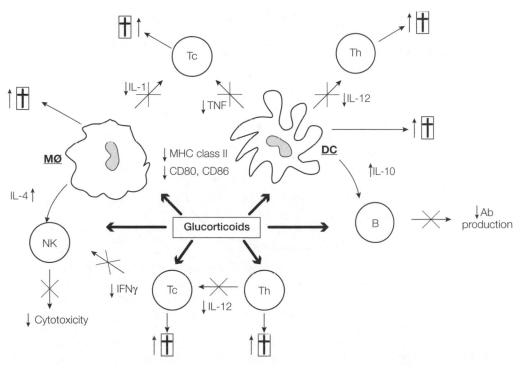

Figure 2. The multiple inhibitory effects of glucocorticoids on immune cells. Glucocorticoids have direct effects on the production of pro-inflammatory cytokines from APC-macrophages (MØ) and dendritic cells (DCs). Inhibition of IL-12 release from DCs suppresses the production of Th1 cells in particular, and reduced TNF from DCs and IL-1 from macrophages reduces Tc development. Both CD86/CD80 and MHC class II are also downregulated, reducing the stimulatory effect on T cells. Reduced IFNγ produced by Tc (and Th1 cells, not shown) also suppresses NK-cell-mediated cytotoxicity. Reduced IL-2 from Th cells also suppresses Tc development. Increased IL10 production by DCs (and macrophages, not shown) also suppresses B-cell production of antibody. Apoptosis (programmed cell death) is enhanced in all these cell populations as an effect of the glucocorticoids.

parasympathetic system, where the fibers release acetylcholine, which also inhibits pro-inflammatory cytokines.

## Sensory peripheral nerve control of immunity

Sensory peripheral nerves involved in pain, touch, and sensory perception release a number of neuropeptides, including CRH, CGRP, and substance P which are generally pro-inflammatory and therefore help in removal of pathogens. These neurotransmitters mediate their effects through receptors on a variety of immune cells, including dendritic cells, macrophages, T cells, and B cells. In contrast to the pro-inflammatory effect of these neurotransmitters, vasoactive intestinal peptide (VIP) released from peripheral nerves inhibits the production of pro-inflammatory peptides and suppresses the expression of some TLRs by DC and macrophages. Figure 3 is a summary of the effects of different neurotransmitters on immune responses.

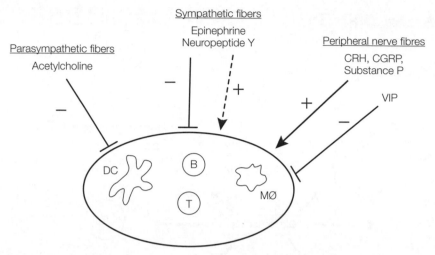

Figure 3. Summary of the effects on immune cells of neurotransmitters released from different types of neurons. Acetylcholine from parasympathetic fibers inhibits cytokine production and immune cell function, as do epinephrine and neuropeptide Y, released from sympathetic fibers. There is also evidence that these can enhance production of some chemokines and therefore be pro-inflammatory. Finally, corticotropin-releasing hormone (CRH), calcitonin gene-related peptide (CGRP) and substance P released from sympathetic fibers are pro-inflammatory, with the exception of vasoactive intestinal peptide (VIP), which is inhibitory.

## Lymphoid organ innervation

It is well documented that central and peripheral lymphoid organs are innervated. The role that neurotransmitter release has on development, function, and cell trafficking is currently unclear. However, thymic atrophy during aging (Section P) is mediated through glucocorticoids. There is an age-associated increase in acetylcholinesterase-positive structures in the human thymus and in rodent models, with older mice having increased noradrenergic sympathetic nerves and a 15-fold increase in the concentration of norepinephrine (noradrenaline). The significance of this finding is presently unclear but does point to a role for the nervous system in thymic function, especially in old age.

# H1 The microbial cosmos

## Key Notes

**Infection and its consequences**

In the past, epidemics caused by plague and influenza have caused the deaths of large numbers of people as well as causing changes in social structures and behavior. Today, 45% of deaths in the developing world are caused by infectious diseases, including malaria, diarrheal diseases, diseases caused by *Mycobacterium tuberculosis* (TB) and human immunodeficiency virus (HIV-1, HIV-2). In the industrial world, HIV/AIDS as well as multidrug-resistant TB represent major public health problems. The emergence of new diseases caused by *Legionella*, *Helicobacter pylori*, *Clostridium difficile*, and the severe acute respiratory syndrome (SARS) virus, as well as Lyme disease, present new challenges to the immune system and man's inventiveness.

**Microbe habitat and immune defense**

Extracellular microbes (e.g., many bacteria and fungi) are usually destroyed by phagocytes, complement, and specific antibodies. Those having an intracellular habitat (e.g., viruses, some bacteria, and protozoa) are mainly controlled or eradicated by cytotoxic T cells, NK cells, and NKT cells.

**Damage caused by pathogens**

Pathogens can cause tissue damage directly by production of toxins but also through an overzealous immune response. In this case, the host cells, in an effort to combat an infection, release many inflammatory mediators such as cytokines (cytokine storm), nitric oxide, superoxides, and other free radicals into the surrounding tissue (friendly fire). In some instances this results in morbidity or mortality. Other mechanisms of immune-mediated damage to the host include anaphylaxis, immune complex disease, necrosis, and apoptosis.

**Related topics**

(B1) Cells of the innate immune system
(B2) Molecules of the innate immune system
(D8) Antibody functions
(F5) Clonal expansion and the development of memory and effector function

## Infection and its consequences

The survival of the human species is dependent on its ability to defend itself against the constant threat from microbial infections. Infectious diseases cause approximately 25% of all deaths in the world but in low-income countries this can rise to 45% (Tables 1 and 2). The major causes of death in the developing world are diarrheal diseases, HIV/AIDS, malaria, measles, pneumonia, and tuberculosis (TB)—in the developing world a young child dies every 3 seconds from an infectious disease. Malaria kills more than one million people a year and accounts for one in five children's deaths. The high annual incidence (over 275 million cases occur globally) imposes a huge economic burden. Approximately 40 million people are infected with HIV/AIDS, with approximately 2.5 million new cases added each year. Tuberculosis, once thought to be under control, kills 2 million people each year and even more if complicated with HIV/AIDS. It is estimated that approximately one-third of the world population has latent TB and that one person becomes infected every second. In the industrialized world, while TB is not as rampant, it is making progressive inroads into society and requires constant vigilance, especially with the increase of infections with multidrug-resistant strains. In the past, plagues and large epidemics have changed social structures and behaviors. It is estimated that the 1918 flu pandemic killed between 40 and 100 million people. This was an unusually severe form of flu caused by a virus subtype of H1N1. It had an infection rate greater than 50% with severe symptoms, in part, suspected to be attributable to the immune system (cytokine storms). Between 2 and 20% of those infected died, in contrast to normal flu epidemics where the mortality rates are about 0.1%. As well as the infectious diseases listed above, new and emerging infectious diseases such as severe acute respiratory syndrome (SARS), *Clostridium difficile* infection, Legionnaires' disease, Lyme disease, and disease caused by hantavirus, present new challenges to the immune system.

Tables 1 and 2 show the infectious diseases in the developing and industrial worlds respectively.

### Table 1. Major infectious diseases in the developing world

| Disease | Numbers infected | Deaths per annum |
| --- | --- | --- |
| Hepatitis B | 2 billion | 600 000 |
| TB | 2 billion | 1.5 million |
| Malaria | 500 million | 1 million |
| Whooping cough, pertussis | 20–40 million | 200 000–300 000 |
| HIV/AIDS | 40 million | 2–3 million |
| Measles | 30 million | >500 000 (mainly children) |
| Diarrhea | 4 billion | 2.2 million |

### Table 2. Major infectious diseases in the industrialized world

| | |
| --- | --- |
| Viruses | Colds, flu, measles, rubella, pneumonia, HIV/AIDS |
| Bacteria | Strep throat, wound infections, septicemia, pneumonia |
| Mycobacteria | TB |
| Fungi | Thrush (*Candida albicans*) |

## Microbe habitat and immune defense

Microbes can invade the host through mucosal surfaces, skin, bites, and wounds. Such invasion is usually countered by innate defense mechanisms that act rapidly, including lytic proteins (complement) and phagocytes that can recognize microbial signatures, for example pathogen-associated molecular patterns (PAMPs) (Section B2). In the event that the infectious agent survives these first lines of defense, the adaptive immune system responds more specifically, but more slowly, in an effort to eliminate the pathogen. The final pathway of defense usually results in immunological memory to the offending pathogen, as is the case with infectious agents such as smallpox and measles, whereby repeated infections with these microbes are uncommon.

The immune responses to bacteria, viruses, fungi, protozoa, and worms differ in the variety of defensive mechanisms used. In general, microbes (e.g., many bacteria and fungi) living outside the cells of the body are more likely to be opsonized by specific antibodies and engulfed by phagocytes or destroyed by the alternative or classical complement pathway, whereas those having an intracellular habitat (e.g., viruses, some bacteria, and protozoa) may require cytotoxic T cells, NK cells, or NKT cells to provide effective protection. An effective immune response to fungi is dependent on the innate and adaptive arms of the immune system working together. Eradication or control of a fungal infection requires an effective Th1 response. Th2 responses correlate with persistence and severity of the disease. Both humoral and cellular responses are required for protection against protozoa, which are difficult to immunize against. Immune protection against helminths (worms) is difficult to achieve because of their size and complexity. The major response mechanisms include the production of antibodies, especially immunoglobulin (IgE), and a cellular response involving eosinophils, mast cells, macrophages, and CD4$^+$ T cells. Both mast cells and basophils degranulate in the presence of IgE antigen complexes; IgA complexes also cause eosinophils to degranulate. Mast cells release histamine, which causes gut spasms, while eosinophils release cationic protein and neurotoxins. Helminth antigens direct the immune system to develop a Th2 response that results in the preferential production of IgE.

## Damage caused by pathogens

Pathogenic organisms can cause tissue damage and disease directly through the production of toxins. For example, bacteria and protozoa produce exotoxins and endotoxins. In addition, most viruses have a lytic stage resulting in tissue damage. On the other hand, the immune response may be more destructive than the infectious agent itself, especially in persistent states, An over-zealous immune response may cause the release of a cytokine storm (IL-1, IL-6, TNF, and IL-10: hypercytokinemia), oxygen free radicals, and coagulation factors as well as other reactive molecules. In the lung this may lead to edema and death, as has happened in some individuals with the recent H1N1 flu or in the past with the Spanish flu. Hypercytokinemia can occur in acute respiratory distress syndrome (ARDS) and was probably responsible for the disproportionate number of deaths of young adults in the 1918 flu pandemic. In this scenario a healthy immune system may be a liability rather than an asset (Section K). Other mechanisms of immune-mediated damage to the host include anaphylaxis, immune complex disease, necrosis, and apoptosis (Table 3).

## Table 3. Pathogens and hypersensitivity

| | |
|---|---|
| Type I | *Echinococcus* hydatid cyst, when it bursts, produces an anaphylactic response |
| Type II | Cross-reactions of antibodies to shared antigens, e.g., streptococci and heart tissues in rheumatic fever |
| Type III | Immune complex deposition in kidney, lung, blood vessel, or joint causing glomerulonephritis (e.g., streptococcal infection), bronchiectasis, vasculitis, or arthritis, respectively |
| Type IV | Granuloma formation, e.g., TB and leprosy |

# H2 Immunity to different organisms

---

**Key Notes**

**Immunity to bacteria**

Phagocytic cells of the innate immune system are able to recognize microbial signatures (PAMPs) and as a consequence ingest, kill, and eradicate many microbial intruders. Extracellular bacteria may also be killed directly through the alternative complement pathway or after activation by antibody binding to the microbe, through the classical complement pathway. Antibodies and complement also act as opsonins facilitating engulfment and killing by phagocytes. For intracellular bacteria, for example TB bacilli that evade the immune system by surviving in host cells such as monocytes and macrophages, a cell-mediated immune (CMI) response is required. This results in the release of cytokines such as IL-12 and IFNγ that enhance monocyte/ macrophage killing of intracellular bacteria.

**Immunity to virus**

The innate immune system inhibitors of viral infection are IFNα and β. However, when viruses replicate in host cells, a cytotoxic lymphocyte response (CTL) is required for their eradication. After infection, viral-specific peptides become expressed on the cell surface in MHC molecules and become targets for CTLs. Antibodies can neutralize extracellular viruses (prevent their attachment to, and infection of, target cells) and enhance phagocytosis of the virus.

**Immunity to fungi**

The immune response to fungal infections (mycoses) is dependent on both the innate and adaptive immune system working together. Th1-type responses are associated with eradication or containment of fungal infections whereas Th2-type responses are associated with severity of disease. Both the innate and adaptive arms of the immune system are involved; immunity principally involves T cells, macrophages, and NKT cells.

**Immunity to protozoa**

Protozoa infections such as malaria, trypanosomiasis, leishmaniasis, and toxoplasmosis are a major threat to health in the tropics and in the developing world. Protozoa are difficult to immunize against and protection is thought to require both cellular and humoral immunity; however, protozoa have evolved different strategies for evading immunity.

**Immunity to worms**

Immune protection against helminths (worms) is difficult to achieve because of their size and complexity. The response mechanisms include the production of antibodies, especially IgE, and a cellular response involving eosinophils, mast cells, macrophages, and CD4+ T cells. Degranulation of mast cells and eosinophils through IgE-antigen and IgA-antigen complexes results in acute inflammation and the release of cationic proteins and neurotoxins.

| **Related topics** | (B1) | Cells of the innate immune system | (D8) | Antibody functions |
|---|---|---|---|---|
| | (B2) | Molecules of the innate immune system | (F5) | Clonal expansion and the development of memory and effector function |
| | (B4) | Innate immunity and inflammation | (H1) | The microbial cosmos |
| | | | (J1) | Deficiencies in the immune system |

## Immunity to bacteria

The primary defenses against microbial infection are the natural physical barriers such as the skin, mucous membranes, tears, earwax, mucus, and stomach acid. The immunological defenses are provided by the innate and adaptive immune systems. Defense against bacterial infections requires many and varied strategies. These include circulating proteins such as complement that can directly destroy some bacteria through lytic activity or engulfment and killing by phagocytic cells that can recognize PAMPs through their pattern recognition receptors (PRRs). Alternatively bacteria may be opsonized by acute phase reactants C-reactive protein and serum amyloid P, or by specific antibodies and engulfed by phagocytes expressing receptors for the Fc region of these antibodies. Both PMNs and macrophages express receptors for IgG as well as IgA. Inflammatory cytokines such as IFNγ can dramatically upregulate expression of these receptors and improve the efficiency of killing by these effector cells. In addition, the TLRs expressed by macrophages and dendritic cells are important in cytokine production and initiating responses against bacteria (and other microbes) by the adaptive immune system, whose primary aim is to produce appropriate specific antibody responses and to provide memory against future infections. (Section B3). A summary of the main effector defense mechanisms against extracellular bacteria is shown in Figure 1.

Some bacteria invade host cells and survive in them, including TB bacilli, *Listeria monocytogenes*, *Salmonella typhi*, and *Brucella* species. These **intracellular** bacteria evade the immune system's surveillance by surviving in host cells such as monocytes and macrophages. The immune system counteracts them by mounting a cell-mediated immune (CMI) response to the infection. Cells involved in the CMI response include Th1 and Th2 CD4+ T cells, CD8+ T cells, monocytes/macrophages, and NK cells. Th1 cells release IFNγ, which makes monocytes/macrophages more potent at killing intracellular bacteria and also enhances their antigen presenting capabilities (Figure 2). The CMI response is important in the protection against not only diseases such as TB but also some viral and fungal infections.

## Immunity to virus

Natural immunity to viral infections is associated with interferons (especially IFNα and IFNβ), so called because of their interference with viral replication (Section B2). IFNγ is probably most effective in protecting against extracellular bacteria through its ability to

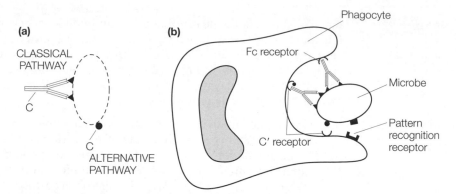

Figure 1.  Defense mechanisms against extracellular bacteria. Bacteria can be killed by (a) complement-dependent lysis with antibody (classical pathway) or without (alternative pathway), and (b) intracellularly following phagocytosis induced through some pattern recognition receptors or receptors for antibody (Fc) and/or complement (C') that recognize opsonized microbes.

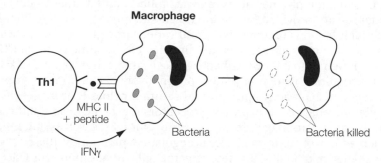

Figure 2.  Defense mechanisms against intracellular bacteria. Macrophages containing intracellular bacteria present microbial peptides on their MHC class II molecules to specific Th1 cells which produce IFNγ. This "activates" the macrophage resulting in enhanced intracellular killing.

enhance immune-mediated mechanisms. Since viruses require attachment to host cells before they can replicate and cause infection, antibodies to the virus that prevent attachment represent an important mechanism that protects against viral infection. These protective antibodies may be IgG or IgA as in the case of polio prevention.

Because viruses replicate in cells where they are no longer exposed to circulating antibodies, their eradication depends upon killing the infected host cells. This of course would require a CTL response. As virally infected cells usually express viral peptides in MHC class I on their surface, they become targets for destruction by cytotoxic CD8⁺ T cells. Cells infected by a virus also become susceptible to killing by NK cells and NKT cells (Section B1). In this way viral replication is prevented and the viral infection eliminated (Figure 3).

## Immunity to fungi

Most fungi, such as *Histoplasma capsulatum*, *Aspergillus* species, *Cryptococcus neoformans*, and *Candida albicans*, are ubiquitous in the environment. Fungal diseases

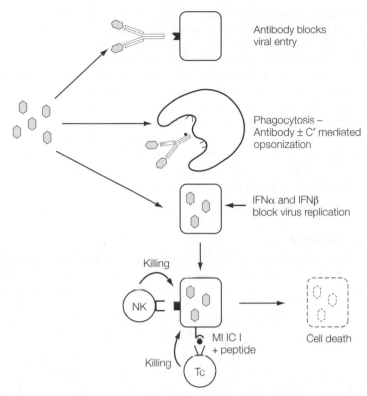

Figure 3. Defense mechanisms against viruses. Antibodies attach to viruses preventing their entry into cells and opsonize them for phagocytosis. IFNα and IFNβ block viral replication in infected cells. NK cells can kill virus-infected cells if they have little or no expression of MHC class I. Tc cells kill virus-infected cells expressing viral peptides in MHC class I.

(mycoses) are common but are most problematic when associated with immunocompromised individuals (Section J1).

In the hospital setting, blood-borne fungal infections are the fourth most important cause of septicemia. Although the immune response to fungal infections is not fully understood both the innate and adaptive arms of the immune system are important to eradicate or keep these infections under control. It also seems clear that protective immunity is principally cellular, especially in those infections deep within the body. This is particularly evidenced by studies of acquired immune deficiency syndrome (AIDS) patients, where low T-cell counts are commonly associated with fungal infections. A number of cell wall components of fungi may act through distinct TLRs on macrophages and dendritic cells activating essential components of the immune system. Dendritic cells are uniquely efficient at presenting fungal antigens to the adaptive immune system as are NKT cells acting through the CD1 family of glycoprotein receptors in responses to glycolipids expressed by fungi. In the past, the immunopathogenesis of fungal infections was explained primarily on the basis of Th1/Th2 responses. Th1 responses are associated with protection and eradication while increased Th2-type cell responses are associated with severity of disease. Although the Th1 axis driven by IL-12 and IFNγ is still central to protection against fungi, patients with inborn deficits of the IL-12/IFNγ loop do not have increased susceptibility to fungal infections. More recently Th17 cells have been shown to have important implications for failure to remove pathogens and resolve chronic inflammation in fungal disease. Inflammation previously attributed to Th1 cells would appear to be a product of dysregulation of these cells. Although IL-17, a Th17-cell product, is important for neutrophil mobilization and eradication of some fungal infections such as candidiasis, in other fungal infections Th17 cells were associated with defective pathogen clearance and failure to resolve inflammation. Moreover, in murine studies of pulmonary aspergillosis, neutralization of IL-17 greatly ameliorates the disease. Thus, Th17 cells appear to have important connections between the failure to resolve inflammation, lack of fungal immune resistance, and susceptibility to chronic fungal infections. Finally, fungal infection represents a paradox to the immune system: whereas a severe inflammatory response may be detrimental to the host, corticosteroids may be ameliorating and protective.

## Immunity to protozoa

Protozoa infections such as malaria, trypanosomiasis, toxoplasmosis, leishmaniasis, and amoebiasis are a major threat to health in the tropics and in particular in the developing world. Protozoa cause chronic and persistent infections because natural immunity against them is weak. They are difficult to immunize against and protection is thought to require both the innate and adaptive immune systems although the humoral response, and in particular the IgG response, may be the most important.

In malaria, antibodies appear to protect against infection by preventing the merozoites (blood stage) from gaining entrance to red cells. However, there are several different strains of malaria and immunity to one strain or species may not be protective against others. Other innate or nonadaptive immune mechanisms may also be involved in protection against certain malaria infections. For example, individuals lacking the Duffy blood group antigen Fy (a–b–) are immune to *Plasmodium vivax* infection. Also, the hemoglobin structure associated with sickle-cell anemia appears to be inhibitory to the intracellular growth of *P. falciparum*.

Trypanosomes continuously challenge the immune system by producing progeny with different antigens. Thus, as the immune system develops a response to antigens on these microbes, they change the structure of some of their surface proteins (switch antigenic coats) so that the antibodies produced in the initial response are no longer reactive or effective in mediating protection against this modified trypanosome. This leads to wave after wave of infection and response.

*Toxoplasma* acquires protection from the immune system by coating itself with laminin, an extracellular matrix protein, which prevents phagocytosis and oxidative damage. The cellular response to *Toxoplasma* appears to be most effective in combating infection, since patients with low T-cell counts, as in HIV infection, are more at risk from infection with *Toxoplasma*. Other protozoan diseases such as leishmaniasis have a predilection for infecting macrophages and require a cellular response for eradication. Moreover, a Th1 response seems to be essential for protection, since IFNγ appears to be the most important cytokine for parasite killing.

## Immunity to worms

An immune response to worms (helminths) is difficult to achieve and not very effective, probably as a result of the size and complexity of these microbes. Thus, diseases such as those caused by *Schistosoma mansoni* (schistosomiasis) and *Wuchereria bancrofti* (lymphatic filariasis, elephantiasis) represent major problems, especially in the developing world. Although PMNs, macrophages, and NK cells may be involved, the main protective mechanism against helminths appears to be mediated by eosinophils and mast cells. While worms are too large to be phagocytosed, they can be opsonized with IgE, IgA, and IgG antibodies. In the event that this happens, the major phagocytic cells, as well as eosinophils and mast cells, will bind to the parasite's surface through their Fc receptors for these molecules and release their toxic cellular contents. Both mast cells and eosinophils degranulate in the presence of IgE–antigen complexes. When mast cells degranulate they release histamine, serotonin, and leukotrienes. These vasoactive amines are neurotransmitters and cause neurovascular as well as neuromuscular changes resulting in gut spasm, diarrhea, and the expulsion of material from the intestine. Eosinophils also have IgA receptors and have been shown to release their granule contents when these receptors are cross-linked. On degranulation, eosinophils release powerful antagonistic chemicals and proteins, including cationic proteins, neurotoxins, and hydrogen peroxide, which also probably contribute to a hostile environment for worm habitation. Helminth infections usually direct the immune system towards a Th2 response with the production of IgE, IgA, and the chemokine eotaxin. The Th2 cytokines IL-3, IL-4, and IL-5 as well as the chemokine eotaxin are chemotactic for eosinophils and mast cells. Figure 4 summarizes the major immune mechanisms for removal of helminths.

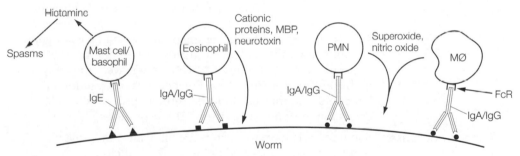

Figure 4. Defense mechanisms against worms. Worms are usually too large to phagocytose, but coated with specific antibodies they can activate a number of "effector" cells via their Fc receptor (FcR). IgE-mediated degranulation of mast cells/basophils results in production of histamine, which causes spasms in the intestine, where these worms are often found. Eosinophils attach to the worm via IgG/IgA antibodies and release cationic proteins, major basic protein (MBP), and neurotoxin. PMNs and macrophages attach via IgG or IgA antibodies and release superoxide, nitric oxide, and enzymes which kill the worm.

# H3 Pathogen defense strategies

## Key Notes

**The battle to stay ahead**

Over time, microbes have developed strategies to circumvent and/or inactivate host immune defense mechanisms. Some pathogens avoid immune recognition by intracellular habitat, mimicking of self-antigens, encapsulation, or changing their surface antigens (antigenic variation). Other pathogens compromise effector mechanisms within phagocytic cells by subverting critical killing mechanisms and in the extravascular spaces by inhibiting complement activation, phagocytosis, cytokine production, or superantigen-induced immunosuppression.

**Avoidance of recognition**

The intracellular habitat of viruses, some bacteria, and protozoa prevents recognition by innate and adaptive immune systems. Other microbes can change their antigens by mutation (drift), by nucleic acid recombination (antigenic shift), and by switching genes encoding cell-surface antigens. Still other microbes express antigens very similar to self antigens (molecular mimicry), while others wear the antigens of the host they are infecting. Furthermore, distraction of lymphoid cells may be achieved through poly- or oligoclonal activation of T and B cells by microbial products.

**Bacterial defense against phagocytosis and killing**

Some bacteria have protective mechanisms that prevent them from being ingested or killed. These include protective coats containing M protein from group A streptococci, protein A from *Staphylococcus aureus*, or polysaccharides associated with LPS of *E. coli*. Other bacterial protective mechanisms are the release of toxins, powerful enzymes, or other chemical substances that kill phagocytic cells.

**Inactivation of other immune defense strategies**

Viruses such as hepatitis B inhibit the production of IFNα by infected cells. Low-affinity antibodies are produced to some organisms, for example treponemes, while others inactivate antibodies by production of proteases. Viruses and bacteria can also block complement activation. HIV inactivates CD4+ T cells while other viruses decrease expression of MHC class I molecules on infected cells, interfering with antigen processing and presentation, blocking the activity of cytotoxic T cells, and inhibiting antibody production. Other microbes produce cytokines that channel the immune response toward one that is ineffective, for example, inducing a humoral response when a cellular response is required for protection.

| **Related topics** | (H1) The microbial cosmos | (L3) Autoimmune diseases – |
|---|---|---|
| | (H2) Immunity to different organisms | mechanisms of development |

## The battle to stay ahead

Immune defense against pathogens is dependent, firstly, on being able to recognize the intruder as a threat, and secondly, being able to eliminate it. While the physical and mechanical barriers as well as the innate and adaptive immune systems are powerful in the prevention of infection, microbes have developed ways both of avoiding recognition and of inactivating components used for their elimination (Table 1). Some pathogens avoid immune recognition by intracellular habitat, mimicking of self-antigens, encapsulation, or changing their surface antigens (antigenic variation). Other pathogens compromise effector mechanisms of immunity by inhibiting complement activation, phagocytosis, and appropriate cytokine production. They can release soluble neutralizing antigens, produce enzymes capable of destroying antibodies or complement, produce superantigens, or induce overall immunosuppression.

## Avoidance of recognition

### Intracellular habitat

Viruses, some bacteria (e.g., mycobacteria, *Listeria*, *Salmonella typhi*, and *Brucella* species) and certain protozoa (i.e., malaria-causing *Plasmodium falciparum*, *P. malariae*, *P. ovale*, and *P. vivax*) are obligate intracellular organisms, thus evading direct recognition by, and the effects of, the innate and adaptive immune systems.

### Antigenic variation

Some viruses, for example influenza, cause alterations of cell-surface antigens through mutation (**antigenic drift**). This makes it very difficult for the immune system to provide protection, as a primary response would need to be generated. The recombination of nucleic acids from human and animal viruses can lead to major **antigenic shifts**, and is known to be responsible for viral pandemics. Other organisms can produce continuous changes in their antigenic coat, distracting the immune system, for example trypanosomes and *Borrelia recurrentis*. In the case of trypanosomes, at least 100 different surface coats can be expressed in sequence.

### Disguise

Some microbes use antigens common or cross-reactive with self (**molecular mimicry**) to avoid immune recognition; for example, the hyaluronic acid capsule of some streptococcal species is the same as that of host connective tissue. While this seems an excellent strategy for the bacteria, it can lead to the development of autoimmune disease (Section L3). *Schistosoma* wears the antigens of the host that it infects, again trying to look like self. In other cases, specific T cells and B cells can be "distracted" through poly- or oligoclonal activation by microbial products. The enterotoxin of *Staphylococcus* (a "superantigen"), for example, activates large numbers of T cells independent of their specificity. Similarly, Epstein–Barr virus activates most B cells but they only produce low-affinity IgM antibodies and few are directed to the virus.

## Table 1.  Pathogen defense strategies

**Avoidance of recognition**

| | |
|---|---|
|    Intracellular habitat | Viruses, mycobacteria, *Brucella*, *Legionella* |
| **  Antigenic variation** | |
|     Drift | Viruses undergo mutation to alter antigens, e.g., influenza, HIV |
|     Shift | Recombination with animal viral nucleic acids, e.g., influenza pandemics |
|     Gene switching | Expression of a sequence of different surface antigens, e.g., *Borellia*, *Trypanosoma* |
| **  Disguise** | |
|     Molecular mimicry | Microbes have antigens in common with self, e.g., *Streptococcus*, *Bacteroides* |
|     Coating with self proteins | Covering of surface with serum proteins, e.g., *Schistosoma*, *Toxoplasma* |
|     Immune distraction | Some microbes, e.g., *Staphylococcus*, produce superantigens which stimulate many different B and T cells, diluting the effects of specific antigens |
| **Inactivation of host immune effector mechanisms** | |
| **  Phagocytosis** | Encapsulation of some bacteria inhibits phagocytosis, e.g., pneumococcus, *Haemophilus influenzae*, and *E. coli*. |
| **  Cytokines** | Inhibition of interferon production, e.g., hepatitis B |
| **  Antibodies** | Low-affinity antibody production, e.g. treponemes |
| | Neutralization of antibody by large amounts of soluble antigens, e.g., *Streptococcus pneumoniae*, *Candida* spp. |
| | Release of proteases that cleave IgA, e.g., *Pseudomonas* spp., *Neisseria gonorrhoeae*, *H. influenzae* |
| | Production of proteins that bind to the IgG Fc region and prevent opsonization, e.g., *Staphylococcus* protein A |
| **  Complement activation** (classical pathway) | Inhibition by incorporation of host complement regulatory proteins into microbial cell wall, e.g., HIV. Inhibition of C3 convertase and C5 by staphylococcal proteins |
| **  T cells** | CD4 T cells infected and killed, e.g., HIV |
| **  Antigen processing** | Inhibition of antigen processing and presentation, e.g., mycobacteria, herpes simplex, HIV, human cytomegalovirus (CMV) and presentation |

## Bacterial defense against phagocytosis and killing

### Inhibition of chemotaxis and engulfment

Phagocytes play a critical role in the killing of extracellular microbes and presenting microbial antigens to the adaptive immune system. Microbes use different strategies to prevent this and ensure their survival. They may release bacterial products, for example streptococcal streptolysin, *Clostridium* toxin, or fractions of mycobacteria, to prevent migration (chemotaxis) to the site of infection. Some pathogenic bacteria have polysaccharide capsules that prevent them from being ingested; for example, *Streptococcus pneumoniae*, *Klebsiella pneumoniae*, and *Haemophilus influenzae* prevent phagocytosis. Others have membrane antigens that inhibit phagocytosis; for example, protein A from *Staphylococcus aureus* binds to the Fc region of IgG and IgA thereby preventing opsonization of the staphylococcal bacteria; M proteins found in fimbriae of streptococci bind to fibrinogen in the serum and prevent complement activation thus allowing the microbe to survive phagocytosis and killing.

## Survival within the phagocytic cell

Once engulfed, the intracellular killing of microorganisms in phagocytes involves the fusion of lysosomes containing bactericidal factors with phagosomes, through oxygen-dependent and oxygen-independent means and other lytic enzymes. Staphylococci produce enzymes such as superoxide dismutase and catalase that can inhibit the oxygen burst. TB resists intracellular killing because they have a waxy coat containing mycolic acids and other lipids and are not easily attacked by lysomal enzymes. The outer membrane and capsular components of *Salmonella*, *Escherichia coli*, *Brucella*, and *Yersinia* can protect against the lytic activity of lysozyme. Bacteria that survive within the cell are called intracellular parasites. Survival also allows these bacteria protection from other arms of the immune system and allows the development of full virulence.

## Killing of phagocytic cells

Many Gram-positive bacteria release substances that cause damage to or kill professional phagocytes. These are extracellular enzymes or toxins usually referred to as "agressins". They include streptolysin O, produced by pathogenic strains of streptococci, which binds to cholesterol in phagocytic cell membranes causing lysosomal granules to burst thereby releasing their toxic products into the cell. Staphylococci produce leukocidin, which can destroy the phagocyte cell membrane and discharge lysosomal granules into the cell's cytoplasm. Exotoxins released by bacteria such as anthrax, *Pseudomonas* species, and pertussis, may also prevent phagocytosis by killing phagocytic cells.

## Inactivation of other immune defense strategies

### Inhibition of cytokines

Some viruses inhibit the production of IFNα by infected cells. This is seen in hepatitis B infection of hepatocytes. TB bacilli interact with the mannose-binding receptor on macrophages inducing an IL-10 response that dampens the pro-inflammatory response essential for protection against mycobacterial infections and favors TB replication. Some *Salmonella* species, after invasion of epithelial cells, downregulate IL-8 production and inhibit NFκB-dependent gene expression. Microbes such as *Salmonella typhi* produce endotoxins that predispose the immune system to develop a Th2 response and antibodies to the pathogen although eradication of these organisms requires a Th1 response.

### Antibodies

Some organisms, for example treponemes, induce low-affinity antibodies, while others, such as *S. pneumoniae* and *Candida*, release large amounts of soluble antigens that bind to antibodies and block their binding to the microbe. In another strategy, microbes produce proteases that destroy antibodies. Bacteria associated with mucosal infections such as *Neisseria gonorrhoeae* and *Pseudomonas* species produce protease enzymes that can destroy IgA, the antibody associated with mucosal protection. *Pseudomonas* also produces an elastase that inactivates C3b and C5a, inhibiting opsonization and chemotaxis.

## Inhibition of the complement system

Microbial products have been shown to inhibit the production of some complement components, modulate the activation process, and degrade other components. For example, two different staphylococcal proteins inhibit C5 and C3 convertase (Section D8). Another example is that HIV incorporates host complement regulatory proteins in its outer membranes to counteract the activation of complement.

## Inhibition of antigen processing and presentation to T cells

Microbes have several strategies for avoiding adaptive immunity by targeting both HLA class I and II processing and presentation pathways. Examples of some microbes that affect antigen processing and presentation are shown in Table 2.

**Table 2. Microbes that impair antigen processing and presentation**

| | Bacteria | Viruses |
| --- | --- | --- |
| Modulation of HLA class II synthesis, trafficking, and expression | Mycobacteria, *Chlamydia*, *Escherichia coli*, *Vibrio cholerae* | Herpes simplex, HIV, human CMV, human rhinovirus 14, RSV, EBV |
| Inhibition of antigen processing for HLA class II presentation | Mycobacteria, *Salmonella* | Herpes simplex |
| Modulation of HLA class I synthesis, trafficking, and expression | *Salmonella*, *Yersinia*, *Chlamydia* | Human CMV |
| Inhibition of antigen processing for HLA class I presentation | | EBV, human CMV, KSHV, herpes simplex and HIV adenovirus |

Certain microbial products can inhibit the proteolytic breakdown of peptides in both pathways (Section F2), the assembly of HLA molecules and their trafficking, and the loading of peptides for presentation to T cells. All of these strategies lead to impairment in appropriate T-cell activation resulting in poor antibody and cytotoxic responses to the microbes. Note that HIV not only impairs antigen processing but also infects and kills CD4 T cells effectively disarming the immune system and leaving the host immunodeficient (Section J3).

## Directing regulation of immune responses

Some microbes produce cytokines that channel the immune response towards one that is ineffective, for example inducing a humoral response when a cellular response is required for protection. The Epstein–Barr virus (EBV) genome codes for an IL-10 homolog that favors a Th2 response by suppressing a Th1 response and thus enhancing an ineffective antibody response to EBV (Section G).

# H4 Pathophysiology of infectious diseases

---

## Key Notes

| | |
|---|---|
| **Entry of microbes into the body** | Microbes can gain access to the body through the respiratory tract, gastrointestinal tract, skin, pharynx, urinogenital tract, and conjunctiva. The most common mode of entry is via the respiratory system. Once the microbe gains entry into the body and avoids innate and adaptive immune responses, infection will ensue. |
| **Events occurring subsequent to infection** | The primary requisite for a bacterium, virus, or fungus to establish an infection is to gain access across the epithelial layer and through the basement membrane and, in the case of obligate intracellular parasites, to access appropriate cells in order to survive and proliferate. Inflammation is the body's primary response to microbial proliferation with the nervous, innate, and adaptive immune systems providing the different defense strategies. |
| **Examples of modes of infection and immunity to specific microbes** | Three known pathogens (*Staphylococcus aureus*, influenza A virus, and *Plasmodium falciparum*) have different modes of entry into the body. The subsequent immune responses described are not effective in all cases. |
| **Related topics** | (A2) External defenses (B2) Molecules of the innate immune system     (C3) Mucosa-associated lymphoid tissue (H3) Pathogen defense strategies |

---

## Entry of microbes into the body

### Respiratory tract

The most common port of entry of microbes into the body is through the respiratory tract; rhinoviruses (colds), influenza, TB, streptococcal, and staphylococcal infections are usually through this mode. Some fungi such as *Aspergillus* or *Candida* species are ubiquitous in nature and may infect through the respiratory system or the skin. It is estimated that we inhale between 5000 and 15 000 microorganisms a day depending on the environment, although these are mainly nonpathogenic bacteria. Efficient cleansing mechanisms help remove inhaled particles keeping the respiratory tract clean. A mucous ciliary lining consisting of mucus-secreting cells and ciliated cells perform the housekeeping duties moving trapped particles back upwards to be expelled via the mucociliary escalator (Section A2). The upper respiratory tract has a similar mucociliary lining and continuous mucous secretions from the nasal cavity each day are responsible for removing harmful bacteria. The alveoli are at the terminal end of the respiratory tract. Here there are no cilia and little mucus, but ready in waiting are large numbers of macrophages. All particles whether viral, bacterial, fungal, or inert particles are dealt with the same way. Large particles are usually trapped in the hairs lining the nostrils, and microbes that escape mucous entrapment are phagocytosed by macrophages. Epithelial lining cells also produce antimicrobial peptides, for example defensins (Section B2). These cationic peptides kill bacteria by causing cell lysis.

## The skin

The body's outer surface is covered with skin, which is a horny layer and usually impermeable. It is the largest organ in the body with a weight of approximately 3 kg. It provides protection against most bacteria and other microorganisms, helps regulate the body's temperature, and allows secretion. The skin is made up of two main layers—the epidermis and the dermis. The outer layer of dead epidermal cells is continuously sloughed off making it difficult for microbes to attach and penetrate. The skin is also slightly acidic and produces antimicrobial substances such as fatty acids, making it a hostile environment for bacteria and providing protection from infection. However, damage to the skin allows the entry of microbes and in some cases insects inject microbes, for example *Plasmodium* (malaria parasite) and trypanosomes, directly through the skin, bypassing the protective skin layer.

## The intestinal tract

The intestinal tract must handle large amounts of organic material each day, including large amounts of nonpathogenic bacteria. There are no specific cleansing mechanisms apart from digestive enzymes in the small intestine, highly acidic fluids in the stomach, and the general flow of intestinal contents. In addition to presenting a physical barrier to microbial penetration, the gut epithelium plays a more active role by producing and secreting large quantities of antimicrobial peptides (Section B2). As much as 20% of the large intestinal gut contents are composed of commensal bacteria which, *in situ*, perform many housekeeping functions such as further degradation of organic material, production of vitamins, synthesis of antifungal and other antimicrobial products, as well as producing an anaerobic environment hostile to aerobic bacteria.

## The pharynx

The throat, mouth, tonsils, and nasal passage are common sites for microbial habitation and further entry into the body. The microbial flora content and numbers differ amongst individuals. Bacteria are the most numerous, but some fungi (*Candida albicans*) and certain protozoa (*Entamoeba gingivalis*) may be found in some individuals. Oral bacteria include streptococci, micrococci, *Neisseria* species, and diphtheroids. Dental plaque is another source of microbial colonization; plaque is a composition of minerals and bacteria.

## Urinogenital tract

Urine is generally sterile because constant flushing makes the urethral environment inhospitable to many microbes. The male urethra is approximately four times longer than the female and normally sterile. As the female's urethra is shorter (approximately 5 cm) and much closer to the anus it is more readily contaminated by microorganisms from the anal canal. This probably accounts for the fact that urinary tract infections are much more common in females than males.

## Conjunctiva

The conjunctiva is kept moist and healthy by continuous washing with tears. Tears contain lysozyme and other antimicrobial products (Section A2). The blinking action of the eyelid serves to keep the eye moist, sweep away foreign particles, and keep the eye lubricated. Eye infections usually resort from injuries, dry eye syndrome, or diseases of the secretory glands.

## Events occurring subsequent to infection

To establish an infection the priority of the invading microbe virus or bacteria is to traverse the epithelial layer and basement membrane. Bacterial infections such as those caused by gonococci and streptococci are usually confined to epithelial cells, as are most *Salmonella* infections of the intestine. Obligate intracellular parasites such as TB, brucella, and viruses require intracellular habitation before they can cause infection. Extracellular bacteria convey a serious disadvantage as they are exposed to all antimicrobial forces that the body can muster. Indeed bacteria and other microbes that are capable of extracellular replication generally advertise their presence by releasing products into the surrounding fluids that may cause immunological notice (e.g., heat shock proteins (HSPs) and lipopolysaccharide (LPS)) and inflammation (exotoxins and endotoxins). During inflammation changes take place in inflamed tissue, vessel dilation is increased, gaps appear between endothelial cells and tight junctions, osmotic changes take place, and there is increased leakage of tissue fluids and plasma proteins from capillary vessels. These changes cause symptoms such as heat, swelling, and pain, all of which will have alerted the nervous system. Cell adhesion molecules will have been expressed at the site of inflammation (Section B4).

Inflammation may also cause epithelial cell and basement membrane barrier destruction. Once across the basement membrane the invading microbe will encounter a hostile environment. This will consist of antimicrobial fluids containing complement components, previously formed antibodies and other antimicrobial proteins such as C-reactive protein, lactoferrin and $\alpha_2$-macroglobulin. Phagocytes will already have been alerted to the site of action through chemotactic processes (also through the nervous system) and are ready to access the site of inflammation by way of the adhesion molecules expressed at the site of attrition. There is a constant flow of plasma proteins from the capillaries and back again into the blood as well as into the lymphatics draining these tissues and then into the draining lymph nodes. In the lymph nodes any microbes that have been phagocytosed, killed, and digested will be presented as antigenic molecules to the T- and B-cell compartments of the immune system. If the invading microbe survives the body's initial defenses, it may spread systemically through the blood stream, the lymphatic system, or by cell-to-cell contact to other areas to form depots or reservoirs of infection.

## Examples of modes of infection and immunity to specific microbes

### Staphylococcus aureus

*Staphylococcus aureus* is a Gram-positive coccus that is present in the skin and mucous membranes of the gastrointestinal and genitourinary tracts of humans. It is a very common microbe found in the anterior nasal passages of 15–30% of adults and attaches to skin and mucosal surfaces through several adhesion molecules.

It is only when there is damage to these epithelia through burns, traumatic wounds, surgical incisions, ulceration, viral skin lesions, and so forth, that infection can take place. Infection is more common in individuals with reduced host resistance, that is, patients with cystic fibrosis or diabetes mellitus and those undergoing immunosuppression. Normally invasion by *S. aureus* is prevented through an intact barrier and the production of antimicrobial peptides such as α-defensins and cathelicidin by the epithelial cells. Once through the epithelial barrier, neutrophils also produce these antimicrobials but the microbe secretes substances to inactivate them (Section H3). In addition, complement-mediated neutrophil chemotaxis and phagocytosis is suppressed. In fact, multiple mechanisms are used by *S. aureus* to evade the immune system (Section H3). It is envisaged that tissue damage through exotoxins released by *S. aureus* results in acute inflammation and that macrophages and dendritic cells process and present microbial antigen to T cells and B cells in draining lymph nodes, giving rise to some neutralizing antibodies. Even the IgG antibodies are blocked by staphylococcus A protein released by the microbe, resulting in the inhibition of opsonization. Entry of *S. aureus* into the blood stream can lead to seeding and infection of many organs, including the heart, lungs, bones, and joints. Pneumonia may result if bacteria reach the lungs, while acute and chronic osteomyelitis or septic arthritis can also occur. Eventually, sufficient neutralizing antibody is produced but generally antibiotic therapy is required for systemic infections. The development of methicillin-resistant *S. aureus* (MRSA) is still a major public health problem.

### Influenza A virus

Influenza A is an RNA virus that is associated with seasonal outbreaks of infection. Up to 10% of the population can be infected, leading to increases in hospital admissions and influenza-related deaths. It enters the body via the nasal or oral mucosal epithelium and through its hemagglutinin (HA) binds to the surface of the respiratory epithelial cells through cell-surface receptors. The virus enters the host cell, uncoats, and replicates and new virions are released by the infected cells through budding at the plasma membrane of the host cell. Damage to the respiratory epithelial surface occurs due to the cytolytic interaction of the virus and the host cell. The exact sequence of host immune responses to influenza is largely unknown. However, it is envisaged that virus-induced damage to the epithelial cells results in acute inflammation mediated by macrophages (probably through pattern recognition receptors) and possibly mast cells in the lamina propria. NK cells probably also play a role early in infection through killing virus-infected cells. Because the virus effectively removes the inner lining of the respiratory tract, poor mucous secretion and damage to the mucociliary escalator make it more likely that other microbes (especially bacteria) can infect the lungs, resulting in secondary infections.

The virus rarely enters the systemic circulation but systemic manifestations do occur due to the release of pro-inflammatory cytokines such as interferon-α (IFNα) and interleukin 6 (IL-6), giving rise to headaches, fever, muscle aches and pains, and severe malaise.

It is likely that macrophages, and particularly dendritic cells in the lamina propria, pick up dead cell fragments containing the virus and migrate to draining lymph nodes where the adaptive immune response is stimulated. Th1 and Th2 cells both help specific B cells and precursor Tc cells to develop. Some neutralizing antibodies to HA on the virus appear after a few days, which may prevent new infections. It is likely that recovery from the infection is mediated by Tc cells (usually directed specifically against internal viral proteins), which take some time to develop. Influenza virus undergoes antigenic drift, meaning that we have little memory against mutating viruses. However, the internal proteins, for example nucleoproteins of the virus, are relatively conserved such that T cells probably do offer some protection against new viruses emerging annually to infect populations. The major problem with new infections (leading to pandemics) comes with antigenic shift resulting in many new antigens derived from animal viruses (Section H3).

### *Plasmodium falciparum* (malaria parasite)

This protozoan is the most virulent species of malaria in humans. A female mosquito taking a meal of human blood by sticking its proboscis through the skin into the subepithelial layer picks up the parasite, in the form of gametocytes, from an infected individual. Following several stages of sexual reproduction in the insect, sporozoites are produced and a small number injected into another unlucky human while the mosquito is feeding. The sporozoites travel via the bloodstream to the liver where they attach to and enter hepatocytes. Rupture of the hepatocytes results in release of merozoites that invade and destroy erythrocytes. During this phase some of the merozoites develop into trophozoites. These undergo sexual reproduction to produce schizonts that are released into the bloodstream as merozoites which infect further erythrocytes. Some schizonts develop into male and female gametocytes that are passed into a mosquito while taking a blood meal and the cycle starts again. The reason for describing the life cycle is that each morphological form carries a different set of parasite antigens, making it very difficult for the host to mount an effective immune response (Section H3). Our knowledge of immune responses to the parasite is far from complete but a number of mechanisms are believed to be important. Some of the sporozoites entering the skin pass into draining lymph nodes where antigen-presenting cells stimulate T and B cells. This results in the production of antibody and probably some cytotoxic T cells that are stimulated to directly kill the hepatocytes and produce a number of cytokines such as TNFα, IFNγ, and IL-1. Merozoites stimulate the production of antibodies, reactive oxygen and nitrogen intermediates, and TNF. Antibodies against the male and female gametes are also produced and may have some protective effect. All of these immune mechanisms are believed to be generated, but the rather weak response is clearly insufficient to prevent the cycle from being completed within the human host. A more complete understanding of the immune responses should help in the development of an effective vaccine against malaria.

# I1 Principles of vaccination

---

**Key Notes**

| | |
|---|---|
| **Principles of vaccination** | The primary goal in vaccination is to provide protective immunity by inducing a memory response to an infectious microbe using a nontoxic antigen preparation. It is important to produce immunity of the appropriate kind: antibody and/or cellular immunity. |
| **Antibody-mediated protection against microbes** | Antibodies produced as a result of immunization are effective primarily against extracellular organisms and their products, for example toxins. Passively administered antibodies have the same effect as induced antibodies. |
| **Cell-mediated immune protection against microbes** | Cell-mediated immunity (T cells, macrophages) induced by vaccination is important particularly in prevention of intracellular bacterial, viral, and fungal infections. |
| **Related topics** | (E3) The cellular basis of the antibody response    (F1) The role of T cells in immune responses |

---

## Principles of vaccination

Edward Jenner, a country physician in England, noticed that dairymaids who frequently contracted cowpox were often immune to the ravages of smallpox, leading him to develop an approach whereby cowpox was used to vaccinate people against smallpox. The term vaccination is derived from the Latin word *vaccinus* meaning "from cows." Vaccination eventually resulted in the complete eradication of smallpox (in 1980) and has been generalized as a reliable method of protection against many pathogens.

The aims of vaccination are to induce memory in T and/or B lymphocytes through the injection of a nonvirulent antigen preparation. Thus, in the event of an actual infection, the infectious agent and/or its toxin are met by a secondary rather than a primary response. The ideal vaccine would protect the individual and ultimately eliminate the disease, but most vaccines simply protect the individual. A standard set of vaccines is now in use worldwide, some of which are (or should be) given to everyone, others to those particularly at risk (Table 1). The timing of vaccination depends on the likelihood of infection, vaccines against common infections being given as early as possible, allowing for the fact that some vaccines do not work properly in very young infants.

## Herd Immunity

Herd immunity results from the vaccination of a significant proportion of the population, which probably causes disruption to the chain of infection. Protecting individuals reduces the number of infectious organisms and decreases the chances of non-vaccinated individuals being infected.

**Table 1. Vaccine recommendations**

| Recommended for | Vaccine | When given/to whom |
| --- | --- | --- |
| All | Measles | From 1 year (6 months in tropics), boost at 10–14 years |
| | Mumps | |
| | Rubella | |
| | Diphtheria | From 2 to 3 months old |
| | Tetanus | Dip/tet boost at 5 years |
| | Pertussis | |
| | Polio (Sabin) | 2–6 months (oral) |
| | (or Salk*) | Parenterally |
| | BCG** | Varies, see footnote |
| Those at risk | Hepatitis B | Travellers, children (some countries) at risk |
| | Hepatitis A | Travelers, people with chronic liver disease, institutions etc. |
| | Influenza | Aged, doctors, nurses etc., annual boost needed |
| | Rabies | Travelers, post-exposure: vaccine + antibody |
| | Meningococcal | Children: 2–10 years |
| | Pneumococcal | Elderly |
| | Haemophilus influenza type b | Infants 2–15 months |
| | Varicella-zoster | Children with leukemia |
| At risk (travel) | Typhoid | Travelers |
| | Cholera | Travelers |
| | Yellow fever | Travelers: boost at 10-year intervals |

*Salk is the polio vaccine of choice in the Netherlands and Scandinavia.
**BCG (bacille Calmette–Guérin) vaccination programs for protection against TB vary worldwide. In some countries, for example the USA and the UK, the immunization schedule is based on risk assessment, that is, where the incidence of TB is significant (greater than 40 per 100000) or if close relatives come from a country with high incidence. In many other countries BCG vaccination is mandatory. Where vaccination is performed, immunizations should be given to neonates while still in hospital. Vaccinations may be given to children up to 16 years of age if warranted and if the child is tuberculin-negative (Mantoux). BCG vaccination is most effective against tuberculous meningitis and miliary TB and lasts for about 15 years. It is contraindicated in patients with immune deficiencies, who are receiving immunosuppressive therapy, or who live where active TB cases have been confirmed.

## Antibody-mediated protection against microbes

Antibodies either produced as a result of immunization or passively introduced into the host are very effective in preventing infection. They will be ready and able to bind the infectious agent at the time of infection instead of waiting for the host's immune system to respond. Antibodies can block viruses or bacteria from entering and infecting host cells and can mediate their killing. They can also prevent damaging effects on other cells by neutralizing toxins such as those produced by *Diphtheria* or *Clostridium* species (Section

H2). IgG antibodies are primarily effective in the blood, whereas IgA plays an important role at mucosal surfaces, where it helps to prevent viral or bacterial access to the mucosa lining cells—the mechanism by which polio vaccination works.

During pregnancy, maternal antibodies are transferred across the placenta to the fetus, providing **passive immunity** that protects the newborn during the first months of life (Section C5). This passive transfer can, in some instances, also be a disadvantage. For instance, the presence of maternal antibody in the newborn inhibits effective immunization. Thus, immunization against some antigens may be delayed until after most of the maternal antibodies have been catabolized. On the other hand, if the mother has an autoimmune disease, for example idiopathic thrombocytopenia, the baby's platelet count will need to be monitored as the mother's antibody can significantly reduce the baby's platelets, causing life-threatening bleeding.

In pre-antibiotic days, it was common to treat or prevent infection by injecting antibody preformed in another animal, usually a horse or a recently recovered patient. This principle is still in use for certain acute conditions where it is too late to induce active immunity by vaccinating the patient (Section J4).

## Cell-mediated immune protection against microbes

This involves helper and cytotoxic T cells, and their activation of other cell populations, including macrophages, PMNs, and NK cells, as well as the release of various cytokines. While antibodies play a major role in combating extracellular infections, cell-mediated immunity is essential for eradicating viral infections as well as certain bacterial, fungal, and protozoan intracellular infections (Sections F1 and H1). Effective vaccination should therefore be aimed at inducing both cellular and humoral responses to the infectious agent.

In certain instances not only are CD4+ and CD8+ T-lymphocyte responses desired, but also those from NK and NKT cells. In other instances it may be more advantageous to specifically target Th1 or Th2 responses; for example, infections with helminths might require Th2-type immunity via induction of IgE antibodies, whereas protection against mycobacterial and fungal infections may be better obtained by a Th1 response that results in production of macrophage activation factors (e.g., IFNγ). CD8+ cytotoxic T cells find and kill infected cells that express protein components of pathogens. The cell that is targeted is determined by the presence of the foreign protein in association with MHC class I molecules. CD8+ T cells lyse the infected cell; hopefully before the progeny of the infectious organisms are fully developed.

CD4+ T cells are basically the directors of the immune response. They interact with foreign antigen expressed in association with MHC class II molecules and then provide soluble or membrane-bound signals for B cells, macrophages, or CD8+ T cells to help them obtain their full effector cell functions: antibody production by B cells, killing by macrophages, and CD8+ T cells. Some diseases only require an antibody response for protection or clearance, while others require a cell-mediated immune response. Yet other diseases are only resolved if both forms of protection are present.

# I2 Immunization

---

**Key Notes**

| | |
|---|---|
| **Passive immunization** | Passive immunization is the administration of preformed antibodies either intravenously or intramuscularly. It is used to provide rapid protection against certain infections such as diphtheria or tetanus, or in the event of accidental exposure to certain pathogens such as hepatitis B. It is also used to provide protection in immunocompromised individuals. |
| **Active immunization** | Active immunization is the administration of vaccines containing microbial products with or without adjuvants in order to obtain long-term immunological protection against the offending microbe. There are two routes of immunization: systemic and mucosal immunization. Systemic immunization, either intramuscularly or subcutaneously, is the normal route of vaccination at present. Mucosal immunization is the method of choice for polio and *Salmonella typhi* vaccines which are given orally. There is increasing awareness that this route may be the best for most immunizations, because many infectious agents enter through mucosal surfaces. |
| **Related topics** | (B2) Molecules of the innate immune system<br>(C3) Mucosa-associated lymphoid tissues <td>(I3) Antigen preparations<br>(K4) Immune-complex-mediated (type III) hypersensitivity</td> |

---

## Passive immunization

Passive immunization is the administration of preformed antibodies, usually IgG, either intravenously or intramuscularly. These may be derived from individuals who have high antibody titers to particular microbes and can thus provide rapid protection against infections such as *Diphtheria*, rabies, or those caused by *Clostridium* species. They are also very useful in the event of accidental exposure to certain pathogens such as hepatitis B. Passive immunization is also used to provide protection in immunocompromised individuals who are unable to make appropriate antibody responses or in some instances incapable of making any antibody at all, that is, in severe combined immunodeficiency. Antibodies given to immune-deficient patients are usually of the IgG class and are derived from pooled normal plasma. These antibodies must be given on a continuous basis, ideally every three weeks, because they are continuously catabolized and are only effective for a short period

Antibodies preformed in animals, notably horses, are also administered for some diseases, but with repeated injections there is the danger of immune-complex formation and serum sickness (Section K4). Antisera are usually injected intramuscularly, but can be given intravenously in extremely acute conditions. Indications for the use of passive immunization by the injection of preformed antibody are shown in Table 1.

## Table 1. Passive immunization

| Disease | Source of antiserum | Indications |
|---|---|---|
| **Infection** | | |
| Tetanus | Immune human; horse | Post-exposure (plus vaccine) |
| Diphtheria | Horse | Post-exposure |
| Gas gangrene | Horse | Post-exposure |
| Botulism | Horse | Post-exposure |
| Varicella-zoster | Immune human | Post-exposure in immunodeficiency |
| Rabies | Immune human | Post-exposure (plus vaccine) |
| Hepatitis B | Immune human | Post-exposure |
| **Prophylactic treatment** | | |
| Hepatitis A | Pooled human Ig | Prophylaxis |
| Measles | Immune human | Post-exposure in infants |
| Snakebite | Horse | Post-bite |
| Some autoimmune diseases | Pooled human Ig | Acute thrombocytopenia |
| | | Autoimmune neutropenia |

## Active immunization

Administration of vaccines containing microbial products with or without adjuvants in order to obtain long-term immunological protection against the offending microbe is termed active immunization. Immunization can be given via two different routes.

## Systemic immunization

This is the method of choice at present for most vaccinations and is usually carried out by injecting the vaccine subcutaneously or intramuscularly into the deltoid muscle. Ideally all vaccines would be given soon after birth, but some are deliberately delayed, for various reasons. The common systemic vaccines for measles, mumps, and rubella are usually given at 1 year of age because, if given earlier, maternal antibody could decrease their effectiveness. The carbohydrate vaccines for *Pneumococcus*, *Meningococcus*, and *Haemophilus* infections are usually given at about 2 years of age, as before this age they respond poorly to polysaccharides unless they are associated with protein components that can act to recruit T-cell help for the development of anti-polysaccharide antibody, for example hen egg albumin (Section I3).

## Mucosal immunization

Most infectious agents gain entry to the body through mucosal surfaces, sites that are also associated with the largest source of lymphoid tissue. Thus, current vaccination approaches have focused on immunization either orally or through nasal-associated immune tissue (NALT: Section C3). This approach has become more efficient and available for public use, obviating the need for painful injection, and permitting self-administration of certain vaccines (e.g., for immunization against influenza). Adjuvant vaccines and live vectors have also been successfully used to target the mucosal immune system. Moreover, in animal models attenuated strains of *Salmonella* act as a powerful immune stimulus and, as carriers of foreign antigen, can be used to immunize mucosal surfaces against herpes simplex virus and human papillomavirus. Furthermore, bacterial toxins, for example those derived from *Vibrio cholerae*, *Escherichia coli*, and *Bordetella pertussis*, have immunomodulatory properties that are being exploited in the development of mucosally active adjuvants. Pertussis toxin augments the co-stimulatory molecules CD86 on B cells and CD28 on T cells (Section F4) and increases IFNγ production.

# I3 Antigen preparations

## Key Notes

| | |
|---|---|
| **Antigen preparations** | Protection against pathogenic microorganisms requires the generation of effective immunity, and thus the induction of appropriate cellular and/or humoral immune responses. Most vaccines consist of attenuated live organisms, killed organisms, inactivated toxins, or subcellular fragments thereof, and, more recently, genes for antigens in viral "vectors," and DNA itself. |
| **Adjuvants** | Nonliving vaccines, especially those consisting of small molecules, require the inclusion of adjuvants to enhance their effectiveness. These include microbial, synthetic, and endogenous preparations having adjuvant activity; at present only aluminum and calcium salts are used in humans. Adjuvants should preserve antigen integrity, enable antigens to be slowly released, and permit effective targeting of antigen-presenting cells. |
| **DNA vaccines** | The use of DNA encoding antigens as vaccines has shown potential. Intramuscular injection of circular DNA results in DNA uptake by muscle cells, expression of the encoded protein and induction of both humoral and cell-mediated immunity. |
| **Recombinant vaccines** | Using molecular genetics, selective recombinant proteins with defined epitopes can be prepared that can induce protective immunity to pathogens. This approach overcomes the problem of disease complications that may occur with modified live vaccines. |
| **Nanoparticle vaccines** | Nanoparticles consisting of synthetic resins, polystyrenes, carbon particles, silica, iron, or polyethylenes are currently being investigated for suitability as adjuvants in vaccine delivery. They offer potential advantages over current vaccines in that they are easily manipulated and may be given through the mucosal route or systemically. |
| **Cytokines** | Cytokines can be added at the time of immunization to skew the immune response to a Th1- or Th2-type response depending on the kind of response needed for protection against a particular pathogen. Cytokines that might be useful are IL-12 or IFNγ, which favor a Th1 response, or IL-4 and IL-10, which favor a Th2 response. |
| **Related topics** | (B2) Molecules of the innate immune system     (E3) The cellular basis of the antibody response |

## Antigen preparations

The protective immune response to pathogenic microorganisms requires the generation of specific T- and B-cell responses and appropriate effector mechanisms. In order to do this, vaccines must be capable of targeting the immune system appropriately. In principle anything from whole organisms to small peptides can be used, but in practice most vaccines consist of attenuated live organisms, killed organisms, inactivated toxins, or subcellular fragments (Table 1).

### Table 1. Antigen preparations used in vaccines

| Type of antigen | Viruses | Bacteria |
| --- | --- | --- |
| Normal heterologous organism | Vaccinia (cowpox) | |
| Living attenuated organism | Measles | BCG |
| | Mumps | Typhoid (new) |
| | Rubella | |
| | Polio: Sabin | |
| | Yellow fever | |
| | Varicella-zoster | |
| Whole killed organism | Rabies | Pertussis |
| | Polio: Salk | Typhoid |
| | Influenza | Cholera |
| **Subcellular fragment** | | |
| Inactivated toxin (toxoid) | | Diphtheria |
| | | Tetanus |
| | | Cholera (new) |
| Capsular polysaccharide | | *Meningococcus* |
| | | *Pneumococcus* |
| | | *Haemophilus* |
| | | Typhoid (new) |
| Surface antigen | Hepatitis B | |

There is also a fundamental distinction between live and dead vaccines. Living and nonliving vaccines differ in many important respects, notably safety and effectiveness. Many live vaccine organisms (mostly viruses) have been attenuated by growth under conditions that result in mutants that have lost virulence but retain antigenicity. Currently, mutation is usually "site-directed" using recombinant DNA technology. These mutant (modified-live) organisms are essentially new strains. They often induce stronger and better localized immunity, do not often require adjuvants or "booster" injections and provide the possibility of "herd" immunity in that mutated nonvirulent virus could be transferred to nonimmunized individuals in a local community. Moreover, the immunity induced, for example Th1 versus Th2 responses, is usually more appropriate for protection against the pathogenic strain of the organism. On the other hand, live attenuated vaccines can sometimes regain virulence by *back-mutation*. Moreover, in immunocompromised individuals even attenuated vaccines that have not regained virulence can cause severe disease.

Killed organisms or molecules derived from these organisms are used when stable attenuated organisms cannot be produced. These antigens may, however, induce weak and/or inappropriate (e.g., antibody vs. CTL) responses. Immune memory may be variable or poor, but they are usually safe if properly inactivated. In only one case (polio) is there a choice between effective live and killed vaccines. Recently, it has been shown that genes for one or more antigens can be inserted into a living vaccine (see DNA vaccines below), thus permitting the development of vaccines that can potentially protect against multiple pathogens.

## Adjuvants

Nonliving vaccines do not induce strong immune responses unless injected along with some other substance such as aluminum hydroxide, aluminum phosphate, calcium phosphate, or hen egg albumin; such substances are called adjuvants. The properties of adjuvants should: (i) include the ability to enable antigens to be slowly released so as to prolong antigen exposure time to the immune system; (ii) preserve antigen integrity; (iii) target antigen-presenting cells; (iv) activate both helper and cytotoxic T cells; (v) produce high-affinity immune responses; and (vi) have the capacity for selective immune intervention. A variety of microbial, synthetic, and endogenous preparations have adjuvant activity, but at present only aluminum and calcium salts are approved for general use in man.

Combinations of macromolecules (oils and bacterial macromolecules) are commonly used as adjuvants in animals to promote an immune response. Oil in the adjuvants increases antigen retention and aggregation (promoting immunogenicity), and inflammation at the site of inoculation. Inflammation increases the macrophage and dendritic cell response and causes local cytokine production, which modulates co-stimulatory molecules needed for T-cell activation. Microparticles, including latex beads and poly(lactide-co-glycolide), have also been used as adjuvants in experimental models. Adjuvants are now being designed to selectively drive Th1 or Th2 responses (Table 2).

### Table 2. Experimental adjuvants currently undergoing assessment

**Experimental, but likely to be approved**

Liposomes (small synthetic lipid vesicles)

Muramyl dipeptide, an active component of mycobacterial cell walls

Immune-stimulating complexes (ISCOMs) (e.g., involving cholesterol or phospholipids)

Bacterial toxins (*Escherichia coli*, pertussis, cholera)

Nanoparticles

**Experimental only**

Cytokines: IL-1, IL-2, IFNγ

Slow-release devices; Freund's adjuvant

Immune complexes

## DNA vaccines

"Naked" cDNA encoding influenza virus hemagglutinin, when inoculated into muscle tissue, stimulates both antibody production and a CTL response that are specific for the influenza protein. The potential for this is still unknown, but if this becomes a routine method of immunization, then the cost of generating and transporting vaccines would be very low. Other uses of recombinant DNA technology are the cloning of defined epitopes into viral or bacterial hosts. Typically, well-characterized infectious agents such as vaccinia virus, poliovirus, or *Salmonella* are used. DNA sequences are cloned into the genome of these agents and are expressed in target structures that are known to be immunogenic for the host. Inclusion of cytokines with the vaccine vectors may prove to be an efficient method for ensuring the correct cytokine environment to steer the immune response accordingly. DNA vaccines have a number of potential advantages over traditional methods of vaccination, including specificity and the induction of potent Th1 and cytotoxic T lymphocyte responses similar to those observed with attenuated vaccines, but without the potential to revert to virulence.

## Recombinant vaccines

Advances in molecular virology and bacteriology have provided many new targets for vaccine development. The last 20 years of study of viral and bacterial pathogenesis have identified not only the components of the immune system that are protective for many infectious agents, but also the epitopes on the pathogen that must be targeted for a vaccine to be effective. That is, certain parts of an infectious disease-causing organism, for example herpes virus glycoprotein D peptide (glyD), stimulate CTL that are protective. If the host is inoculated with glyD, they develop CTL responses to the epitope, without the potential for disease from vaccination with a modified live vaccine. This approach is also possible for antibody-mediated protection from infectious agents. In this scenario, both a B-cell epitope (the site that antibody binds to on the infectious agent) and a T-cell epitope (the peptide that binds to MHC class II to stimulate CD4+ helper cells) must be present, so as to select the appropriate B cells, and to stimulate specific T-cell help.

## Nanoparticle vaccines

Nanoparticles for vaccine production are currently being investigated for suitability as adjuvants in vaccine delivery. Nanoparticles are solid particles measuring from 10 to 1000 nm, to which the vaccine antigen is absorbed, attached, or bound. Common nanoparticles used for vaccine preparations include the polyesters poly (lactic acid) and poly (glycolic acid). Some biopolymers such as poly(lactide-co-glycolide) (PLGA) have natural adjuvant properties and have been extensively studied as immune stimulants. PLGA appears to promote dendritic cell maturation and nanoparticles coated with mannan have been used to target the mannose receptor on antigen-presenting cells. In other studies, monoclonal antibodies have been absorbed to the particles and used to locate specific targets on dendritic cells. Nanoparticle vaccines are currently being studied for application through the musosal and systemic routes of vaccination.

## Cytokines

Cytokines can significantly influence the function of professional antigen-presenting cells (APCs). Thus, IFNγ and IL-4 induce increased expression of class II molecules on APCs, thereby enhancing their antigen-presenting abilities. The use of such effector cytokines is considered a useful adjunct in vaccination, as polarization of the immune system to a Th1 or Th2 response may be preferable in some instances; for example, a Th1 response is the preferred response in tuberculosis, whereas a Th2 response is important in protecting against polio.

Thus, manipulation of Th1 and Th2 responses permits selective activation of the immune pathways important to protection against particular pathogens.

# I4 Vaccines to pathogens and tumors

## Key Notes

| | |
|---|---|
| **Bacterial vaccines** | Vaccines have been developed to many different types of bacteria, including, *Mycobacterium tuberculosis, Escherichia coli, Haemophilus*, pneumococcus, *Vibrio, Helicobacter* (ulcer-causing bacteria), and the Lyme disease spirochete. Especially notable is the diphtheria, pertussis, and tetanus (DPT) vaccine that protects children from these often fatal childhood diseases. |
| **Viral vaccines** | Vaccines have been developed to viruses that infect the respiratory tract (influenza, adenovirus), the gastrointestinal tract (polio, rotavirus), the skin (yellow fever, La Crosse fever) and the reproductive tract (herpes). Viral vaccines, like those for bacteria, are either modified-live, killed, or subunit. |
| **Vaccines to other infectious agents** | Protozoan parasites, such as those that cause malaria (*Plasmodium*), African sleeping sickness (*Trypanosoma*), and schistosomiasis are major diseases, mostly in the Third World. The ability to vaccinate people and animals to protozoan diseases will allow people to live in areas endemic (where the organism is always present) for the disease. |
| **Tumor vaccines** | Vaccination strategies for cancer therapy are being actively pursued. Prophylactic vaccines induce immunity to viruses associated with tumor development. Other approaches are designed to enhance/induce effective immunity in tumor-bearing patients against tumor antigens, including those associated with prostate, breast, colon, brain, and ovarian cancers, one of which (Provenge for prostate cancer) has been approved. |
| **Vaccines to addictive drugs** | Addictive behavior to drugs of abuse such as heroin/morphine and cocaine is a major problem in many countries. Current pharmaceutical therapies have shown limitations in the prevention of relapse and the continuous abuse of these drugs. Newer strategies involving vaccines to nullify the addictive properties are under investigation. |
| **Related topics** | (E2) B-cell activation      (H2) Immunity to<br>(F4) T-cell activation                  different organisms<br>                                         (N7) Tumor vaccines |

## Bacterial vaccines

Modified-live, killed, and subunit vaccines have been developed to many different types of bacteria, including *Mycobacterium tuberculosis*, *Escherichia coli*, *Haemophilus*, pneumococcus, *Vibrio*, *Helicobacter* (ulcer-causing bacteria), and the Lyme disease spirochete. Most familiar is the diphtheria, pertussis, and tetanus (DPT) vaccine that many young children receive to protect them from often fatal childhood diseases. Some bacterial vaccines induce antibodies specific for proteins on the bacteria required for their attachment and subsequent invasion of the host; others induce immunity to endo- or exotoxins. BCG, which protects against tuberculosis, is most effective against the miliary form of this disease as well as tuberculous meningitis. T-independent vaccines to carbohydrates such as the capsule of pneumococcus or *Haemophilus* are effective but have limitations because T-cell help is not provided for affinity maturation and isotype switching. However, a "conjugate" vaccine with pneumococcal carbohydrates attached to diptheria toxoid is available that allows the development of IgG antibodies of high affinity.

## Viral vaccines

Vaccines to viruses that infect the respiratory tract (influenza, adenovirus), the gastrointestinal tract (polio, rotavirus), the skin (yellow fever, La Crosse fever) and the reproductive tract (herpes) have been developed. As with bacteria, viral vaccines are modified-live, killed, or subunit vaccines. The recent emergence of AIDS as a worldwide health hazard has focused attention on viral vaccine development. In fact, effective viral vaccines have been developed for viruses that are in the same genetic classification group as HIV. However, none of the vaccines developed for immunization against HIV have been effective, due to the ability of HIV to modify its immunogenic epitopes faster than the immune system can produce protective immune responses to these new epitopes.

This highlights the most important question in vaccine development—what are the properties of a good vaccine? It must be safe, effective, cheap to make and distribute, stable for long-term storage or transport, and insensitive to major changes in temperature, and it should provoke an immune response that lasts for a long period of time.

## Vaccines to other infectious agents

Protozoan parasites, such as those that cause malaria (*Plasmodium*), African sleeping sickness (*Trypanosoma*), and schistosomiasis are very important diseases, mostly in the Third World. The ability to vaccinate against protozoan diseases would dramatically affect the life style, health, longevity, and productivity of those living in areas where the organism is always present (endemic). Parasites express many antigens that are usually immunogenic, but most do not consistently stimulate protective responses. As with HIV, parasites have evolved defense mechanisms that allow a continual evolution of their immunogenic epitopes. This is best typified by *Plasmodium*, which continually and rapidly develops variants with different surface proteins so that the current immune response is no longer effective. Parasites have also developed mechanisms to shift the focus of the immune response by altering the cytokine profile during the induction phase to one that is not protective, for example from Th1 to Th2 (Section H3) as in the case of *Mycobacterium lepri*.

## Tumor vaccines

Several prophylactic vaccines that induce immunity to viruses that cause tumors have been developed, are quite effective, and have been approved (Sections N2 and N7). In particular, hepatitis B vaccines prevent infection by this virus and reduce the incidence of liver cancer, and the human papillomavirus (HPV) vaccine reduces cervical carcinoma.

Other approaches are designed to enhance/induce effective immunity in tumor-bearing patients. In principle, the immune system, through immune surveillance, should be able to recognize and kill tumors that express foreign antigens. Tumors that are induced by chemicals or irradiation are more likely to have new or neo-antigens that are immunogenic and characteristic of the individual tumors. However, most tumor-specific antigens are either absent or weakly immunogenic through being expressed at low levels. Moreover, such antigens have been difficult to find and may be patient-specific.

Most approaches to both direct therapy and vaccines are through targeting the over-expressed products of proto-oncogenes found in a variety of tumors, for example the HER2/neu antigen overexpressed by many prostate and breast tumors. However, almost all of these antigens are also expressed on normal cells and thus tolerance to these antigens has already developed. Thus, a successful vaccine must break this tolerance.

The major challenge for immunologists is to optimize the routes of delivery of these antigens to maximize induction of protective immunity (Section N7). CTLs are clearly of great importance in immunity to tumors. Studies have shown that immunogenic peptides isolated from class I molecules expressed on myeloma tumor cells are effective at inducing tumor-specific immunity, and in studies using peptide vaccines, objective responses have been seen in patients with either advanced glioma or hormone-refractory prostate cancer, where there was a marked increase in the CTL response.

A very active area of tumor vaccine research involves loading patient dendritic cells *in vitro* with tumor-associated antigens and reinjection of these cells into the patient. The most successful of these approaches and a breakthrough in the development of tumor vaccines is Provenge, a therapy that has now been approved by the US FDA for treatment of prostate cancer.

## Vaccines to addictive drugs

Addictive behavior to drugs such as opiates (morphine/heroin), cocaine, or nicotine from smoking tobacco lead to health issues as well as social and economic concerns due to abuse of these drugs. Current therapies for addiction include withdrawal, substitution, or removal of the drug from society, and the possibility that vaccines to addictive drugs could be another very useful approach is being explored.

Drugs of abuse are usually haptens, small nonpeptide molecules (~200–400 Da) that are not immunogenic (Section A4). They can be made immunogenic by linking them to known antigens, for example tetanus toxoid. For a vaccine approach to work requires that the vaccine induce an IgG response to the drug that would prevent the drug from crossing the blood–brain barrier. Human and animal studies, some of which are in various stages of clinical trials, have shown that vaccines directed against nicotine, heroin/morphine, and cocaine could block the addictive properties of these drugs.

# J1 Deficiencies in the immune system

---

## Key Notes

**Components of the immune system**
Each of the four components of the immune system (T cells, B cells, phagocytes, and complement) has its domain of function important to protection against certain pathogens. These components are intimately integrated into a program of immune defense that could be severely compromised if even one were absent or deficient.

**Defects in specific immune components**
The occurrence of repeated or unusual infections in a patient is a primary indication of immunodeficiency. Although a deficiency may compromise several components of the immune system, in most instances the deficiency is more restricted and results in susceptibility to infection by some but not all microbes. Defects in T cells tend to result in infections with intracellular microbes, whereas those involving other components result in extracellular infections.

**Classification of immuno-deficiencies**
Immunodeficiencies are either primary (congenital/inherited), or secondary (acquired as the consequence of diseases and their treatments or other factors, e.g., malnutrition). These can be defined on the basis of the specific immune component that is abnormal.

**Related topics**

| | |
|---|---|
| (A5) Hemopoiesis – development of blood cells | (D8) Antibody functions |
| | (F5) Clonal expansion and the development of memory and effector function |
| (B1) Cells of the innate immune system | |
| (B2) Molecules of the innate immune system | |

---

## Components of the immune system

The multiple interactive cellular and molecular components making up the immune response usually provide sufficient protection against bacterial, viral, or fungal infections. However, any situation that results in impaired immune function may contribute to a spectrum of disorders referred to as immunodeficiency diseases. In particular, immunodeficiency is defined as an increased susceptibility to infection.

It is evident from a consideration of the disorders and infections in individuals with selective immunodeficiency that each component of the immune response (T cells, B cells, phagocytes, and complement) has its domain of function. These four systems, although somewhat independent, are intimately integrated into a program of immune defense that could be severely compromised if even one were absent or deficient. In particular, the requirements for cell cooperation, the importance of chemotactic stimuli, and activating factors emphasize the interdependence of these systems and the potential consequences of an abnormality in any of them. However, although the absence of, or an abnormality in, one domain may compromise the individual, they need not be life threatening if other components of the immune system can compensate for this deficiency.

## Defects in specific immune components

The occurrence of repeated or unusual infections in a patient is a primary indication of abnormalities in immune function and of immunodeficiency. A variety of circumstances may be involved in this impairment of immune function, including genetics, cancer, chemotherapy and irradiation, malnutrition, and aging. Although it is possible that the deficiency could affect several components of the immune system (e.g., as in the case of severe combined immunodeficiency, SCID), in most instances the deficiency is restricted to a single component. Such deficiencies result in susceptibility to infection by some but not all microbes. For example, diseases involving defects in T cells predispose to infections with intracellular organisms, including mycobacteria, some fungi, and viruses, whereas those involving other components of the immune response tend to result in infections by bacteria that have an extracellular habitat. In other words, infections with particular microbes are a reflection of which components of the immune system are defective. Moreover, it is often possible to define the abnormal immune component in an immune deficiency disease and, in the process, discover a considerable amount about the importance of that component in normal immune defense and in its interrelationships with the other components of the immune system. Furthermore, it is important to recognize such abnormalities and to pinpoint them as accurately as possible, since correction, if possible, must be tailored to the specific abnormality.

## Classification of immunodeficiencies

The immunodeficiency diseases can be classified as either primary—usually **congenital** (the result of improper development of the innate, humoral, and/or cellular immune systems)—or secondary, that is, **acquired** (the consequences of other diseases, their treatment, or other factors). A large number of specific congenital and acquired abnormalities in the immune system have been identified that contribute to patient susceptibility to recurrent infections. These abnormalities range from those that affect the immune system very early in life, and thus compromise the immune response to many antigens, to those that affect the final stages of differentiation of particular immune cells and hence lead to very selective abnormalities. The primary diseases are very rare while the secondary diseases are relatively common. A more pathophysiological description characterizes the specific immune component that is abnormal by defining quantitative or qualitative abnormalities of the cells (lymphocytes, phagocytes) and/or molecules (antibodies, cytokines, complement components) of the immune system.

# J2 Primary/congenital (inherited) immunodeficiency

## Key Notes

| | |
|---|---|
| **Complement** | Patients deficient in certain complement components (especially C3) are prone to recurrent infections with encapsulated organisms (*Pneumococcus* and *Streptococcus*) and *Neisseria*. Opsonization of these pathogens by C3b is important for their removal by phagocytosis. Deficiencies in membrane attack complex (MAC) components and in complement regulatory molecules also result in increased susceptibility to certain infections or to inflammation. |
| **Phagocytosis** | Intrinsic defects include those associated with differentiation, chemoattraction, and intracellular killing of the microbe. Extrinsic or secondary defects (not an inherent phagocytic defect) may result from antibody or complement deficiency or suppression of phagocytic activity. |
| **Humoral Immunity** | Primary antibody deficiency may result from abnormal development of B cells or from lack of T-helper activity. Patients suffer from recurrent extracellular bacterial infections. Those with severe combined immunodeficiency (SCID) and Bruton's disease have few or no B lymphocytes and no antibodies. In hyper-IgM syndrome, CD40 signaling is defective and there is no class switch from IgM. Common variable immunodeficiency (CVID) may result from lack of B-cell terminal differentiation or absence of T-cell help. |
| **Cellular immunity** | T-cell deficiencies result in severely compromised humoral as well as cellular immunity. Patients have recurrent life-threatening viral, fungal, mycobacterial, and protozoan infections. In DiGeorge syndrome, thymus embryogenesis is defective and few T cells develop. SCID may result from defects in the common cytokine receptor γ chain or from adenosine deaminase or purine nucleoside phosphorylase deficiency. |
| **Related topics** | (A5) Hemopoiesis – development of blood cells<br>(B2) Molecules of the innate immune system<br>(B4) Innate immunity and inflammation<br>(D8) Antibody functions<br>(E2) B-cell activation<br>(F5) Clonal expansion and development of memory and effector function |

## Complement

Primary immune deficiencies of the complement system have been described for many of the complement components and their inhibitors, some in terms of specific gene mutations. Patients with a deficiency of certain of these complement components (especially C3) are prone to recurrent infections both with encapsulated organisms such as *Pneumococcus* and *Streptococcus* and with *Neisseria* (Table 1). The attachment of complement to the surface of some of these organisms is clearly important for their removal by phagocytic cells. Deficiencies in the later complement components and in the regulatory molecules of the complement system also result in increased susceptibility to infections by meningococci (e.g., *Neisseria*) or to inflammation, respectively.

### Table 1. Complement deficiencies

| Component deficient | Disease caused/infections |
| --- | --- |
| **Regulatory components** | |
| C1q inhibitor | Hereditary angioedema (continuous complement activation and consumption) |
| Decay accelerating factor, DAF, (CD55) | Paroxysmal nocturnal hemoglobulinuria (lysis of red blood cells) |
| **Complement components** | |
| C1, C2, or C4 | Immune complex disease, Ag–Ab complexes not removed; C2 deficiency associated with SLE |
| C3 | Most serious; infection with pyogenic bacteria |
| **MAC (membrane attack complex)** | |
| C5–8 | Meningococcal infections, e.g., *Neisseria* |
| C9 | Asymptomatic |

## Phagocytosis

Defects in phagocytic function can be classified as either intrinsic (related to the inherent properties of the phagocyte) or extrinsic (not the result of an inherent phagocytic defect). Intrinsic disorders related to different stages of phagocyte differentiation and function have been identified, including those associated with stem cell differentiation, chemoattraction to the site of microbial assault, and intracellular killing of the microbe (Table 2). Extrinsic defects may result from (i) deficiency of antibody or complement, that is, other primary defects, or (ii) suppression of phagocytic activity (e.g., by glucocorticoids or autoantibodies), that is, secondary defects, to be discussed later.

## Humoral immunity

Primary antibody deficiency mainly results from abnormal development of the B-cell system. Any of the steps involved in B-cell maturation may be blocked or abnormal (Table 3 and Figure 1). Patients with severe combined immunodeficiency (SCID) due to deficiencies in ADA (adenosine deaminase), PNP (purine nucleoside phosphorylase), or *RAG1* and *RAG2* (recombinase-activating genes), or those with Bruton's disease, have few or no B lymphocytes and therefore few if any antibodies in their circulation. Thus, they are unable to coat the surface of (opsonize) bacteria for which phagocytosis is the primary defense. The overall lack of antibodies means that the patients suffer from recurrent bacterial infections, predominately by *Pneumococcus*, *Streptococcus*, and *Haemophilus*.

In particular, deficiencies in ADA or PNP (enzymes important in the removal of toxic levels of dATP and dGTP, Figure 2) compromise very early B-cell (and T-cell) development,

## Table 2. Phagocytic defects

| Defect in | Disease/mechanism |
| --- | --- |
| Stem cell differentiation | Neutropenia: too few neutrophils to adequately remove microbes |
| Adhesion to endothelium for margination | Leukocyte adhesion deficiency (LAD): defective CD18 (due to gene mutation), a critical leukocyte function-associated (LFA) molecule* |
| Phagocytosis | Chediak–Higashi syndrome: failure of phagolysosome formation and lysosome degranulation caused by a mutation of the LYST gene that encodes a lysosomal trafficking regulator |
| Intracellular killing | Chronic granulomatous disease (CGD): defect in genes encoding the proteins making up the NADPH (nicotinamide adenine dinucleotide phosphate) oxidase system involved in oxygen-dependent killing within the phagolysosome |
| IFNγ or IL-12 receptors | Mycobacterial infections: failure to activate NADPH oxidase |

*CD18, with CD11, form the C3bi receptor (CR3) responsible for binding C3b-opsonized microbes, a critical step in engulfment of a microbe. LFA molecules are present on all effector cell populations (including cytotoxic T cells) and are important in linking effector and target cells as an initial step in cytotoxicity or phagocytosis. Thus, the function of more than just phagocytes is affected by this defect.

## Table 3. B-cell deficiencies

| Stage of differentiation/maturation | Disease |
| --- | --- |
| Lack of development of stem cells into pre- or pro-B cells | Severe combined immunodeficiency (SCID): due to ADA/PNP or *RAG* deficiency; also affects T-cell development |
| B cells fail to develop from B-cell precursors | Bruton's agammaglobulinemia: congenital, mostly X-linked; due to a defective gene encoding a tyrosine kinase (Btk) involved in activation of pre-B cells to immature B cells (Section E2); T cells normal |
| B cells do not class switch from IgM | Hyper-IgM syndrome: increased IgM but little or no IgG in the circulation; due to defects in genes encoding CD40 on B cells or CD40L on activated T cells (Section F4) |
| IgG/IgA deficiency | Common variable immunodeficiency (CVID): defects include inability of B cells to mature into plasma cells and defective T-cell help; IgA deficiency is most common (1 in 700 people) |
| Antibody deficiency early in life | Transient hypogammaglobulinemia: thought to be the result of an intrinsic maturational B-cell defect and/or one involving T-helper cells |

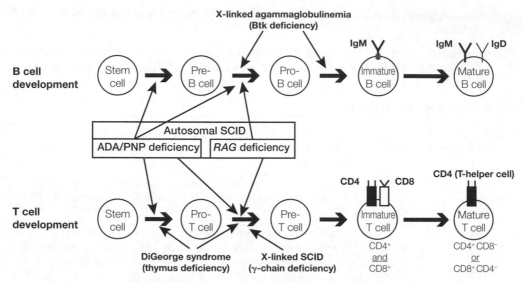

Figure 1. ADA and PNP deficiencies have some effect on B-cell development, but far less than on T-cell development. Thus, humoral immune deficiency resulting from ADA or PNP deficiency is due primarily to lack of T-cell help. Mutations in *RAG* genes affect equally both T and B cell development, whereas mutations in the B-cell tyrosine kinase (Btk) would only affect B cells and prevent their maturation to immature B cells.

to a large extent at the stem-cell level. Deficiencies in *RAG1* and *RAG2*, which encode proteins critical to V and J, as well as V, D, J, gene rearrangement (Section D3) result in the lack of B-cell (and T-cell) generation of diversity, and thus antigen specificity, and cannot progress past the pre-B-cell stage. In Btk (Bruton's tyrosine kinase) deficiency, signals necessary for maturation of B cells past the pre/pro-B-cell stage are not properly provided and thus few if any mature B cells develop.

Mutations that result in defects in genes for CD40 or CD40 ligand result in hyper-IgM syndrome, an X-linked disease in which CD40 signaling is defective and there is no class switch from IgM. Common variable immunodeficiency (CVID) and **transient hypogammaglobulinemia** may result from lack of B-cell terminal differentiation or absence of T-cell help.

Although some of these disorders are related to basic biochemical abnormalities of the B-cell lineage, others are the result of defective regulation by T cells. Thus, humoral immune deficiency may result from the absence of T-helper activity. This is seen as one form of CVID. Another form of CVID involves B cells that do not respond to signals from other cells. It is also possible that abnormalities in monocyte or dendritic cell presentation, and/or IL-1 (or other cytokine) production, may contribute to, or be responsible for, some of these disorders. Moreover, since different classes of immunoglobulin are regulated by different T-helper cell subpopulations (e.g., Th1 cells help IgG1 and IgG3 responses; Th2 cells help IgA and IgE responses) selective antibody class (IgA or IgG) deficiencies may result from abnormalities in these T-cell subpopulations.

## Cellular immunity

Deficiencies caused only by the loss of cellular immunity are very rare, as most T-cell deficiencies (Table 4 and Figure 1) result in severely compromised humoral immunity as

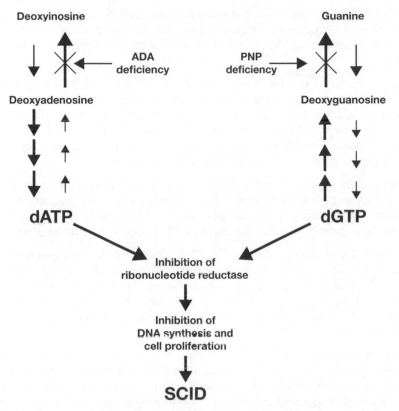

**Figure 2.** Role of ADA and PNP in SCID. Deficiencies in these components result in reversal of the normal pathway of removal of toxic purines and cause the buildup of dATP and dGTP, eventually resulting in inhibition of DNA synthesis and thus cell proliferation.

## Table 4. T-cell deficiencies during development

| Deficiency | Disease |
| --- | --- |
| Lack of a thymus | DiGeorge syndrome: a defect in thymus embryogenesis leading to incomplete development of the thymus and thus decreased T-cell numbers |
| Stem cell defect | SCID: 50% have a defect in the γ chain used by many cytokine receptors, including the IL-7 and IL-2 receptors |
| Death of developing thymocytes | SCID: 25% have an ADA or PNP deficiency; toxicity is due to buildup of purine metabolites that inhibit DNA synthesis |
| Defective development (maturation) | SCID: due to mutations in the *RAG1* and *RAG2* genes critical to TCR rearrangement, maturation past the pro-T-cell stage does not occur |
| Variable T-cell function | Hereditary ataxia telangiectasia (AT); severe sinus/lung infections due to mutations in a DNA-repair gene and TCR chromosome breaks |
| T cells don't help B cells | Wiscot–Aldrich syndrome (WAS); X-linked mutation in a gene encoding adapter proteins |

well. One of the most frequent is DiGeorge syndrome, a defect in thymus embryogenesis leading to irregular development of the third and fourth branchial pouches that results in incomplete development of the thymus, compromised T-cell maturation, and decreased T-cell numbers.

As with primary antibody deficiency, primary T-cell deficiency mainly results from abnormal development and maturation. Because of a lack of, or deficiency in, ADA or PNP (Figure 2), patients with SCID are compromised very early in the development of T cells, to a large extent at the stem cell level. SCID due to deficiencies in *RAG1* and *RAG2* results in the lack of T-cell generation of diversity, and thus of antigen specificity, and T-cell development cannot progress past the pro-T-cell stage.

X-linked SCID is the result of a deficiency of a common γ chain that is a signaling subunit of several important cytokine receptors (IL-2, IL-4, and IL-7, among others). This deficiency results in an inability of developing T cells to proliferate past the pro-T-cell stage in response to these cytokines. Lack of response to IL-7 seems to be particularly important.

Other primary T-cell deficiencies include hereditary ataxia telangiectasia (AT) and Wiscott-Aldrich syndrome (WAS). AT is a variable T-cell deficiency that is due to mutations in a DNA repair gene. In its absence, chromosomal breaks in the genes for the T-cell receptor are not repaired, resulting in increased sinus and lung infections. WAS is an X-linked disease in which T cells are abnormal and do not interact effectively with B cells, resulting in increased infections.

Note that some deficiencies in cellular immunity may relate to T-effector cells (e.g., cytotoxic T cells), whereas the T-helper population may be normal. Children have also been described with an inability to produce or respond to IFNγ. Overall, patients (mostly children) with T-cell deficiencies have recurrent viral, fungal, mycobacterial, and protozoal infections.

# J3 Secondary (acquired) immunodeficiency

## Key Notes

| | |
|---|---|
| **Factors causing acquired immunodeficiency** | Secondary, or acquired, immunodeficiencies are by far the most common and may result from infection (e.g., HIV), malnutrition, aging, cytotoxic drugs, and trauma, among others. |
| **Infection: HIV and AIDS** | Acquired immune deficiency syndrome (AIDS) is caused by the human immunodeficiency viruses HIV-1 or HIV-2. They enter the body via infected body fluids and infect monocytes/macrophages (the primary reservoir for the virus) and helper T cells, through CD4 on these cells. Chemokine receptors are also involved in HIV gp120 binding to these cells and are critical to infection. Loss of CD4$^+$ T cells due to HIV infection eventually compromises the ability of the immune system to combat infections. |
| **Malnutrition** | Worldwide, the major predisposing factor for secondary immunodeficiency is malnutrition and, in particular, protein–calorie malnutrition (PCM). Lack of certain dietary elements (e.g., iron, zinc) and metabolic derangements are also significant contributors to malnutrition-mediated immune dysfunction, e.g. glucocorticoids. |
| **Tumors** | Tumors can release factors, including IL-10, TGFβ, VEGF, and prostaglandins, that are immunosuppressive. Moreover, they can recruit host regulatory cells, including Tregs and myeloid suppressor cells, to compromise host immunity to the tumor. Thus, anti-tumor therapies must not only focus on tumor killing, but also on eliminating these immunosuppressive molecules and cells. |
| **Therapy using cytotoxic drugs and irradiation** | Cytotoxic drugs and irradiation that especially target dividing cells are widely used for tumor therapy. Cytotoxic drugs are also used to suppress allogeneic graft rejection in patients receiving kidney or other grafts. These therapies also kill normal immune cells, including stem cells, neutrophil progenitors, and rapidly dividing precursors for all leukocytes, and thus compromise immune defense against pathogens. |
| **Immune senescence: consequences of aging** | With aging, memory T cells increase but become less able to expand. Fewer new (naive) T cells enter the pool due to thymic involution, diminishing the immune repertoire and the quality of T- and B-cell responses. B-cell development |

|  |  |  |  |
|---|---|---|---|
|  | in the bone marrow may also decrease. This reduction in immune capability in the elderly results in a decreased response to infection and vaccination. | | |
| **Trauma** | Patients suffering trauma (e.g., associated with burns or major surgery) are less able to deal with pathogens, perhaps as a result of the release of factors that dampen immune responses, e.g. glucocorticoids. | | |
| **Related topics** | (B1) Cells of the innate immune system | (F4) | T-cell activation |
|  | (B2) Molecules of the innate immune system | (F5) | Clonal expansion and development of memory and effector function |
|  | (E3) The cellular basis of the antibody response | (P) | Aging and the immune system (immunosenescence) |
|  | (F3) Shaping of the T-cell repertoire | | |

## Factors causing acquired immunodeficiency

Secondary, or acquired, immunodeficiency is by far the most common immunodeficiency and contributes a significant proportion to hospital admissions. Factors causing secondary immunodeficiency mainly affect phagocytic and lymphocyte functions and include infections (HIV), malnutrition, tumors, cytotoxic drugs, aging, and trauma (Table 1).

**Table 1. Factors causing secondary immunodeficiency**

| Factor | Components involved |
|---|---|
| Infection (HIV) | Immunosuppression by microbes, especially HIV. Other examples include malaria and measles; mechanisms involve decreased T-helper cell function and antigen processing/presentation |
| Malnutrition | Protein–calorie malnutrition is, worldwide, the major predisposing factor for secondary immunodeficiency; lack of dietary elements (e.g., iron, zinc) and metabolic derangements inhibit lymphocyte maturation and function |
| Tumors (Section N) | Tumors release immunosuppressive molecules (e.g., TGFβ) and recruit T-regulatory cells (Tregs) or myeloid suppressor cells to mediate immunosuppression through cell–cell contact or release of soluble molecules |
| Cytotoxic drugs/ irradiation | Widely used for tumor therapy, but also kills normal rapidly dividing cells, including stem cells, neutrophil progenitors, and bone marrow leukocyte precursors, thus compromising immune defense against infection |
| Aging (Section P) | Increased infections; reduced responses to vaccination; T- and B-cell, responses decrease and the quality of the response changes; antibody specificities tend to be less toward foreign and more toward auto-antigens; antibody isotypes transition from IgG to IgM, affinities from high to low and there are fewer naive cells |
| Trauma | Stress tends to increase infections, probably due to release of immunosuppressive molecules such as glucocorticoids; removal of the spleen after an accident results in decreased phagocytosis of microbes |

## Infection: HIV and AIDS

Acquired immune deficiency syndrome (AIDS) is caused mainly by human immunodeficiency virus (HIV)-1 but also by HIV-2. These retroviruses enter the body via infected body fluids and exhibit trophism for T cells, in particular the T-helper population, and for monocytes and macrophages. They bind and gain entry into T cells and monocytes (the primary reservoir for the virus) through the CD4 molecule on these cells. Other accessory receptors (chemokine receptors, Section B2) are involved in viral gp120 binding to T lymphocytes and monocytes, and individuals lacking functioning chemokine receptors do not progress from HIV infection to AIDS. In particular, the chemokine receptors CXCR4 and CCR5 are co-receptors for HIV and are required for productive HIV infection of $CD4^+$ cells, including monocytes, macrophages, and T-helper cells.

The development of AIDS is defined as the occurrence of opportunistic infections (e.g., *Pneumocystis*) or Kaposi's sarcoma (caused by human herpesvirus) in an individual who has been infected with HIV. This is a direct result of the loss of $CD4^+$ helper cells. Damage to the pivotal $CD4^+$ T cell has major effects on the functions of other cells of the immune system (Figure 1). Infection of monocytes and antigen-presenting cells is also likely to be important in the speed of progression of the disease.

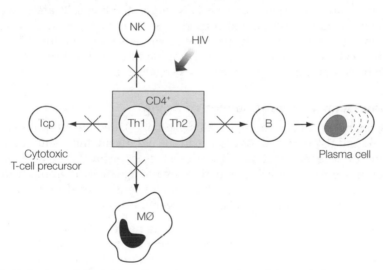

Figure 1. HIV infection of $CD4^+$ Th cells compromises their ability to help other immune cell populations.

## Malnutrition

Worldwide, the major predisposing factor for secondary immunodeficiency is malnutrition. Malnourished individuals are more susceptible to infectious disease. In particular, protein–calorie malnutrition (PCM) is by far the most important factor in secondary immunodeficiency. All biochemical and physiological functions, including those associated with immune defense, require energy. Dietary proteins are also critical, as they supply the essential amino acids for synthesis of all body proteins, again including those of the immune system. Certain vitamins and minerals are also important in immune defense. Lack of dietary elements such as iron and zinc as well as metabolic derangements are contributors to malnutrition-mediated immune dysfunction. Zinc is essential for actively proliferating cells and, if deficient, B and T cells are compromised, and the functions of cells of the innate immune system, including monocytes, neutrophils, and NK cells, are also significantly impaired. Iron deficiency affects many cell processes, especially those associated with red blood cells, but also significantly impairs immune cell proliferation and function.

## Tumors

A variety of immunosuppressive molecules, including IL-10, TGFβ, vascular endothelial growth factor (VEGF), and prostaglandins, are produced by tumors. In addition, cells of the adaptive immune system may be induced by tumors to suppress anti-tumor immune responses. These cells include regulatory T (Treg) cells (CD4⁺/CD25⁺) that can mediate immunosuppression by cell–cell contact or through release of soluble molecules. Cells of the innate immune system, for example myeloid suppressor cells (MSCs), can also be recruited by tumor cells to suppress anti-tumor immunity. Thus, in developing therapeutic approaches to tumor eradication, especially those designed to induce or enhance host immunity to the tumor, consideration must be given to ways of circumventing these tumor-associated immunosuppressive factors.

## Therapy using cytotoxic drugs and irradiation

Cytotoxic drugs and irradiation that especially target dividing cells are widely used for tumor therapy. Cytotoxic drugs are also used to suppress allogeneic graft rejection in patients receiving kidney or other grafts. These therapies also kill normal immune cells, including stem cells, neutrophil progenitors, and rapidly dividing precursors for all leukocytes, and thus compromise immune defense against pathogens.

## Immune senescence: consequences of aging

As one ages, changes occur in immune status (Section P). Reduced responses to vaccination and increased risk of infectious disease in the elderly are the result of reductions in immune function. The most striking of these changes is the involution of the thymus and the subsequent decrease in T-cell production. By the age of 60, thymus tissue is almost completely replaced by fat. Thus, T-cell education is significantly decreased and the host becomes more dependent on the pool of T cells generated earlier in life.

As one ages, memory T cells (CD40RO⁺) increase and naive cells decrease, resulting in an accumulation of activated T cells and fewer naive cells entering the pool. In addition, the ability of T cells from aged individuals to expand is limited, thus further diminishing cell-mediated immune responses.

There are also age-associated reductions in humoral immunity, at least one part of which results in reduced B-cell development in the bone marrow and thus reduced B-cell diversity. This is manifest as a change in the quality of the antibody response, including a decrease in antibody affinity, a diminished response to vaccines, and an increase in auto-antibody production (Section P3). Some of these alterations in humoral immunity may be due to the impaired capacity of T cells to induce the maturation of B cells to produce high-affinity, isotype-switched antibody.

Overall, the immune system appears to shift with age from one dependent primarily on adaptive immune responses to one somewhat more dependent on innate immunity.

## Trauma

After significant trauma, including that associated with burns or major surgery, the immune system seems less able to deal with pathogens. Although the basis for this apparent immunodeficiency is not understood, it is possible that these traumatic events induce release of immunomodulatory factors (e.g., glucocorticoids), which dampen immune responses (Section G6). Removal of the spleen, the primary organ for removing microbes from the circulation (e.g., after an accident), results in decreased phagocytosis of microbes and thus increased susceptibility to blood-borne infections, especially to infections by encapsulated bacteria, for example pneumococci.

# J4 Diagnosis and treatment of immunodeficiency

---

## Key Notes

**Family history**
Because defective genes can be inherited, an investigation into the family history is especially important in the diagnosis of primary immunodeficiencies.

**Evaluation of specific immune components**
Evaluation of the nature of immunodeficient components in a patient is important for determining appropriate treatment. This may be achieved by assay of Ig levels and B-cell numbers for antibody-mediated immunity; T cell and T-cell subset numbers and their cytokine production for cell-mediated immunity; lytic ability and individual components for complement activity; granulocyte and monocyte counts, their ability to phagocytose and kill bacteria, and their response to various chemotactic and activation signals, for the phagocytic system.

**Antibiotics and antibodies**
Antibiotic therapy is the standard treatment for infections. In addition, antibodies from a pool of donors are used for antibody deficiencies.

**Bone marrow transplants and gene therapy**
Replacement of faulty cells/organs, with those from normal individuals or with fetal liver or thymus grafts is used when MHC-compatible donors can be found and has been used to reconstitute normal phagocytic function in CGD and B and T cells in SCID. Treatment for genetic defects includes, in some cases, gene therapy, that is, replacing a defective gene with a normal gene. Although gene therapy has been successfully used for some diseases, it still has difficulties that need to be addressed.

**Related topics**

| | |
|---|---|
| (A5) Hemopoiesis – development of blood cells | (F5) Clonal expansion and the development of memory and effector function |
| (B1) Cells of the innate immune system | (I2) Immunization |
| (D7) Immunoassay | (J2) Primary/congenital (inherited) immunodeficiency |
| (D8) Antibody functions | (M3) Rejection mechanisms |

---

## Family history

Because defective genes can be inherited, for example, defective CD40 and/or CD40 ligand in hyper-IgM syndrome, ADA and PNP in SCID, and Btk in Bruton's agammaglo-bulinemia (Table 3 in Section J2), it is important to establish the history and genetics of family members with similar recurrent episodes of infection. This information on family history is especially important in the diagnosis of primary immunodeficiencies and is valuable for genetic counseling.

## Evaluation of specific immune components

Recognizing and pinpointing immune defects is critical, since correction must be tailored to the abnormality. Although the nature of an infection or disorder provides clues to which immune component is affected, it is not always clear which subcomponents are compromised. It is therefore important to apply a systematic evaluation of immune function to individuals suspected of immune abnormalities (Table 1).

Humoral (antibody) immunity may be initially evaluated by determining the presence and levels of the different antibody classes and subclasses in the serum of a patient using

## Table 1. Evaluation of the different components of the immune system

**Evaluation of antibody-mediated immunity**

Serum immunoelectrophoresis to determine if all classes of antibody are present

Quantitate antibodies in serum and secretions by ELISA or radial immunodiffusion

Assay for specific antibodies:
    By agglutination, for IgM antibodies to blood group substances A and B
    Before and after immunization with killed vaccines

Quantitate circulating B cells by flow cytometry with mAbs to surface Ig

Evaluate induction of B-cell differentiation *in vitro*

Evaluate by biopsy the presence of B cells and plasma cells in lymph nodes

**Evaluation of cell-mediated immunity**

DTH skin tests to common antigens: *Candida*, streptokinase, streptodornase

Determine:
    Total lymphocyte count (60–80% of blood lymphocytes are T cells)
    T-cell number in blood (using mAb to CD3 and flow cytometry)
    T-cell subpopulation percentages (using mAbs to CD4 and to CD8)

Evaluate lymphocyte proliferation to lectins (PHA, Con A) and alloantigens (MLR)

Analyze T-lymphocyte function:
    Lymphokine production: IFNγ, IL-2, etc.
    Helper cell activity and cellular cytotoxicity

**Evaluation of the complement system**

Assay for total hemolytic complement: $CH_{50}$

Quantitate individual complement components by immunoassay

Assay for neutrophil chemotaxis using complement in patient serum as a chemoattractant

**Evaluation of phagocyte function**

Determine total granulocyte and monocyte counts

Assay for:
    Chemotaxis, using a Boyden chamber
    Phagocytosis, using opsonized particles
    Superoxide generation using nitroblue tetrazolium (NBT) reduction
    Bacterial killing
    Individual enzymes and cytokines (e.g., IL-1 and IL-12)
    Integrins important to adhesion and margination (e.g., CD18) and their function
    Response to activation by IFNγ, GM-CSF, etc.

Evaluate ability to process and present antigen

serum immunoelectrophoresis, radial immunodiffusion, and/or radioimmunoassays (Sections D6 and D7). Detection of specific antibodies can be determined using skin tests (Section K5), by agglutination (e.g., for IgM antibodies to blood group substances A and B) and/or enzyme-linked immunosorbent assay (ELISA) (e.g., for specific antibodies after immunization with killed vaccines). It is also important to determine B-cell numbers and functional properties using monoclonal antibodies to surface immunoglobulin and B-cell differentiation assays, respectively. Lymph node biopsy is used to determine the presence and numbers of B cells and plasma cells in tissues.

Cell-mediated immunity is often evaluated by skin tests for delayed-type hypersensitivity, DTH (Section K5), to common antigens (e.g., *Candida*, streptokinase, streptodornase). Total T-cell ($CD3^+$) and T-cell subpopulation ($CD4^+$ or $CD8^+$) numbers are also useful in evaluating the potential for cell-mediated immune responses (Section D7). However, normal numbers of T cells and T-cell subpopulations in a patient do not mean that they function normally. Thus, lymphocyte proliferation to lectins (phytohemagglutinin and concanavalin A) and alloantigens (mixed lymphocyte reaction), lymphokine production (e.g., $IFN\gamma$, IL-2), and helper and killer cell activities may also need to be evaluated.

Both classical and alternative pathways of the complement system can be evaluated for their overall functional activity in red-cell lysis assays that determine total hemolytic complement ($CH_{50}$). Immunoassays can then be used to determine the concentration of individual complement components, including those associated with the alternative pathways. Neutrophil chemotaxis assays, using complement from a patient's serum as a chemoattractant, can be used to evaluate complement chemotactic factors (Sections B2 and D8) such as C5a.

Cells of the phagocyte system are able to respond to chemotactic stimuli and migrate toward a pathogen, recognize it, and mediate its phagocytosis and/or killing. These cells are involved in immune defense both as a result of their own recognition of microbe molecular patterns (Section B3) and as a result of direction by the humoral, cellular, and/or complement systems. Total granulocyte and monocyte blood counts permit determination of whether they are present in normal numbers. Chemotaxis assays (using Boyden chambers) evaluate their response to chemotactic molecules such as C5a. Assays for phagocytosis (using antibody and/or complement opsonized particles), superoxide generation (using the nitroblue tetrazolium, NBT, test), and bacterial killing are important in determining the functional capability of these cells. Assays for specific enzymes and for cytokines (e.g., IL-1 and IL-12) indicate the ability of these cells to produce molecules critical to microbe killing and in recruiting other cells and immune mechanisms. Their response to activation by $IFN\gamma$, GM-CSF, and so forth, indicates their ability to be induced to increased cytotoxic ability. The presence on phagocytes of molecules important to their ability to adhere to endothelium and marginate (CD18) should be determined. Finally, as monocytes, macrophages, and dendritic cells process and present antigen, it may be important to assay their ability to activate T cells and thus to initiate specific immune responses.

One of the best ways to evaluate immune function involves looking at both the afferent (initiation) and efferent (effector) limbs of the immune system of an individual. This can be done by injecting antigen into an individual and determining if a normal humoral and cellular response develops. If it does, all of the T- and B-cell systems are probably intact. Another even more definitive evaluation procedure might be to use a live attenuated vaccine, for example polio virus, as this would permit evaluation of the immune response in a very real setting. However, this would never be done as even an attenuated live pathogen could cause a lethal infection in an immunodeficient individual (Section I3).

## Antibiotics and antibodies

Antibiotic therapy is the standard treatment for infections. Children whose immune system produces no antibodies begin to experience recurrent infections as maternal antibody from placental transfer *in utero* is depleted. These individuals are treated with antibiotics and intravenously with periodic injections of pooled immunoglobulins from normal human serum. Contamination of immunoglobulin preparations with viruses, including HIV and hepatitis B and C, must be excluded.

## Bone marrow transplants and gene therapy

Replacement of faulty cells/organs with cells and organs from normal individuals is now commonly used when MHC-compatible donors can be found. Although this approach has been extensively used primarily for reconstitution of a failed organ system, it has also been applied to therapy of immune disorders, especially after ablative therapy used in the treatment of tumors. In addition, bone marrow transplantation as a source of hemopoietic stem cells has been successfully used for reconstitution of normal phago-cytic function in chronic granulomatous disease (CGD) and of B and T cells in SCID. In most cases, such transplants carry the risk of rejection (Section M3) and require appro-priately regulated immunosuppression for transplant and patient survival. Fetal liver and thymus grafts and transplantation of stem cells enriched from peripheral blood of normal donors and umbilical cord blood are other very promising approaches to the treatment of some of these diseases, and are being aggressively pursued at the present time.

In cases where it is clear that a single gene is deficient or abnormal, gene therapy may be used. As described in Section J2, a number of genes have already been identified as faulty in patients with primary immunodeficiency diseases (Tables 2, 3, and 4 in Section J2). Thus, a definitive treatment for these defects may well be gene replacement therapy, in which a faulty gene in a patient is replaced with a "normal" gene. In particular, a vector, usually a virus modified to carry human DNA, into which a copy of the normal gene has been inserted is injected into the patient, or into patient stem cells that are then injected into the patient. The gene carried by the vector unloads into the targeted cells and a normal functional gene is expressed that properly fulfills a function critical to immune defense.

Although the US Food and Drug Administration (FDA) has not yet approved a gene ther-apy product, clinical trials of several promising candidates are ongoing. This approach has already been tried with success for adenosine deaminase (ADA) deficiency and is currently being tried for other disorders for which a faulty gene has been identified. Gene therapy has also been used successfully to treat patients with diseases of the myeloid system.

Probably the most promising of the gene therapy trials has been that focused on treat-ing X-linked SCID, which is the result of a deficiency of the common $\gamma$ chain, a critical signaling subunit of several cytokines, including IL-2, IL-4, and IL-7 (Section J2). In this trial, the results of which were reported in 2000, all 10 children with X-SCID who had been treated with CD34$^+$ hematopoietic stem cells into which the common $\gamma$ chain was inserted were found to have had their genetic defect corrected; a remarkable result for an otherwise lethal disease. Unfortunately, two years later a few of these patients devel-oped leukemia, which was eventually found to be the result of insertion of the normal gene next to a cancer-causing gene, and the trial was halted. With time other difficulties with gene therapy were identified. These difficulties include (i) appropriately expressing the normal gene; (ii) controlling where the normal gene inserts in the patient's DNA; (iii) immune responses against the vector; and (iv) the complexity of multigene disorders. Nonetheless, it seems likely that in the not-too-distant future the promise of this approach in treating rare but often fatal diseases will be realized.

# K1 Definition and classification

## Introduction

The immune system normally responds to a variety of microbial invaders with little or no damage to host tissues. However, in some situations, immune responses (especially to some antigens) can lead to more severe tissue-damaging reactions (immunopathology). This "inappropriate response" by the immune system to antigens is referred to as hypersensitivity and is by no means restricted to antigens of microbial origin since it also includes both inert and self antigens (autoimmunity). Hypersensitivity reactions are antigen specific and occur after the immune system has already responded to an antigen (i.e., the immune system has been primed). The adverse reactions are therefore mainly the result of antigen-specific memory responses. It is important to note that these responses are part of normal immune defense mechanisms and occur daily as immune cells and molecules come in contact with antigens and/or pathogens that had previously induced immunity. What is unusual about hypersensitivities is that these normal responses become clinically evident because they are localized and/or involve interactions between large amounts of antigen and antibodies or immune cells. Moreover, in certain hypersensitive states, such as allergy, an inappropriate immune response leads to the production of IgE antibodies, which in the presence of specific allergens can cause mast cell and basophil degranulation with the concomitant release of chemical and immune mediators resulting in tissue damage and pathology. Factors leading to the development of hypersensitive states include environment, genetics, and age. Moreover, some antigens can induce more than one type of tissue damaging reaction. For example, penicillin can induce type I, II, III, and IV hypersensitivity reactions.

## Classification of hypersensitivities

Hypersensitivity reactions occur at different times after coming into contact with the offending antigens: within a few minutes (i.e., immediate); minutes to hours (intermediate); or after many hours (delayed). Generally, the delayed responses are mediated by the cellular components of the immune system (i.e., T cells) while the others are the result of the humoral arm of the immune response that includes antibodies and the complement system. The original classification by Gell and Coombs was into four main types; a fifth has since been added. Table 1 summarizes the main immune system components that contribute to tissue damage. It should be stressed that more than one of these mechanisms can contribute to any one particular disease process.

**Table 1. Classification of hypersensitivities**

| Time of appearance | Type | Immune mechanism |
|---|---|---|
| 2–30 min (immediate) | I | IgE antibodies (acute inflammatory response) |
| 5–8 h (cytotoxic) | II | Antibody and complement |
| 2–8 h (immune complex) | III | Antibody–antigen complexes |
| 24–72 h (delayed) | IV | T-cell mediated (can be granulomatous) |
| Chronic | V | Antibody-mediated stimulation |

# K2 IgE-mediated (type I) hypersensitivity: allergy

## Key Notes

| | |
|---|---|
| **Introduction** | This most common type of hypersensitivity (Type I) is mediated by IgE and causes mild (e.g., hay fever) to life-threatening (e.g., anaphylactic shock, which may be caused by insect venoms or food allergens) situations. Individuals who make high levels of IgE antibodies have a genetic predisposition to do this and are said to be atopic. The incidence of allergy varies throughout the world, as high as 30% in the industrial world and less common in the developing world. |
| **Clinical manifestations and their incidence** | Allergic responses can be relatively mild or result in life-threatening anaphylactic reactions that lead to vascular collapse and death. Allergic rhinitis is the most prevalent allergic disease. Asthma affects about 300 million people worldwide. Atopic dermatitis has a prevalence that varies from 1% to 30% depending on the country. Mortality is mainly associated with asthma and anaphylaxis. The increasing incidence of allergic diseases may be due to more pollution and, paradoxically, to better hygiene. Late-phase responses occur to many allergens. |
| **Diagnosis** | Clinical diagnosis is dependent on clinical history and skin testing to a range of allergens. A positive skin test usually occurs within 30 minutes in the form of a wheal (fluid accumulation) and flare (redness) at the site of injection. Laboratory diagnosis is based on measuring serum IgE levels and allergen-specific IgE. |
| **Common allergens** | The most common allergens in the USA are probably grass and tree pollens, but insect venoms, nuts, drugs, and animal dander are also common. Fungal and worm antigens can also induce type I hypersensitivity. |
| **Sensitization phase** | Sensitization to a particular antigen is dependent on stimulation of IgE antibody production. This requires $CD4^+$ Th2 cells to secrete IL-4 for B-cell growth and differentiation and to induce class switching of antigen-specific B cells to IgE. |
| **Effector phase – IgE-mediated mast cell degranulation** | IgE antibodies produced following initial contact with the specific antigen bind to IgE receptors on mast cells, basophils, and eosinophils. Cross-linking, by antigen, of the IgE and the receptors with which it is associated results |

|  |  |
|---|---|
|  | in rapid degranulation and release of pharmacological mediators (e.g., histamine) causing local inflammation (anaphylaxis). In the case of systemic anaphylactic reactions, adrenaline (epinephrine) treatment is required to restore blood pressure. |
| **Treatment of allergy** | Drugs used to inhibit the production, release, or action of inflammatory mediators and which then relieve symptoms include diphenhydramine, and glucocorticoids. Epinephrine (adrenaline) is used to counteract mediator effects such as low blood pressure and bronchospasms. Desensitization results in inducing an IgG immune response that diverts the immune response away from production of IgE. This approach has been used successfully for only a few allergens (e.g., bee venom). Immunotherapy using monoclonal antibodies against IgE is now used in the treatment of severe allergies. |
| **Related topics** | (B1) Cells of the innate immune system (B4) Innate immunity and inflammation (D2) Antibody classes (D7) Immunoassay (G5) Regulation by T cells and antibody |

## Introduction

This most common type of hypersensitivity is mediated by IgE, an antibody normally found in very small amounts in the circulation (Section D2) that probably evolved to protect us against worm infestations (Sections B4 and H2). Some individuals in the population are genetically predisposed to respond to certain antigens by producing IgE to these antigens and are said to be atopic. An atopic individual is frequently allergic to more than one allergen. Allergic reactions can occur to normally harmless allergens (such as pollen or foodstuffs) and microbial antigens (fungi, worms). An allergen may be a protein or carbohydrate and is defined by its ability to stimulate an adaptive Th2-type response and IgE antibody production. The main cellular types involved in allergic reactions are, in addition to Th2 cells, mast cells, basophils, and eosinophils.

## Clinical manifestations and their incidence

The main clinical manifestations of allergy are shown in Table 1. In some individuals severe allergies to dietary products such as peanuts may result in life-threatening anaphylactic

## Table 1. Clinical manifestations of allergic diseases

Allergic asthma
Angiodema
Anaphylaxis
Rhinitis (hay fever)
Atopic dermatitis
Urticaria
Hives

reactions that lead to vascular collapse and death. The incidence of allergies varies throughout the world, with levels as high as 30% of the population in the industrial world but less common in developing nations. The incidence has increased significantly over the last 20 years and is becoming more common in emerging nations. Allergic rhinitis is the most prevalent allergic disease and affects approximately 20% of the population. Asthma affects about 300 million people worldwide and is more prevalent in children than adults at a ratio 2.5:1. Similar to other allergic conditions, atopic dermatitis may vary from country to country with prevalence values varying from 1% in some countries to 30% in others. Mortality is mainly associated with asthma and anaphylaxis. Although death from asthma has significantly declined due to better clinical intervention and understanding of the disease, anaphylaxis still accounts for approximately 500 deaths annually in the United States.

The explanation for the increasing incidence of allergic diseases is unknown and although the incidence of asthma, allergic rhinitis, and atopic dermatitis appears to be stabilizing in the UK, other allergic conditions such as urticaria, food allergy, and angiodema have seen major increases in hospital admissions. The increased incidences in allergic conditions have highlighted the **hygiene hypothesis** whereby individuals exposed to poor hygiene may have a lower prevalence of allergic diseases; this is thought to be due to increased Th1 responses in these individuals (see "sensitization phase" below). This is supported by studies showing that children placed in day-care centers have a lower incidence of atopic diseases and further supported by studies showing that children from large families are less likely to develop allergies than children from families with one child. Another possibility often discussed to explain increased incidence of many allergic diseases is increased pollution.

Allergic symptoms following exposure to an allergen occur very rapidly and usually soon subside. However, many allergens, for example peanuts, can induce a "late phase" response (2–24 hours later) mediated by Th2 lymphocytes and other cells, including eosinophils, which result in a burst of inflammatory cytokines and further clinical symptoms. Inappropriate responses to a number of "allergens," in the absence of specific IgE responses, can also produce "anaphyactoid" responses, probably mediated directly through contact of mast cells with the "allergen," which induces release of histamine (Section K4).

## Diagnosis

The clinical diagnosis of allergy is based on the patients clinical history and skin testing to a range of allergens (Prausnitz–Kustner test), which involves injection of the allergen intradermally or its application to the skin and "scratching in" (scratch test). A positive skin test usually occurs within 30 minutes in the form of a wheal (fluid accumulation) and flare (redness) at the site of injection. Laboratory diagnosis is based on measuring serum IgE levels and allergen-specific IgE. This is carried out using RAST, but more frequently now by ELISA (Section D7).

In addition to allergens, other substances may cause irritation in both allergic and nonallergic individuals. This "intolerance" is not to be confused with allergies. An example of this is a food intolerance—lactose intolerance. Cow's milk contains lactose and individuals deficient in the enzyme lactase cannot break down the lactose, which is therefore not absorbed. The correct diagnosis of this condition can be made by clinical and laboratory observations.

## Common allergens

The most common activators of type 1 allergies in the US (Table 2) are grass and tree pollens, which frequently cause allergic rhinitis (hay fever). Other important allergens

**Table 2. Commonest initiators of allergic reactions**

| Allergens | Source |
|---|---|
| 1. Pollens | Plants and trees |
| 2. Animal hair | Animal dander |
| 3. House dust mite | Fecal waste products |
| 4. Moulds | Spores |
| 5. Insect bites | Venoms, wasps, bees, etc. |
| 6. Food | Shellfish, peanuts, milk, etc. |
| 7. Latex | Proteins from sap rubber |
| 8. Medicines | Penicillin, salicylates, anesthetics, etc. |

include insect venoms, nuts, drugs, and animal dander. Allergens from some invading organisms (fungal spores, viruses, and worms) can also give rise to allergic reactions. Systemic release of worm (*Echinococcus*) antigens from hydatid cysts binding to receptor-bound IgE can cause anaphylaxis (Section H1). Allergic asthma is an important disease that can be triggered by a number of different environmental antigens and, in its early stages, is mediated by IgE.

## Sensitization phase

Sensitization to a particular allergen is dependent on stimulation of IgE antibody production. Thus, B-cell antigen receptors specific for the allergen bind, internalize, process, and present the antigen in MHC class II molecules. CD4[+] Th2 cells recognize the antigen presented by these B cells, secrete IL-4 important for B-cell growth and differentiation, and induce class switch of antigen-specific B cells to IgE production (Sections B2, E3, and F5) (Figure 1). Recent data also suggest that mast cells themselves may contribute to the sensitization phase. In this model, allergens can directly induce mast cell release of histamine and cytokines such as TNFα, IL-4, and IL-10 into the local environment. Dendritic cells, which have taken up the innocuous allergen, would become "conditioned" and activated to direct naive T cells towards a Th2 phenotype (Figure 1).

Why certain individuals become sensitized to particular antigens by producing IgE is unclear, but the possibilities include (i) the genetics of the individual (identical twins have 70% concordance for the same allergies, whereas non-identical twins have 40% concordance); (ii) environmental factors (i.e., pollution) that condition mucosal tissues of the immune system to produce IL-4 which then predisposes a Th2 response; (iii) defective regulation of the response through Th1 cells and/or Tregs (Sections F5 and G5).

## Effector phase – IgE-mediated mast cell degranulation

Specific IgE antibodies produced as a result of previous contact with antigen (allergen) diffuse throughout the body, eventually coming in contact with mast cells and basophils. These cells have high-affinity receptors for the Fc region of IgE (FcεR) and therefore bind to these antibodies. This has no effect on the mast cell until the specific antigen (allergen) is reintroduced into the body and comes in contact with these IgE-coated mast cells. Cross-linking of these antibodies on the mast cell surface (Figure 2) causes immediate release of granules (degranulation) that contain large amounts of pharmacological mediators (Table 3). These substances have a direct effect on nearby blood vessels, causing

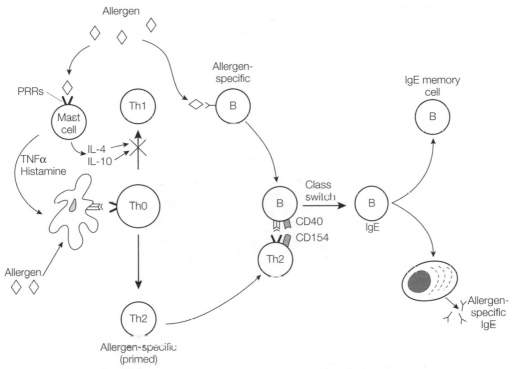

Figure 1. Model for the induction of allergen-specific IgE responses. Mast cells are activated through allergens via PRRs to produce histamine and cytokines, including TNFα, IL-10, and IL-4. Histamine "conditions" dendritic cells, which have processed and presented the allergen, to induce Th0 cells to become Th2 cells, while IL-10 and IL-4 directly inhibit the development of Th1 cells. Interaction of allergen-specific B cells with primed allergen-specific Th2 cells, and the subsequent ligation of B cell CD40 with CD154 on T cells, result in class switch of the B cells to IgE. These B cells then develop into memory cells and plasma cells that produce allergen-specific IgE.

vasodilation and an influx of eosinophils, which also release mediators. Systemic release of histamine and other substances by mast cells can lead to severe vasodilation and vascular collapse resulting in life-threatening systemic anaphylactic reactions, which require treatment with epinephrine (adrenaline) to restore blood pressure. In particular, leukotrienes, histamine, prostaglandins, and platelet activating factor released from mast cells are key mediators of type I hypersensitivity and can be classified by their effects on target cells and tissues (Table 3). In addition, these mediators, together with T-cell cytokines, cause a prolonged "late-phase" reaction that is often seen 2–12 hours after the initial reaction. Locally (e.g., in the nose), mediator release results in the symptoms of redness, itching and increased secretions by mucosal epithelial cells leading to a runny nose.

## Treatment of allergy

Drugs used to treat immediate hypersensitivity act at one of two levels: (1) Inhibitors of the production or release of inflammatory mediators; these include nonsteroidal anti-inflammatory drugs (NSAIDs), such as aspirin and indomethacin, synthetic steroids (glucocorticoids) such as dexamethasone and prednisolone, and the inhibitor of histamine release, cromolyn. (2) Inhibitors of mediator action such as histamine receptor antagonists. Diphenhydramine,

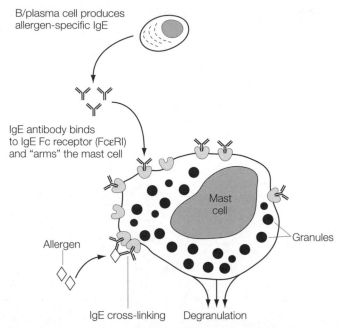

B/plasma cell produces
allergen-specific IgE

IgE antibody binds
to IgE Fc receptor (FcεRI)
and "arms" the mast cell

Mast
cell

Granules

Allergen

IgE cross-linking    Degranulation

Figure 2. IgE-mediated mast cell degranulation. Allergen binds to and cross-links cytophilic (cell-bound) IgE, which signals through the IgE receptor to trigger mast cell activation and degranulation with release of histamine, leukotrienes, and so forth.

chlorphenamine, and cetirizine are representative $H_1$-blocking agents, which are most useful for relief of the sneezing, rhinorrhea (runny nose) and itching eyes associated with hay fever. They are not useful for bronchial asthma or systemic anaphylaxis, where mediators other than histamine play a more important role. Glucocorticoids also inhibit some of the actions of inflammatory mediators. Other drugs such as epinephrine (adrenaline) and theophyline are used to counteract mediator effects such as low blood pressure and bronchospasm.

More specific treatment involves desensitization that is used to divert the immune response away from a predominantly Th2-driven IgE antibody response and towards a Th1-driven IgG response. This involves injection or ingestion of allergen in low and increasing amounts. Success has been achieved with only a few allergens, for example bee venom. A Th1-induced IgG response could have two significant effects: (1) Larger amounts of IgG would be produced than IgE and this excess IgG antibody would bind and remove the antigen before it could bind IgE on the mast cells or basophils and trigger degranulation. (2) IgG would also remove antigen before it could bind to and stimulate Th2-driven IgE-producing B cells, thus decreasing the amount of antigen-specific IgE produced (Figure 3).

Immunotherapy using a monoclonal anti-IgE antibody (omalizumab) has become available for the treatment of asthma and other severe allergic conditions. This mAb prevents IgE binding to its high-affinity receptor (FcεR) on mast cells (Section Q3). Omalizumab is also used in helping to desensitize individuals to insect venoms by allowing increasing concentrations of venom to be used and decreasing the chance of anaphylaxis during desensitizing. Other treatments targeting IL-4 (important for Th2 maturation) or binding IL-4 *in situ* with recombinant soluble IL-4 receptor (sIL-4R) have shown too little efficacy for patient use.

### Table 3. Inflammatory mediators classified by their effects on target cells

**Mediators with pharmacologic effects on smooth muscle and mucous glands**

1. *Histamine.* Histamine binds to two types of receptors on target cells, $H_1$ and $H_2$. On binding to $H_1$ receptors, histamine induces contraction of smooth muscle (e.g., in airways), increases vascular permeability and mucous secretion by goblet cells. Via H2 receptors, histamine increases gastric secretion, decreases mediator release by basophils and mast cells and stimulates Th2 cells by polarizing DCs into Th2 potentiating cells
2. *Slow-reacting substance of anaphylaxis (SRS-A).* These cysteinyl-leukotrienes ($LTC_4$, $LTD_4$, $LTE_4$), are potent constrictors of peripheral airways (i.e., bronchoconstrictors) and also cause leakage of post-capillary venules, leading to edema. Leukotrienes are derived from the membrane fatty acids of mast cells, neutrophils, and macrophages
3. *Prostaglandins.* A variety of effects are manifested by this large family of related compounds. Prostaglandin $D_2$ is produced by mast cells and causes bronchial constriction. Prostaglandin $I_2$ (prostacyclin) is produced by endothelial cells and probably synergizes with $LTB_4$ to cause edema
4. *Platelet activating factor (PAF).* A low-molecular-weight lipid which causes platelet aggregation with release of vasoactive mediators (serotonin) and smooth muscle contraction
5. *Kinins.* Bradykinin (a nonapeptide) and lysyl-bradykinin (a decapeptide) cause increased vascular permeability, decreased blood pressure, and contraction of smooth muscle

**Mediators that are pro-inflammatory and/or have chemotactic properties**

1. *Eosinophil chemotactic factors of anaphylaxis (ECF-A).* Includes histamine and tetrapeptides from mast cell granules
2. *Neutrophil chemotactic factor of anaphylaxis (IL-8).* A granule-derived protein of mast cells which attracts and activates neutrophils
3. *Late-phase reactants of anaphylaxis.* Mediators that cause delayed inflammatory cell infiltration
4. *Leukotriene $B_4$ ($LTB_4$).* Derived from membrane fatty acids, a potent chemotactic factor for PMNs, eosinophils, and macrophages, LTB induces adhesion of leukocytes to post-capillary venules, degranulation and edema

**Mediators that cause tissue destruction**

1. *Toxic oxygen and nitrogen radicals* (e.g., superoxide and nitric oxide). Released from polymononuclear cells, macrophages and mast cells
2. *Acid hydrolases.* From mast cells
3. *Major basic protein.* A very destructive protein from eosinophil granules
4. *Tryptase.* A major protease released from mast cells that is a good indicator of mast cell activation. Tryptase can cleave C3 and C5a as well as causing tissue destruction

Future approaches in the treatment of allergy include the involvement of T-regulatory cells. Tregs have been shown to be important in the homeostatic mechanisms of the immune system and in the development or prevention of autoimmune disease (Section L3). Regulation of the immune response is thought in part to be under the control of thymus-derived CD4$^+$CD25$^+$ Treg cells or CD4$^+$CD25$^+$ Treg cells that can be induced peripherally from naive T cells. The role of Treg cells in allergy is not understood but would seem important. In experimental models of allergy, CD4$^+$CD25$^+$ T cells downregulate IgE responses to allergens and, *in vitro*, suppress the differentiation of Th2 cells from naive CD4 cells. In other *in vitro* studies of allergy, peripheral blood from atopic individuals has been shown to have reduced suppressive ability in allergen-stimulated T-cell cultures.

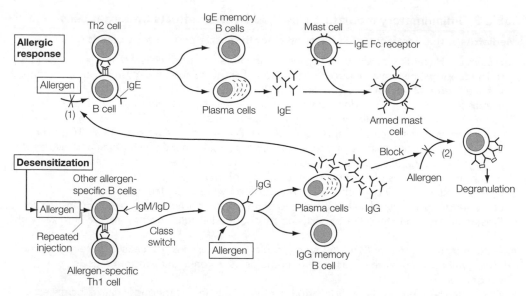

Figure 3. Desensitization. In an individual with IgE antibody to an allergen, there are memory cells which respond to allergen by differentiating into plasma cells which produce allergen-specific IgE (Figure 1 in Section D2). This IgE binds to IgE Fc receptors on mast cells that degranulate when allergen is reintroduced and cross-links IgE on these cells. Repeated injections of allergen are intended to induce an IgG response by stimulating allergen-specific B cells which have not yet undergone a class switch. In particular, allergen-specific Th1 cells would provide help to these B cells inducing class switch to IgG. This IgG would be produced in larger quantity than IgE and compete effectively for the allergen when it is reintroduced, preventing the allergen from stimulating IgE memory B cells (1) and removing the allergen before it can bind IgE on mast cells (2).

In allergy, it is thought that pervading conditions may lead to an inappropriate Treg response in which production of IL-4 cytokines results in a Th2-type expansion. Thus, it is possible that induction of appropriate Treg responses could prevent the development of allergen-specific IgE responses that are mediated through Th2 cells.

# K3 IgG- and IgM-mediated (type II) hypersensitivity

## Key Notes

| | |
|---|---|
| **Introduction** | In type II hypersensitivity (also called cytotoxic hypersensitivity) antibody (IgM or IgG) directed mainly to cellular antigens (e.g., on erythrocytes) or cell-surface autoantigens causes damage through opsonization, lysis, or antibody-dependent cellular cytotoxicity. |
| **Rhesus incompatibility** | Pregnant mothers who are rhesus D (RhD) antigen-negative can develop antibodies to RhD antigen inherited from the father. Sensitization occurs either through prior blood transfusion with RhD+ erythrocytes, or at parturition, when fetal erythrocytes pass into the maternal circulation. During subsequent pregnancies, small numbers of fetal erythrocytes that pass across the placenta stimulate a memory response with the result that IgG antibodies to RhD antigen pass back across the placenta and destroy the fetal erythrocytes (hemolytic disease of the newborn). |
| **Transfusion reactions** | Natural antibodies (isohemagglutinins) to major blood group antigens (A, B) bind to transfused erythrocytes expressing the target antigens resulting in massive hemolysis. This is now rare due to blood group typing before transfusion. |
| **Autoantigens** | Antibodies to a variety of self antigens such as basement membranes of lung and kidney (Goodpasture's syndrome), the acetylcholine receptor (myasthenia gravis), and erythrocytes (hemolytic anemia) can result in tissue-damaging reactions. |
| **Drugs** | Antibiotics such as penicillin can attach to erythrocytes and cause IgG-mediated damage to erythrocytes. |
| **Stimulatory hypersensitivity** | A variant of type II hypersensitivity (sometimes called type V), results when stimulatory, rather than cytolytic, antibodies to self antigens develop. In this case, the antibodies bind to a receptor and act like the natural ligand. In Graves' disease, autoantibodies are present which react with the thyroid-stimulating receptor, stimulating hyperthyroidism. |
| **Related topics** | (D8) Antibody functions<br>(L2) Factors contributing to the development of autoimmune disease<br><br>(L4) Disease pathogenesis – effector mechanisms<br>(M2) Transplantation antigens |

## Introduction

Antibody, alone or together with complement, can cause hypersensitivity reactions that usually result in the direct lysis or removal of cells, hence the alternative name of cytotoxic hypersensitivity. Diseases caused by this type of hypersensitivity often involve erythrocytes (anemias) and self cells (autoimmune diseases). Cell death (or lysis) is mediated through normal mechanisms by which antibodies and complement carry out their function including phagocytosis, lysis, and antibody-dependent cellular cytotoxicity (Section D8).

## Rhesus incompatibility

Rhesus D (RhD) antigen is carried by erythrocytes. Children born to RhD– mothers and RhD+ fathers may express RhD on their erythrocytes. The mother can become sensitized to RhD antigen prior to pregnancy, through blood transfusion, or during pregnancy, and especially at birth, by the baby's RhD+ erythrocytes coming into contact with the mother's immune system. Some pass across the placenta but most are released into the maternal circulation during placental shedding. Since RhD is not present in the mother, her immune system responds to it as a foreign antigen and makes antibodies (Figure 1).

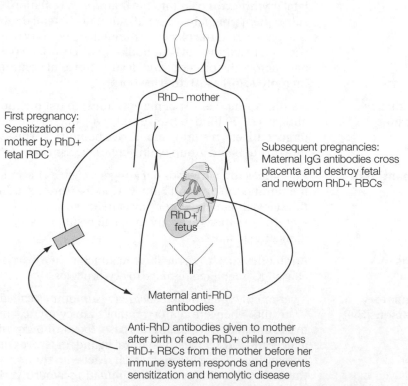

First pregnancy: Sensitization of mother by RhD+ fetal RDC

RhD– mother

Subsequent pregnancies: Maternal IgG antibodies cross placenta and destroy fetal and newborn RhD+ RBCs

RhD+ fetus

Maternal anti-RhD antibodies

Anti-RhD antibodies given to mother after birth of each RhD+ child removes RhD+ RBCs from the mother before her immune system responds and prevents sensitization and hemolytic disease

Figure 1. RhD antigen and hemolytic disease of the newborn. RhD– mothers who give birth to RhD+ infants become immunized at birth with RhD antigen when fetal red blood cells (RBCs) pass into the mother's circulation. This results in IgG antibodies to RhD which cross the placenta during subsequent pregnancies and destroy fetal and newborn RhD+ RBCs. This can be prevented by giving the mother anti-RhD antibodies immediately after the birth of each RhD+ infant or during pregnancy in order to destroy RhD+ RBCs before they stimulate an active immune response in the mother.

This is often not a problem during the first pregnancy but in subsequent pregnancies small amounts of erythrocytes passing across the placenta stimulate a memory response leading to specific anti-RhD antibody production. IgG antibodies pass across the placenta and bind to the fetal erythrocytes leading to their opsonization and lysis. If not prevented, this results in hemolytic disease of the newborn (HDN). Generally, mothers at risk are detected during pregnancy at early routine checks and monitored thereafter. Following each pregnancy with an RhD+ fetus, RhD– mothers are given antibodies to RhD to remove the fetal erythrocytes from the blood stream and suppress the development of a subsequent anti-RhD immune response.

## Transfusion reactions

It is common practice to give blood transfusions in cases of severe blood loss. Most individuals have IgM antibodies (isohemagglutinins, Section M2) to the major blood group antigens A and B that are expressed on the surface of erythrocytes. Individuals who are blood group A have antibodies to B antigens, those who are blood group B will have anti-A antibodies, and those who are AB will have neither. Those who are blood group O will have both antibodies. It is therefore important to do blood group typing on transfusion donors and recipients. In most cases, this is done accurately, but occasionally accidents occur whereby blood is given to a recipient who has the reactive isohemagglutinins. This can result in a transfusion reaction which manifests itself as a complement-mediated massive intravascular life-threatening hemolysis.

## Autoantigens

Antibodies can be made to self antigens when there is breakdown of tolerance to self (Sections M3 and L3). These autoantibodies can cause tissue-damaging reactions. In Goodpasture's disease, autoantibodies to the lung and kidney basement membranes cause inflammation and hemorrhage at the site of antibody binding. Antibodies to the acetylcholine receptor cause loss of receptors (Figure 2), reducing conduction of nerve impulses across the neuromuscular junctions (myasthenia gravis). Autoantibodies to erythrocytes result in their lysis and/or removal, leading to autoimmune hemolytic anemia.

## Drugs

Penicillin, as well as inducing immediate-type hypersensitivity through IgE, can also stimulate an IgG response. This IgG binds to penicillin attached to erythrocytes and, in the presence of complement, induces hemolysis. This disappears on drug removal.

## Stimulatory hypersensitivity

This antibody-mediated hypersensitivity is a variant of type II hypersensitivity but is not cytotoxic. In this case, the autoantibodies are directed to hormone receptor molecules and function in a stimulatory fashion, like the natural ligand, that is, the hormone itself. The classical example is Graves' disease, where antibodies to the thyroid-stimulating receptor result in overactivity of the thyroid (Figure 3).

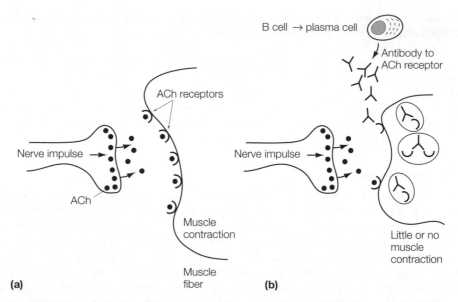

Figure 2. Myasthenia gravis. (a) Normal stimulation of muscle contraction. Nerve impulses trigger release of acetylcholine (ACh) from the nerve ending. The ACh then binds to ACh receptors on muscle cells triggering their contraction. (b) Autoantibodies to the ACh receptor bind to these receptors on muscle cells and cause their internalization and degradation, so that when ACh is released as the result of a nerve impulse, there are few ACh receptors with which to bind; thus, muscle contraction does not occur or is diminished.

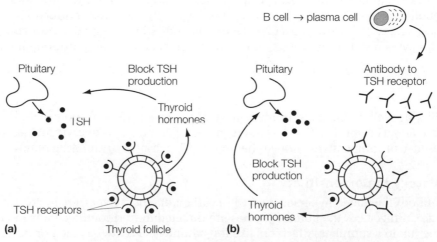

Figure 3. Graves' disease. (a) The pituitary makes thyroid-stimulating hormone (TSH) which binds to TSH receptors on cells of the thyroid follicle and triggers them to make thyroid hormones. In turn, these thyroid hormones inhibit production of TSH by the pituitary as a form of normal feedback regulation of TSH production by thyroid hormones. (b) Autoantibodies to TSH receptor bind TSH receptors and trigger the thyroid follicle cells to release thyroid hormones which stop the pituitary from making TSH. However, they have no effect on production of the autoantibody, which continues to stimulate thyroid follicle cells to make the thyroid hormones that cause hyperthyroidism.

# K4 Immune-complex-mediated (type III) hypersensitivity

---

**Key Notes**

| | |
|---|---|
| **Introduction** | Immune complexes that result from interaction of antibodies (IgM or IgG) with foreign serum products, for example immunoglobulins, as well as microbial and self antigens, either in local sites or systemically, may lead to phagocytic and complement-mediated damage. |
| **Mechanisms of type III hypersensitivity** | Tissue damage is caused mainly by antibody-mediated complement activation and release of lytic enzymes from neutrophils. Local damage (Arthus reaction) can be seen in pulmonary disease resulting from inhaled antigen. Systemic antibody complexes with microbial or autoantigens result in immune complex deposition in blood vessels (vasculitis) or in the renal vessels (glomeruli) of the kidneys leading to glomerulonephritis. |
| **Diseases associated with type III hypersensitivity** | Pulmonary diseases result from inhalation of bacterial spores (farmer's lung) or avian serum/fecal proteins (bird fancier's disease). Systemic disease can occur from streptococcal infections (streptococcal nephritis), autoimmune complexes (e.g., systemic lupus erythematosus, SLE), drugs (e.g., penicillin), or antisera made in animals. |
| **Related topics** | (B4) Innate immunity and inflammation (D2) Antibody classes (D6) Antigen–antibody complexes (immune complexes)      (L4) Disease pathogenesis – effector mechanisms |

---

## Introduction

Normally, antigen–antibody immune complexes are removed by phagocytic cells (or by erythrocytes, Section D) and there is no tissue damage. However, when there are large amounts of immune complexes and they persist in tissues, they can cause damage which may be localized within tissues (Arthus reaction) or systemic. This type of hypersensitivity can be induced by microbial antigens, autoantigens, and foreign serum components.

## Mechanisms of type III hypersensitivity

Much of the tissue damage is the result of antibody-mediated complement activation leading to neutrophil chemoattraction and release of lytic enzymes by the degranulating neutrophils (Section B1). Local deposition of immune complexes results in an Arthus reaction (Figure 1). Immune complexes (usually small) can also cause systemic effects such as fever, weakness, vasculitis, arthritis, edema, and glomerulonephritis. An example of this is when passive antibodies are given to patients to protect them against microbial toxins such as tetanus toxin (Section I4). For example, an antibody response can develop (serum sickness) against horse anti-tetanus toxin antibodies that results in immune complex formation. Serum immune complexes can deposit in blood vessels (vasculitis) or can become trapped in the blood vessels of the kidneys leading to glomerulonephritis.

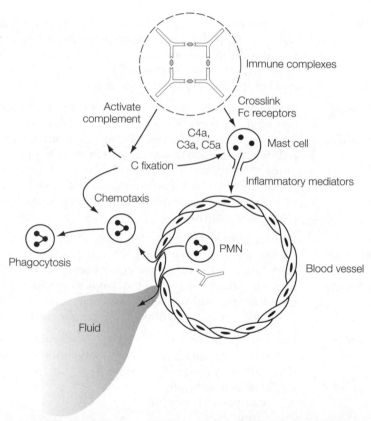

Figure 1. The Arthus reaction. Small immune complexes in the skin directly trigger Fc receptors and activate complement, resulting in an acute inflammatory response mediated through degranulation of mast cells induced by C3a and C5a. Small immune complexes can also lodge in blood vessels and induce vasculitis or glomerulonephritis in the kidney.

## Diseases associated with type III hypersensitivity

A list of some diseases mediated by type III hypersensitivity is shown in Table 1. IgG antibodies complexed with inhaled antigens cause local damage in the airways of the lung (including pneumonitis and alveolitis). Immune complexes made against antigens encountered systemically cause a variety of symptoms and in particular kidney damage through immune complex deposition.

**Table 1. Type III hypersensitivity diseases**

| Site of reaction | Antigens | Disease |
|---|---|---|
| Localized (inhaled) | Bacterial spores | Farmer's lung |
| | Fungal spores | |
| | Pigeon serum/fecal proteins | Bird fancier's disease |
| Systemic | Microbes including: | |
| | *Streptococcus* | Streptococcal nephritis |
| | Hepatitis B virus | |
| | Epstein–Barr virus | |
| | *Plasmodium* (malaria) | |
| | Autoantigens, e.g., DNA | Systemic lupus erythematosus |
| | Drugs: penicillin, sulfonamides | Drug allergy |

# K5 Delayed (type IV) hypersensitivity

## Key Notes

**Introduction**

Type IV hypersensitivity develops 24 h after contact with an antigen and is mediated by T cells together with dendritic cells, macrophages, and cytokines. The persistent presence of the antigen, for example chronic mycobacterial infections, results in granulomas. Skin contact with a number of small molecules (chemicals and plant molecules) can also result in delayed (contact) hypersensitivity.

**The tuberculin reaction**

This is a "recall" response to purified mycobacterial antigens that is used as the basis of a skin test to determine if a patient has a memory immune response (not necessarily curative) to TB that would indicate prior infection or vaccination.

**The production of granulomas**

The inability of T cells to kill all mycobacteria in macrophages often results in chronic stimulation. In particular, mycobacterial-specific T cells produce cytokines responsible for "walling off" the macrophages containing mycobacteria, which in turn results in the production of granulomas. This also occurs in response to schistosome worms, and is seen in other clinical conditions with, as yet, undefined antigens.

**Contact sensitivity**

Contact with a number of small-molecular-weight chemicals (e.g., nickel in a watch strap buckle) and molecules from some plants (poison ivy) can result in them penetrating the skin, binding to self proteins, and inducing a specific $CD4^+$ T-cell response. The resulting cytokines induce a local redness and swelling which usually disappears on removal of the antigen.

**Related topics**

(B1) Cells of the innate immune system

(F1) The role of T cells in immune responses

(F5) Clonal expansion and development of effector function

(G5) Regulation by T cells and antibody

(G6) Neuroendocrine regulation of immune responses

(H2) Immunity to different organisms

## Introduction

Unlike type 1 (immediate) hypersensitivity, which manifests itself in seconds or minutes, this hypersensitivity reaction begins at least 24 h after contact with the eliciting antigen. It was first associated with T-cell-mediated immune responses to *Mycobacterium tuberculosis* (MTb) and was therefore initially termed "bacterial hypersensitivity." Such responses often lead to the production of granulomas some weeks later. This delayed type of hypersensitivity (DTH) now covers a range of T-cell-mediated responses, including those induced by small molecules coming into contact with the skin—contact hypersensitivity. In addition to T cells, the key players in this type of sensitivity are dendritic cells, macrophages, and cytokines. This type of hypersensitivity also plays a role in several clinical situations where there is persistence of antigen which the immune system is unable to remove, leading to chronic inflammation.

## The tuberculin reaction

Initial experiments by Koch showed that patients with tuberculosis (TB) given subcutaneous injection of mycobacterial antigens derived from MTb developed fever and sickness. This "tuberculin reaction" is now the basis of a "recall" test to determine if individuals have T-cell-mediated reactivity against TB. In this test (Mantoux test) small amounts of the purified protein derivative (PPD) of tuberculin derived from MTb organisms are injected into the skin and the site examined up to 72 h later. A positive skin test shows up as a firm red swelling which is maximal at 48–72 h after injection and is mediated by dendritic cells and an influx of macrophages and tuberculin-specific memory T cells, into the site of injection (Figure 1).

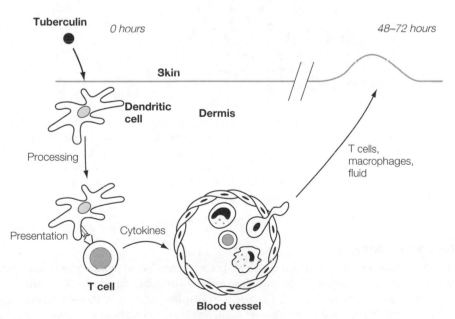

Figure 1. The tuberculin reaction (delayed-type hypersensitivity). Tuberculin protein introduced into the dermis is processed and presented by dendritic cells to T cells via MHC class II molecules. Cytokines produced by the memory T cells with specificity for MTb antigens alter local endothelial cell-adhesion molecules, allowing monocytes to enter the site of injection and develop into macrophages. T cells and macrophage products induce edema (fluid) and swelling. A positive skin test shows up as a firm red swelling which is maximal at 48–72 h after injection.

## The production of granulomas

CD4+ T cells control intracellular microbial infections such as those by mycobacteria and some fungi (Section F5). However, mycobacteria, in addition to some other intracellular infections, have escape mechanisms to prevent their elimination (Section H3). Thus, macrophage activation factors produced by CD4+ T cells are not always effective (Figure 2). Antigen therefore persists and leads to the "chronic" stimulation of CD4+ T cells and continuous production of cytokines. These mediate fusion of the macrophages containing the microbes, and fibroblast proliferation, which results in "walling off" the offending microbes in the form of a granuloma. This chronic inflammatory state is seen both in TB and in the tuberculoid type of leprosy caused by *Mycobacterium leprae* (Section H3). Granulomatous reactions also occur with schistosoma infections and in some clinical situations where the antigens have not yet been defined (e.g., sarcoidosis and Crohn's disease). Nonimmune granulomas are produced by persistent particles such as asbestos that cannot easily be removed from the body by phagocytosis.

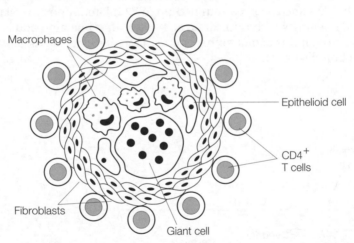

Figure 2. Granulomas. Immune granulomas are formed in response to chronic stimulation of CD4+ T cells by persistent nondegradable antigens, including mycobacteria. They consist of epithelioid cells, macrophages, and giant cells which are "walled off" by fibroblasts surrounded by an outer layer of CD4+ T cells. Cytokines produced by the different cells all contribute to the granuloma formation, which is the immune system's way of isolating the nondegradable microbes from the rest of the body.

## Contact sensitivity

Certain small molecules penetrating the skin can give rise to contact sensitivity, seen clinically as dermatitis. Some of the chemical agents and plant products that induce contact sensitivity are listed in Table 1. Classical examples include reactions against metal fasteners on watch straps and rashes seen in response to poison ivy and poison oak. Removal of the agent usually results in resolution of the hypersensitivity.

Sensitization against these molecules is thought to be mediated through binding to skin proteins and through the powerful antigen-presenting properties of skin dendritic cells, Langerhans cells, which present antigen on MHC class II molecules to CD4+ Th1 cells (Figure 3). The subsequent contact sensitivity reaction involves presentation of the antigens to memory CD4+ T cells that release cytokines causing vasodilation, traffic into

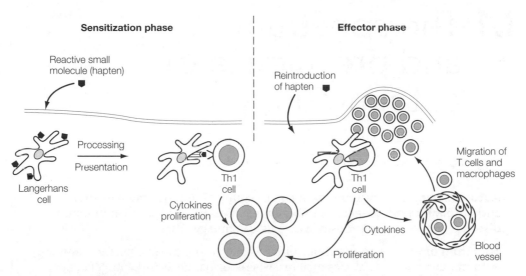

**Figure 3.** Contact sensitivity mediated through Langerhans cells. In the sensitization phase, reactive small molecules, haptens (e.g., pentadecacatechol associated with poison ivy), which come in contact with the skin bind to self proteins (including those on Langerhans cells) and are internalized, processed, and presented by Langerhans cells to T cells. These proliferate to form clones of Th1 cells specific for hapten-modified self peptide. When hapten is reintroduced, the modified self peptide is again presented on Langerhans cells in MHC class II. Memory T cells eventually find and respond to these antigens by releasing cytokines (e.g., IFNγ) which attract primarily Th1 cells and monocytes to this area and upregulate expression of adhesion molecules on endothelial cells that result in passage of Langerhans cells into the tissues.

the site of nonspecific CD4$^+$ T cells and activated macrophages, and localized pustule formation.

## Table 1.  Agents causing contact sensitivity

| | |
|---|---|
| Chemicals | Nickel, turpentine, some cosmetics, formaldehyde |
| Plants | Poison ivy, poison oak |

# L1 The spectrum and prevalence of autoimmunity

---

**Key Notes**

| | |
|---|---|
| **Autoimmunity and autoimmune disease** | Autoimmunity is an acquired immune reactivity to self antigens. Autoimmune diseases occur when autoimmune responses lead to tissue damage. |
| **Spectrum of autoimmune conditions** | Autoimmune diseases may be organ-specific, for example diabetes mellitus, where the pancreas is the target organ, or systemic (non-organ-specific), for example systemic lupus erythematosus (SLE), where multiple organs may be involved. Pathogenesis associated with these diseases may be mediated primarily by antibody, by T cells, or a combination of both. |
| **Prevalence** | Approximately 3.5% of individuals have autoimmune disease, 94% of which is accounted for by Graves' disease/hyperthyroidism, type I diabetes, pernicious anemia, rheumatoid arthritis (RA), thyroiditis, celiac disease (CD), vitiligo, multiple sclerosis (MS), and SLE. Women are more likely than men to develop autoimmune disease. |
| **Related topics** | (G2) Central tolerance    (I3) Antigen preparations |
| | (G3) Peripheral tolerance |

---

## Autoimmunity and autoimmune disease

The immune system has the capacity to mount an immune response to virtually all molecules and/or cells. Although the capacity to respond to self antigen is present in all of us, in most instances we are tolerant or anergic to these antigens (Section G), indicating that mechanisms must exist to prevent or subdue autoimmune responses. Moreover, autoreactive T and B cells as well as autoantibodies are found in people who do not have autoimmune diseases, demonstrating that immunological autoreactivity alone is not sufficient for the development of disease. The mechanisms currently thought to prevent/suppress autoimmune responses include inactivation or deletion of autoreactive T and B cells, active suppression by Treg cells or cytokines, and the immunosuppressive adrenal hormones, the glucocorticoids. When suppressor mechanisms fail or are overridden, a response directed against self antigen can occur, resulting in autoimmune diseases that range from those which are organ-specific (diabetes and thyroiditis) to those which are systemic (non-organ-specific) such as systemic lupus erythematosus (SLE) and rheumatoid arthritis (RA).

Several important co-factors in the development of autoimmune disease have been identified, and include genetics (e.g., HLA associations), gender, and age. Characteristics of the antigen and how it is "presented" to the immune system are also important. For example, injection of animals with chemically modified thyroid protein or with normal protein plus Freund's adjuvant (Section I3) can give rise to severe thyroiditis that is due to immune recognition of normal thyroid proteins. Infection by organisms, including Epstein–Barr virus (EBV) or mycoplasma, can provoke autoantibody production in otherwise normal persons. In addition, some dietary proteins such as those found in wheat and other cereals may cause celiac disease. Other causes of autoimmune pathology include certain drugs, such as procainamide, which are used to treat cardiac arrhythmias, or toxic substances such as mercuric chloride and polyvinyl chloride.

## Spectrum of autoimmune conditions

That autoimmune diseases involve immune recognition of specific antigens is evidenced by organ-specific diseases, including thyroiditis, diabetes mellitus, multiple sclerosis (MS), celiac disease (CD), and inflammatory bowel disease. Antigens shared by multiple tissue sites are apparently involved in systemic autoimmunity in diseases such as SLE, RA, systemic vasculitis, and scleroderma. It is also clear that a given individual may develop autoimmune disease of more than one type. For example, thyroid autoimmune disease is sometimes associated with gastric autoimmunity and the incidence of CD in Sjögren syndrome, type 1 diabetes, and autoimmune thyroiditis is markedly increased. The pathogenesis associated with autoimmune disease may be mediated primarily by antibody (e.g., hemolytic anemia), primarily by cellular immunity (e.g., MS), or by a combination of antibody and cell-mediated immunity (e.g., RA).

## Prevalence

Autoimmune diseases are prevalent in the general population, and it is estimated that approximately 3.5% of individuals are afflicted. The most common are Graves' disease/hyperthyroidism, type I diabetes, pernicious anemia, RA, thyroiditis, CD, vitiligo, MS, and SLE, which together account for 94% of all cases. Overall, women are 2.7 times more likely than men to develop an autoimmune disease, but the female-to-male ratio can be as high as 10:1 in SLE (Section O3).

# L2 Factors contributing to the development of autoimmune disease

## Key Notes

**Autoimmune diseases are multifactorial**

Autoimmune diseases arise as the result of a breakdown in self-tolerance. Factors predisposing and/or contributing to the development of autoimmune diseases include age, genetics, gender, infections, and the nature of the autoantigen. Combinations of these factors are probably important in the development of autoimmune disease.

**Age and gender**

Autoantibodies are more prevalent in older people and women have a greater risk than men for developing an autoimmune disease. In SLE and Graves' disease, there is a female-to-male bias of 10:1 and 7:1, respectively. A higher incidence in female mice of autoimmune diseases is consistent with hormones playing an important role.

**Genetic factors**

Antigen-specific autoimmune phenomena cluster in certain families. Particular HLA genes are associated with certain autoimmune diseases and particular HLA haplotypes predict the relative risk of developing a particular autoimmune disease. Gene polymorphisms and or mutations also play a role, as evidenced by the findings that Fas deficiency leads to autoimmune lymphoproliferative syndrome (ALPS) and mutation of the *AIRE* gene expressed by epithelial cells in thymic medulla leads to the development of a rare condition with a range of autoimmune diseases. Mutations in genes for certain complement components lead to an increased risk of SLE.

**Infections**

Many infectious agents (EBV, mycoplasma, streptococci, *Klebsiella*, malaria, etc.) have been linked to particular autoimmune diseases and may be important in their etiology.

**Nature of autoantigens**

Immune responses to highly conserved proteins are frequently found in autoimmune diseases. Target antigens are often conserved proteins and include heat shock proteins (HSPs) and enzymes. Antibodies to HSPs can be seen in some autoimmune diseases and are probably derived from cross-reactions with microbial HSPs. Autoimmune responses to enzymes include transglutaminase (tTG) in celiac disease and glutamic acid dehydrogenase (GAD) in diabetes.

**Drugs and autoimmune reactions**

Certain drugs can initiate autoimmune reactions by unknown mechanisms. For example, patients receiving procainamide develop SLE-like symptoms and have antinuclear antibodies that disappear following discontinuation of the drug.

| Immunodeficiency | A deficient immune response may allow persistence of infection or inflammation, which can lead to an increased incidence of autoimmune disease. For example, patients deficient in the complement components C2, C4, C5, or C8 have an increased incidence of autoimmune diseases, perhaps because of inefficient clearance of immune complexes. | | |
|---|---|---|---|
| **Related topics** | (G5) Regulation by T cells and antibody | (J2) | Primary/congenital (inherited) immunodeficiency |
| | (G6) Neuroendocrine regulation of immune responses | | |

## Autoimmune diseases are multifactorial

Autoimmune diseases arise as the result of a breakdown in tolerance to self antigens. Moreover, autoimmune diseases are multifactorial in that their development, in most cases, probably results from combinations of predisposing and/or contributing factors. These factors include (Table 1) (i) genetics—inheritance of a particular HLA haplotype increases the risk of developing disease, for example ankylosing spondylitis, Reiter syndrome, and SLE; (ii) gender—more females than males develop disease; (iii) infections— EBV, mycoplasma, streptococci, *Klebsiella*, malaria, and so forth, have been linked to particular autoimmune diseases; (iv) the nature of the autoantigen—highly conserved enzymes and heat shock proteins (HSPs) are often target antigens and may be cross-reactive with microbial antigens; (v) drugs—certain drugs can induce autoimmune-like syndromes; and (vi) age—most autoimmune diseases occur in adults. Autoimmune diseases can also occur through breakdown of tolerance to dietary proteins; for example, ingestion of gluten proteins derived from wheat or other cereals leads to the development of celiac disease.

## Age and gender

Autoantibodies are more prevalent in older people and animals, perhaps due to less stringent immunoregulation by the aging immune system. Few autoimmune diseases occur

## Table 1. Summary of factors contributing to development of autoimmune diseases

| | |
|---|---|
| Age | Higher incidence in the aged population |
| Gender | Females generally more prone than males |
| Genetics | Some diseases are HLA-associated |
| Infections | Some common infections, e.g., EBV, streptococcus, malaria |
| Nature of autoantigen | Often conserved antigens, e.g., heat shock proteins and enzymes |
| Dietary antigens | Ingestion of gluten leads to the development of celiac disease |
| Drugs | Some drugs, e.g., procainamide and hydralazine, induce SLE-like symptoms |

in children, the majority being in adults. Women have a greater risk than men for developing an autoimmune disease. In SLE and Graves' disease, there is a female-to-male bias of 10:1 and 7:1 respectively, whereas ankylosing spondylitis is almost exclusively a male disease. Taken together, these facts suggest that the neuroendocrine system plays an important role in the development of these diseases. This is supported by animal studies where it has been shown that female mice of a particular strain spontaneously develop SLE. This can be prevented by removing their ovaries (estrogen source) or by treating them with testosterone. Similarly, male mice that are more resistant to developing the disease lose this resistance if castrated (Section O3).

## Genetic factors

Antigen-specific autoimmune phenomena have a familiar clustering. For example, thyroid-reactive antibodies are much more common in genetically related family members of a person with autoimmune thyroid disease than in the population at large. The role of the MHC (presumably in presenting autoantigenic peptides) is evidenced by the strong association between HLA type and incidence of certain autoimmune diseases. The possession of particular HLA haplotypes predicts the relative risk of developing a particular autoimmune disease (Table 2).

Polymorphisms and/or mutations of many other genes involved in lymphocyte activation or suppression are also likely to play a crucial role. For example, in MLR lpr mice, an autosomal recessive mutation in the Fas apoptosis gene leads to progressive lymphadenopathy and hypergammaglobulinemia, with production of multiple SLE-like autoantibodies. In man, mutation in the same Fas gene leads to autoimmune lymphoproliferative syndrome (ALPS) with patients having nuclear, red blood cell, and platelet autoantibodies. As another example, patients with a mutation in the *AIRE* gene expressed by epithelial cells in thymic medulla (Section G2) develop autoimmune polyendocrinopathy-candidiasis-ectodermal dystrophy (APECED). In this rare condition, patients develop a range of autoimmune diseases including hypoparathyroidism. Complement deficiency due to mutations in genes for C2, C4, C5, and C8 results in increased risk of SLE, demonstrating the importance of complement in the clearance of immune complexes.

**Table 2. Some autoimmune diseases showing HLA association (Caucasians)**

| Disease | HLA | Risk* |
|---|---|---|
| Ankylosing spondylitis | B27 | 90 |
| Reiter syndrome | B27 | 36 |
| Systemic lupus erythematosus | DR3 | 15 |
| Myasthenia gravis | DR3 | 2.5 |
| Juvenile diabetes mellitus (insulin-dependent) | DR3/DR4 | 25 |
| Psoriasis vulgaris | DR4 | 14 |
| Multiple sclerosis | DR2 | 5 |
| Rheumatoid arthritis | DR4 | 4 |
| Celiac disease | DQ2/DQ8 | 4 |

*For example, an individual who has HLA-B27 is 90 times more likely to develop ankylosing spondylitis than an individual not having HLA-B27.

## Infections

Many infectious agents (EBV, mycoplasma, streptococci, klebsiella, malaria, etc.) have been linked to particular autoimmune diseases. Lyme arthritis, for example, is initiated by chronic infection with spirochetes of the genus *Borrelia* (e.g., *Borrelia burgdorferi*) which are transmitted by deer ticks from deer and rodents to people. Some microbial antigens also have structures similar to self antigens and induce autoimmune responses through "antigenic mimicry" (see below).

## Nature of autoantigens

Target antigens in autoimmune disease may be cell surface, cytoplasmic, nuclear, or secreted (Table 3). They are often highly conserved proteins such as heat shock proteins

### Table 3. Antigens targeted in autoimmune disease

| Organ-specific diseases | | Non-organ-specific diseases | |
|---|---|---|---|
| Disease | Antigen(s) | Disease | Antigen(s) |
| Addison's disease | Adrenal cortical cells* | Ankylosing spondylitis | Vertebral |
| Autoimmune hemolytic anemia | RBC membrane antigens | Chronic active hepatitis | Nuclei, DNA |
| Graves' disease | TSH receptor | Multiple sclerosis | Brain/myelin basic protein |
| Guillain–Barré syndrome | Peripheral nerves (gangliosides) | Rheumatoid arthritis | IgG (rheumatoid factor) connective tissues |
| Hashimoto's thyroiditis | Thyroid peroxidase thyroglobulin/T4 | Scleroderma | Nuclei, elastin, nucleoli centromeres, topoisomerase 1 |
| Insulin-dependent diabetes mellitus | β cells in the pancreas (GAD, tyrosine phosphatase | Sjögren syndrome | Exocrine glands, kidney, liver, thyroid |
| Pemphigus | Desmosal antigens in keratinocytes | Systemic lupus erythematosus | Double-stranded DNA, nuclear antigens |
| Pernicious anemia | Intrinsic factor | Wegener's granulomatosis | Proteinase 3 |
| Polymyositis | Muscle (histidine tRNA synthetase) | Celiac disease | Small intestine |
| Primary biliary cirrhosis | Pyruvate dehydrogenase | | |
| **Several organs affected** | | | |
| Goodpasture syndrome | Basement membrane of kidney and lung (type IV collagen) | | |
| Polyendocrine | Multiple endocrine organs (hepatic cytochrome p450; intestinal tryptophan hydroxylase) | | |

*Adrenocorticotropic hormone receptor; 17α- and 21-hydroxylases.
TSH, thyroid-stimulating hormone; GAD, glutamic acid decarboxylase.

(HSPs) and other stress proteins. Of importance, the primary immune response to microbial infections includes a strong response to HSPs followed by a response to specific microbial components. It is believed that this anti-HSP response might confer on the host the ability to respond generally to other microbial infections. However, microbial and human HSPs have a high sequence homology that could target human HSPs. Immune responses to enzymes and substrates are frequently found in autoimmune diseases (Table 4). For example, IgA antibodies to tTG are found in the sera of patients with celiac disease, IgG antibodies to glutamic acid decarboxylase in diabetes, and thyroid peroxidase in autoimmune thyroiditis.

**Table 4. Enzymes as autoantigens**

| Enzyme | Disease |
| --- | --- |
| Pyruvate dehydrogenase | Primary biliary cirrhosis |
| Glutamic acid decarboxylase | Insulin-dependent diabetes |
| Myeloperoxidase | Glomerulonephritis |
| Thyroid peroxidase | Autoimmune thyroiditis |
| 17α- and 21-hydroxylases | Addison's disease |
| Proteinase 3 | Wegener's granulomatosis |
| Tyrosinase | Vitiligo |
| Tissue transglutaminase | Celiac disease |

## Drugs and autoimmune reactions

Certain drugs can initiate autoimmune reactions by unknown mechanisms. For example, antinuclear antibodies appear in the blood of the vast majority of patients receiving prolonged treatment with procainamide for ventricular arrhythmias, and nearly 10% develop an SLE-like syndrome which resolves following discontinuation of the drug.

## Immunodeficiency

A deficient immune response may allow persistence of infection or inflammation. This possibility is supported by the observation that immune deficiency syndromes are associated with autoimmune abnormalities. For example, patients deficient in the complement components C2, C4, C5, or C8 have an increased incidence of autoimmune diseases (see Genetic factors). There are also diseases where, paradoxically, immunodeficiency and autoimmunity coexist. An example of this is in common variable immune deficiency (Section J2) where autoantibodies to platelets are sometimes found. Autoimmune diseases are also more common in patients with IgA deficiency (Section J2).

# L3 Autoimmune diseases – mechanisms of development

---

**Key Notes**

| | |
|---|---|
| **Breakdown of self-tolerance** | The mechanisms that lead to autoimmunity are unclear but may include molecular mimicry, defective regulation of the anti-self response through Th1, Th2, and Treg cells, polyclonal activation, modification of self antigens through microbes and drugs, and availability of normally sequestered self antigen. |
| **Molecular mimicry and the T-cell bypass** | An immune response may be generated against an epitope that is identical, or nearly identical, in both a microbe and host tissue, resulting in attack on host tissue by the same effector mechanisms activated to eliminate the pathogen. For example, a cross-reactive antigen between heart muscle and group A streptococci predisposes to the development of rheumatic fever as a result of inducing autoantibodies to heart muscle. |
| **Defective regulation mediated via Th cells** | Microbial infection induces primarily either Th1 or Th2 cytokines. The Th1 response leads to the production of the pro-inflammatory cytokines, while the Th2 response is associated with anti-inflammatory cytokines. Predominance of Th1 or Th2 responses occurs in some autoimmune diseases, and changes in the relative contribution of these subsets (e.g., as seen in pregnancy) can influence disease activity in RA and SLE. |
| **Polyclonal activation via microbial antigens** | Some microbes or their products activate lymphocytes independently of their antigenic specificity, that is, are polyclonal activators (e.g., LPS and EBV). Patients with infectious mononucleosis produce only IgM antibodies to several autoantigens, including DNA. Since a switch to production of IgG autoantibodies (which requires Th cells) does not occur, T cells are probably not involved or are inhibited in their action. |
| **Modification of cell surfaces by microbes and drugs** | Foreign antigens, for example viruses and drugs, may become adsorbed onto the surfaces of cells or react chemically with surface antigens in a hapten-like manner to alter their immunogenicity. For example, thrombocytopenia and anemia are relatively common in drug-induced autoimmune disease. Thrombocytopenia is also common in children following viral infections, and may involve |

|  | association of viral antigens or immune complexes with the surface of platelets. |
|---|---|
| **Availability of normally sequestered self antigen** | Because tolerance induction occurs mainly during embryonic development, antigens which are absent or anatomically separated (sequestered) from the immune system during this period are not recognized as self. Such antigens include the lens proteins of the eye, and molecules associated with the central nervous system, the thyroid, and testes. |
| **Defective regulatory T cells (Tregs)** | Tregs play an important role in the maintenance of self-tolerance. Decreased Treg numbers and/or function have been described in several human autoimmune diseases, suggesting a breakdown in tolerance. |

| **Related topics** | (A4) Antigens | (G6) Neuroendocrine |
|---|---|---|
| | (G2) Central tolerance | regulation of immune |
| | (G3) Peripheral tolerance | responses |
| | (G5) Regulation by T cells and antibody | (H3) Pathogen defense strategies |

## Breakdown of self-tolerance

The mechanisms that lead to autoimmunity are unclear and involve many factors. In an ideal immune response, danger signals from only foreign antigens would activate both innate and adaptive effector mechanisms, with the foreign antigens being cleared without damage to the host and with immune effector mechanisms being turned off when they are no longer needed. However, foreign antigens, in the context of microbial infections, usually cause inflammation and cell destruction. The debris and apoptotic cells would normally be cleared by phagocytic cells. In SLE there is evidence for a defective phagocytosis of apoptotic cells resulting in accessibility of multiple cellular antigens to the immune system. Release of autoantigens, including HSPs and nuclear components, in the absence of effective tolerance mechanisms could result in the development of autoimmunity. The various mechanisms which may explain breakdown of tolerance to self and how reactions to autoantigens may be initiated include molecular mimicry, defective regulation of the anti-self response through Th1, Th2, and Treg cells, polyclonal activation, modification of self antigens by microbes and drugs, and availability of normally sequestered self antigen.

## Molecular mimicry and the T-cell bypass

The adaptive immune response continuously monitors microbial infections and responds accordingly. In some cases, however, a response may be generated against an epitope that is identical, or nearly identical, in both a microbe and host tissue, resulting in attack on host tissue by the same effector mechanisms which are activated to eliminate the pathogen. One example is rheumatic heart disease, which is due to an epitope that is common to heart muscle and group A streptococci (Figure 1). In this case, previously anergized anti-self B cells (which also cross-react with streptococci) may be reactivated by receiving co-stimulatory signals from microbe-specific T cells. The B cell interacts with the microbial antigen through its antigen receptor and presents microbial peptides

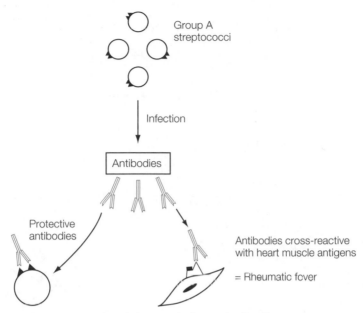

Figure 1. Group A streptococci and rheumatic fever. Antibodies to a streptococcal antigen cross-react with heart muscle antigen leading to damage and rheumatic fever. Disease abates when the bacteria are eliminated and antibody production ceases.

to antimicrobial T cells which then provide help and activate the anti-self B cells (Figure 2). Self-reactive B cells also become activated if the self antigen forms a complex with a microbial antigen. In this event, the self-reactive B cell can endocytose microbial antigens along with the self antigen and present microbial peptides to T cells. The microbe-specific T cell in this instance will provide help to the self-reactive B cell in the form of co-stimulatory molecules and cytokines, leading to breakdown in tolerance (Section G1).

## Defective regulation mediated via Th cells

The initial response to a microbial infection is usually associated with predominantly either Th1 or Th2 cytokines (Section G5). The Th1 response leads to the production of the pro-inflammatory cytokines IFNγ, IL-2, and TNFα, followed by the release of the anti-inflammatory cytokines TGFβ, IL-4, and IL-10 from Th2 cells. Another Th population, Th17 (Section B1), also produces the pro-inflammatory cytokine IL-17. The Th2 response is associated with anti-inflammatory cytokines. That polarized Th1 or Th2 responses may be involved in autoimmune pathogenesis is suggested by the observation that during pregnancy (a time of hormonal change), a period when Th2 cytokines predominate, clinical symptoms associated with the autoimmune disease SLE are exacerbated, whereas in RA, a Th1-type disease, clinical symptoms are ameliorated. These results related to pregnancy give credence to the role that the endocrine system plays in regulating the immune system (Section O).

## Polyclonal activation via microbial antigens

Some microbes or their products activate lymphocytes independently of their antigenic specificity, that is, are polyclonal activators. An example of this in mice is endotoxin or lipopolysaccharide (LPS), which is produced mainly by Gram-negative bacteria. In man,

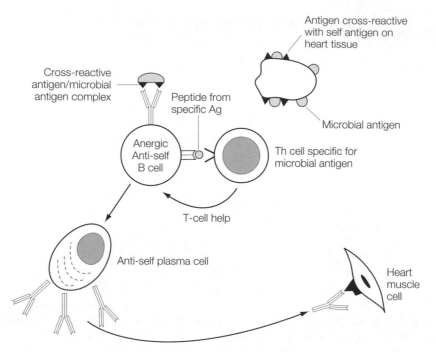

Figure 2.  Activation of anergic anti-self B cells. The BCR on an anti-self B cell binds to self–microbial Ag complex. The B cell presents the microbial component of the complex to a T cell and receives T cell help for activation (second signal). This is also called the "T-cell bypass" mechanism of autoimmunity since prevention of T cell help for self is bypassed by presentation via a non-self antigen.

EBV has been linked to autoimmunity in a small subset of infected individuals. Most patients with infectious mononucleosis, which is caused by EBV, develop only IgM auto-antibodies against several cellular antigens, including DNA (Figure 3). Since a switch to production of IgG autoantibodies, which requires Th cells, does not occur, T cells are probably not involved or are inhibited in their action. Moreover, on recovery, when the strong EBV stimulus is removed, autoantibodies disappear. Clearly, multiple factors are important for maintenance of long-term tolerance to self, and polyclonal activation together with a defect or impairment of immunoregulation following infection could result in activation and expansion of autoreactive clones.

## Modification of cell surfaces by microbes and drugs

Foreign antigens may become adsorbed onto the surfaces of cells or react chemically with surface antigens in a hapten-like manner to alter their immunogenicity. Thrombocyto-penia (low platelet levels) and anemia (low red blood cell levels) are relatively common examples of drug-induced autoimmune disease. Thrombocytopenia is also common in children following viral infections, and may involve association of viral antigens or virus–antibody immune complexes with the surface of platelets. Similarly, an autoimmune-like situation may result when microbial antigens become actively expressed on the surfaces of infected or transformed cells, especially during viral infection. Although the immune response that subsequently develops normally results in removal of these infected cells, in some cases the tissue destruction associated with elimination of these antigens may result in immunologically mediated disease which is much more serious than the

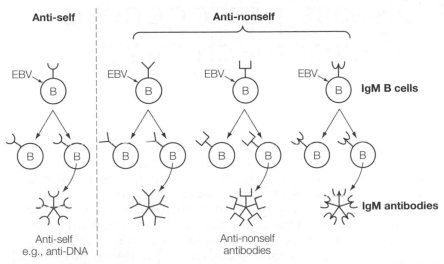

Figure 3. Autoantibodies produced through polyclonal activation of B cells. B cells of all specificities, including self, that have not been eliminated by central tolerance mechanisms may be polyclonally activated (e.g., by EBV infection) to synthesize and release the antibodies they are programmed to produce, perhaps including some autoantibodies. Transient production of the antibodies normally subsides after the microbe is eliminated or controlled.

infection itself. For example, mice infected *in utero* or at birth with lymphocytic chorio-meningitis virus (LCMV) become tolerant to the virus and harbor it for life without overt disease symptoms. However, if normal adult mice are exposed to LCMV, the infection is invariably fatal. In X-irradiated or neonatally thymectomized (i.e., immunosuppressed) mice, the viral infection is not lethal. Thus, lethal neurological damage results, not from the virus itself, but from the immune response to LCMV-infected cells (Section K5).

## Availability of normally sequestered self antigen

Since tolerance induction occurs mainly during embryonic development, antigens which are absent or anatomically separated (sequestered) from the immune system during this period are not recognized as self. These antigens either are present in too low amounts to stimulate autoimmunity or are sequestered in immunologically privileged sites. In later life, these antigens may be released as a result of trauma or infection. They may then stimulate lymphocytes that have escaped tolerance, and induce the development of autoimmune disease. Antigens which fit this model include those found in the lens of the eye, central nervous system, thyroid, and testes. For example, after vasectomy blocks the release of sperm through spermatic ducts, antibodies to spermatozoa are produced. In addition, trauma to the lens in one eye results in autoantibodies that can damage the nontraumatized eye.

## Defective regulatory T cells (Tregs)

Tregs play an important role in the maintenance of self-tolerance (Section G3). Experimental animal models have shown that removal of Tregs results in the development of a number of autoimmune diseases. Decreased Treg numbers and/or function have been described in several human autoimmune diseases. These changes, together with other factors contributing to breakdown in tolerance, could lead to development of autoimmune diseases.

# L4 Disease pathogenesis – effector mechanisms

## Key Notes

| | |
|---|---|
| **Tissue-damaging reactions in autoimmune diseases** | The mechanisms of tissue destruction by autoantibodies are the same as those that lead to protective responses—phagocytosis, complement activation, and interference with molecular function. T and B cells may be involved as well as inflammatory cytokines, immune complexes, phagocytes, and complement components. The main difference between antimicrobial and autoimmune responses is that, in autoimmune disease, autoantigen is always present and in most cases cannot be removed from the body. |
| **Autoantibodies can directly mediate cell destruction** | Autoantibodies can bind to self cells and, either alone or with complement, cause damage mediated mainly through opsonization via Fc and C3 receptors on phagocytic cells. For example, IgG autoantibodies bind to red blood cells in autoimmune hemolytic anemia (AIHA) or to platelets in immune thrombocytopenic purpura (ITP) and mediate phagocytosis of these self cells. |
| **Autoantibodies can modulate cell function** | Antibodies to certain self cell-surface molecules can either interfere with or enhance the functional activity of the cell. For example, antibodies to the acetylcholine receptor in myasthenia gravis block their effective interaction with acetylcholine. In Graves' disease, antibodies to the TSH receptor overstimulate the thyroid. |
| **Autoantibodies can form damaging immune complexes** | Circulating immune complexes, whether composed of autologous or foreign antigens, can result in damage to tissue by complement activation and by triggering release of mediators from Fc receptor-bearing cells (type III hypersensitivity). Immune complexes can deposit in the glomeruli, especially in SLE, or in blood vessels, leading to kidney damage or vasculitis respectively. |
| **Cell-mediated immunity in pathogenesis** | Although autoantibodies have been clearly linked to autoimmune disease, cell-mediated immunity also plays an essential part in pathogenesis in some, if not all, autoimmune disorders. Inflammatory T-cell infiltrates are a hallmark of organ-specific diseases such as diabetes and multiple sclerosis. Their importance is indicated by studies showing that T cells can transfer particular autoimmune diseases in animal models. |

| Related topics | (B1) Cells of the innate immune system | (K4) Immune-complex-mediated (type III) hypersensitivity |
|---|---|---|
| | (D8) Antibody functions | |
| | (K3) IgG- and IgM-mediated (type II) hypersensitivity | (K5) Delayed (type IV) hypersensitivity |

## Tissue-damaging reactions in autoimmune diseases

The inflammatory processes underlying the tissue damage that occurs in autoimmune disease are the same as those responsible for the protective role of the immune system, but become "chronic" because the "antigen" is not cleared. The inflammatory infiltrate usually consists of T cells, macrophages, neutrophils, B cells, mast cells, and in some instances plasma cells. However, the nature of the primary insult, whether microbial or other, and site of the target tissue may influence the type of cellular infiltrate. For example, increased numbers of mast cells, eosinophils, lymphocytes, and plasma cells may be a feature of gastrointestinal-associated autoimmune diseases such as celiac disease and Crohn's disease, whereas, in the pancreas of a person with diabetes the cellular infiltrate may be mainly mononuclear cells, that is, lymphocytes and macrophages. Some autoimmune diseases, such as Goodpasture syndrome, are caused by autoantibodies to lung and kidney basement membranes, which lead to renal failure. Immune complexes become deposited in the kidney, also leading to kidney failure, in SLE (Section K4). Paradoxically, immunodeficiency is often associated with an increased incidence of autoimmune disease. Autoimmune diseases are driven by antigen and when this is removed in experimental animals or man the autoimmune response subsides, for example, removal of the thyroid gland in Hashimoto's thyroiditis removes the source of autoimmune stimulation and autoantibodies are no longer produced.

## Autoantibodies can directly mediate cell destruction

Autoantibodies can bind to self cells and either alone or with complement cause damage. This can be mediated through opsonization via Fc receptors or C3 receptors on phagocytic cells. An example of this is IgG autoantibodies binding to red blood cells in autoimmune hemolytic anemia (AIHA), or to platelets in immune thrombocytopenic purpura (ITP; Figure 1). The Fc-mediated mechanism appears to be more important, since successful therapy (e.g., with immunosuppressive steroid hormones—glucocorticoids or corticosteroids) coincides with decreased Fc receptors on monocytes and macrophages, but not with lower autoantibody titers. Furthermore, the injection of high amounts of non-immune IgG in AIHA decreases cell destruction, an effect partly due to blocking of the Fc receptors on the body's phagocytes. Autoantibodies can also bind directly to cells in tissues. For example, in Goodpasture's syndrome, IgG antibodies bind to the basement membranes of kidney and lungs, attracting phagocytes, which release enzymes that damage these tissues (frustrated phagocytosis).

## Autoantibodies can modulate cell function

Antibodies to certain self cell-surface molecules can either interfere with or enhance the functional activity of the cell. For example, myasthenia gravis (MG) is characterized by weakened and easily tired muscles. Serum antibodies directed against muscle, and in particular antibodies to the acetylcholine receptor, play a key role. These antibodies not only block the acetylcholine-binding sites, but appear to act by cross-linking the receptor

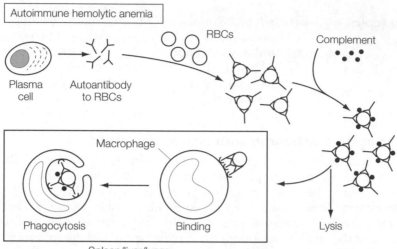

Figure 1. Autoantibody-mediated removal of erythrocytes (in AIHA) or platelets (in ITP). In autoimmune hemolytic anemia (AIHA), autoantibody to red blood cells (RBCs) binds to the RBC. As these antibody-coated cells pass through the spleen, liver, and lungs, they are recognized and bound by Fc receptors for IgG on macrophages in these organs. The RBCs are phagocytosed by these macrophages and destroyed. Similarly, in idiopathic thrombocytopenia (ITP), which is mediated by autoantibody to platelets, the antibody-coated platelets are removed and destroyed. Complement may also play a role in lysing autoantibody-coated RBCs or platelets and/or in opsonizing these self cells for phagocytosis by macrophages.

so that it becomes nonfunctional (Figure 2 in Section K3). This is an example of type II hypersensitivity. The opposite is true in Graves' disease, an autoimmune thyroid disease in which autoantibodies stimulate rather than inhibit receptor function. Both thyroid growth-stimulating immunoglobulin (TGSI; an example of type V hypersensitivity) and thyrotropin-binding inhibitory immunoglobulin (TBII) have been demonstrated. TBII, by binding to receptors for thyroid-stimulating hormone (TSH, thyrotropin), stimulates the thyroid gland to make high levels of thyroid hormone, resulting in hyperthyroidism. IgG autoantibodies can cross the placenta and can cause transient hyperthyroidism in the newborn of women who have Graves' disease and MG in the newborns of mothers with MG. It would appear that in MG and Graves' disease, only B cells specific for a few bodily components are activated. The defect may therefore lie with a very small subset of T or B cells. Since total antibody titer does not correlate well with disease state, antibody class and subclass (e.g., C′ binding or nonbinding) may be a crucial consideration.

## Autoantibodies can form damaging immune complexes

Circulating immune complexes, whether composed of autologous or foreign antigens, can result in damage to tissue by complement activation and triggering release of mediators from Fc-receptor-bearing cells (type III hypersensitivity, Section K4). Immune complexes may also perturb normal immunoregulation, perhaps through triggering Fc receptors on lymphocytes. For example, although SLE may involve some target-cell-specific autoantibodies (e.g., to erythrocytes), the most life-threatening manifestation of SLE is usually kidney damage, which results from the deposition of soluble immune complexes in the glomeruli. Immune complexes may also deposit in blood vessels, leading to vasculitis. Since autoantibodies are produced to many bodily components, there may be a generalized defect in self tolerance similar to the Fas/FasL apoptotic defects seen in certain autoimmune (lpr and gld) strains of mice. Antibodies to T cells are common as well, and may contribute to progression of the disease.

## Cell-mediated immunity in pathogenesis

Although autoantibodies play a significant role in autoimmune disease, cell-mediated immunity also plays an essential part in pathogenesis in some, if not all, autoimmune disorders. In particular, T cells are involved not only in the development of autoimmune disease but also in tissue inflammation. For example, inflammatory T-cell infiltrates are a hallmark of organ-specific diseases, such as diabetes and MS, and are also present in skin lesions in SLE. However, a clear understanding of their involvement in autoimmune pathogenesis has been complicated by the MHC-restricted nature of T-cell recognition and the difficulty in isolating these T cells and in identifying their target antigens. In animal models, using inbred populations, it has been possible to clone autoimmune T cells that are able to transfer the autoimmune disease to other animals. For example, injection of myelin basic protein has been shown to induce experimental allergic encephalomyelitis (EAE) in rats, a disease very like MS in humans. Both encephalogenic and tolerogenic peptides to which T cells bind have been identified and either disease or protection against disease can be transferred to other rats of the same inbred strain with different cloned T cells. In general, clones making Th2 cytokines are protective whereas those making Th1 cytokines elicit disease. Thus, it is clear that T cells play a central role in both pro- and anti-inflammatory aspects of autoimmune disease, and that their MHC restriction, peptide specificity, and Th1/Th2 and Th17 cytokine profiles are important contributors to pathogenesis.

# L5 Diagnosis and treatment of autoimmune disease

---

**Key Notes**

| | |
|---|---|
| **Diagnosis** | Diagnosis of autoimmune disease is through clinical and laboratory criteria that differ for each disease. Autoantibodies to a variety of autoantigens are detected using immunofluorescence on tissue sections and ELISA techniques. These techniques allow the detection of the IgG Abs to double-stranded DNA that are characteristic of SLE, IgA antibodies to tTG that are diagnostic of celiac disease, rheumatoid factor found in RA patients, and a variety of other autoantigens. |
| **Replacement therapy** | In some cases, critical self antigens are compromised by the autoimmune process and may need to be replaced. In thyroid autoimmunity, the patient is treated with thyroid hormones. In myasthenia gravis, inhibitors of enzymes which break down acetylcholine are used. In diabetes, insulin is given to replace that lost by damage to islet cells. |
| **Suppression of the autoimmune process** | Current treatments are aimed at nonspecific suppression of the autoimmune inflammatory process. These include nonspecific aspirin-like drugs (nonsteroidal anti-inflammatory drugs; NSAIDs) or glucocorticoids, used to dampen inflammation, and plasmapheresis to remove autoantibodies. Cytotoxic drugs, cyclosporine, and monoclonal antibodies to T or B cells are also used to modulate or eliminate autoreactive lymphocytes. Drugs targeting cytokines (or their receptors) are used to treat a number of autoimmune diseases. |
| **Related topics** | (D5) Monoclonal antibodies<br>(D6) Antigen–antibody complexes (immune complexes)<br>(D7) Immunoassay<br>(Q3) Antibody-mediated immunotherapy |

---

## Diagnosis

Diagnosis of autoimmune diseases is through clinical and laboratory criteria that differ for each disease. In the clinical laboratory, autoantibodies to a variety of autoantigens are detected using immunofluorescence techniques on tissue sections or ELISA (Section D7). For example, sera containing antinuclear antibodies (ANAs) found in a number of autoimmune diseases can be detected on thyroid tissues, as can antibodies to thyroid peroxidase, which are characteristic of Hashimoto's thyroiditis. Antibodies to neutrophil cytoplasmic antigen (ANCAs) are detected by immunofluorescence on normal neutrophils; their presence indicates a diagnosis of Wegener's granulomatosis. However, ELISA is the most common technique used for detecting autoantibodies, allowing for automation and measurement of many sera for many different autoantibodies. These include measurement of IgG antibodies to (i) double-stranded DNA in the sera of SLE patients; (ii) the acetylcholine receptor in myasthenia gravis; (iii) the TSH receptor in Graves' disease; (iv) pyruvate dehydrogenase in primary biliary cirrhosis; (v) GAD in type 1 diabetes; and (vi) IgA tTG antibodies in patients with celiac disease. Rheumatoid factor (RF), an autoantibody specific for the Fc region of IgG that is found in 70% of patients with RA, can also be measured by ELISA.

## Replacement therapy

In some cases the autoantigen that is being removed either directly by the autoimmune response (e.g., vitamin $B_{12}$ in pernicious anemia or, thyroid hormones in autoimmune thyroiditis) or indirectly by immune damage (e.g., insulin in diabetes) may need to be given back to the patient. This includes platelets in autoimmune thrombocytopenias, thyroid hormones in thyroid autoimmunity, vitamin $B_{12}$ in pernicious anemia, and insulin in insulin-dependent diabetes (Table 1).

## Suppression of the autoimmune process

Current treatment is aimed at non-autoantigen-specific inhibition of the ongoing inflammatory response (Table 1). Nonspecific aspirin-like drugs (nonsteroidal anti-inflammatory drugs; NSAIDs) or glucocorticoids are often used to dampen inflammation. Removal of autoantibodies and immune complexes from the blood and replacement of patient

## Table 1. Therapy of autoimmune diseases

| | |
|---|---|
| Replacement of targeted autoantigen | For example, thyroid hormone for thyroid autoimmune disease, insulin for type 1 diabetes |
| Nonsteroidal anti-inflammatory drugs (NSAIDs), e.g., aspirin, ibuprofen | Inhibit prostaglandins: RA and others |
| Corticosteroids, e.g., prednisone | Anti-inflammatory |
| Cytotoxic drugs | |
|    Azathioprine | Inhibits cell division, suppresses T cells |
|    Cyclophosphamide | Blocks cell division, inhibits antibody production |
| Cyclosporine (ciclosporin) | Inhibits T-cell cytokine IL-2 production |
| Monoclonal antibodies to CD20 | In drug-resistant RA |
| Inhibitors of TNFα | In drug-resistant RA |
| Inhibitors of inflammatory cytokines | Psoriasis, RA |
| Molecules involved in T-lymphocyte traffic | Treatment of MS |
| **Experimental – not yet approved** | |
| Collagen | Treatment of RA |
| Small peptides from myelin basic protein | Treatment of MS |

RA, rheumatoid arthritis; MS, multiple sclerosis

plasma with plasma from normal donors (plasmapheresis) can be useful, but has a short-lived effectiveness. Since TNF plays a major role in pathogenesis of several auto-immune diseases, synthetic receptors for this cytokine or monoclonal antibodies to it are now being used to treat RA, ankylosing spondylitis, psoriatic arthritis, and inflammatory bowel disease. Drugs that inhibit cytokine release by T cells (e.g., cyclosporine) are also used to treat some autoimmune diseases. Cytotoxic drugs used to treat tumors are used in severe cases of autoimmune disease to eliminate the autoantigen-specific T and B cells that are the origin of the disease. Other more specific drugs are also used to eliminate specific T and B cells. For example, monoclonal antibodies to CD20 on B cells reduce autoantibodies and are used to treat RA, SLE, primary Sjögren syndrome, AIHA, and other autoimmune diseases (Section Q3). However, care must be taken, especially with relatively nonspecific immunosuppressive drugs, to avoid elimination of important immune cells that could lead to secondary immunodeficiency (Section J3).

Autologous hematopoietic stem cell transplants (HSCT) have been successfully used in the treatment of some severe autoimmune diseases, including MS, systemic sclerosis, SLE, RA, juvenile idiopathic arthritis, and immune cytopenias. Many patients have experienced long-term disease-free remissions.

Monoclonal antibodies that specifically target immune cells, especially T or B cells, have the capability to eliminate lymphocytes responsible for the disease. However, it is important to avoid nonspecific elimination of too many of these important immune cells as doing so may lead to secondary immunodeficiency (Section J3). Although induction of specific immune tolerance to particular autoantigens would be ideal, more than one autoantigen is often involved and induction of tolerance is very difficult to achieve during an ongoing immune response (Section G3). However, in celiac disease, since the inducer of the disease is known (gluten), its removal from the diet leads to a return to clinical normality and the cessation of autoantibody production to the target enzyme, transglutaminase.

Use of the oral route for induction of systemic tolerance and reinstatement of tolerance to autoantigens has thus far not proven successful in man. However, experimental treatments that interfere with T-cell help during cognate interactions with antigen and/or that enhance Treg function seem very useful avenues to explore. In fact, in animal models of autoimmune disease, introduction of Tregs expanded *ex vivo* has shown decreased inflammation and may be a promising approach for treatment of at least some human autoimmune diseases.

# M1 The transplantation problem

## Historical perspective

Skin grafts were used to treat major wounds acquired during the Second World War, and it was from this experience that the early concept of transplantation rejection was founded. This led to the now widely known fact that transplantation of donor organs/tissues to another individual usually results in rejection, unless histocompatible tissues (based on specific tissue typing) and immunosuppression are used. Early experiments in mice in the 1950s and 1960s defined the role of the major histocompatibility molecules in graft rejection. Transplantation is now common in medical practice and many different organs/tissues are transplanted successfully (Table 1).

**Table 1. Commonly transplanted organs/tissues**

| Allografts | Autografts |
| --- | --- |
| Kidney, pancreas, heart (heart/lung), skin, cornea, bone marrow, liver, blood | Skin, bone marrow |

## Types of grafts

Tissues/organs transplanted from one part of the body to another (autografts) are not rejected, because they are self. Transplants of tissue/organs from an individual within the same species are called allografts (e.g., human to human) and those from one species to another are called xenografts (e.g., pig to human). Human transplants are usually allografts and autografts. Xenografts have for some time been considered as an alternative to allografts, because of inadequate supplies of human donor organs/tissues. However, due to ethical issues, significant likelihood of rejection, and the possible transfer of pathogens (e.g., viruses) to humans, this type of transplantation has been "put on the back burner."

## The major problem of rejection

That the immune system is responsible for the rejection process has been demonstrated in animal models and in humans. The immune mechanisms used for rejection are the same as those used in immune responses to invading microbes and are mainly adaptive immune responses. The cause of the problem is genetic polymorphism, and in particular that the transplantation antigens are predominantly polymorphic gene products, for example blood groups and major histocompatibility complex (MHC) molecules, which vary among different individuals within the same species. Rejection can be minimized by using familial donors, tissue typing, and immunosuppressive drugs. Bone marrow transplantation given as a source of stem cells can result in life-threatening graft-versus-host reactions.

# M2 Transplantation antigens

## Key Notes

| | |
|---|---|
| **The blood group antigens** | The major blood group antigens are those of the ABO system. These carbohydrate antigens are present on erythrocytes and some other tissues. Most individuals have antibodies (isohemagglutinins) which recognize these antigens. Thus, blood group A individuals have antibodies to blood group B, and blood group B individuals antibodies to blood group A. Blood transfused from one group to the other would be rejected. |
| **The major histocompatibility complex antigens** | The main tissue transplantation antigens are encoded by the polymorphic MHC locus (HLA in man). The inheritance of two alleles (out of many possible) at six different loci (A, B, C, DP, DQ, DR) means that the chance of all HLA antigens of two unrelated individuals being exactly the same is extremely low. |
| **Minor histocompatibility antigens** | Minor transplantation antigens include non-ABO blood group alloantigens and antigens associated with the sex chromosomes, for example Y antigen of the male chromosome. These are usually "weaker" than the MHC antigens, and may be the antigens targeted by the immune system in late-onset rejection. |
| **Related topics** | (A4) Antigens<br>(F2) T-cell recognition of antigen<br>(G5) Regulation by T cells and antibody<br><br>(G6) Neuroendocrine regulation of immune responses |

## The blood group antigens

The major blood group ABO antigens are mainly present on the surfaces of erythrocytes (although they are also expressed on the endothelial cells of the kidney) and the genes encoding them are polymorphic, that is, there is more than one allele coding for the gene product. This is in contrast to most proteins, for example albumin, which are coded for by nonpolymorphic genes or genes which lack allelic variation. The major blood group alleles A and B code for enzymes which create different sugars on proteins and lipids on the surface of erythrocytes. Blood group O is a null allele and does not add sugars. These alleles are inherited in a simple Mendelian inheritance pattern and are co-dominantly expressed (i.e., both allelic products are expressed on the erythrocyte surface, Table 1). An individual can be either homozygous (the same) or heterozygous (different) for the inherited alleles.

The major problem with transplanting blood is that all of us (except those having AB blood group) have antibodies (isohemagglutinins) to these blood group antigens (Table 1). The reason for development of these antibodies is unclear, but is probably due to

**Table 1. Blood group antigens and isohemagglutinins**

| Blood group | A | B | AB | O |
|---|---|---|---|---|
| Genotype | AA or AO | BB or BO | AB | OO |
| Isohemagglutinins | Anti-B | Anti-A | None | Anti-A and B |

cross-reactivity of AB antigens with those of certain ubiquitous microbes (Section E3). Transplantation of blood to a recipient who has serum isohemagglutinins can result in a severe transfusion reaction mediated by a type II hypersensitivity reaction (Section K3) that can lead to death.

### The major histocompatibility complex antigens

These are the major barrier to transplantation of nucleated cells. As previously described, MHC molecules are expressed on all nucleated cells of the body and their physiological function is to direct T cells to carry out their function. However, like the locus coding for the major blood group antigens and unlike the majority of other gene products, genes coding for MHC molecules are polymorphic. In contrast to the ABO system, each MHC locus can encode for a very large number of different allelic forms and, to further increase the complexity, there are six different loci. In humans, these loci are found on chromosome 6 (Figure 1); this is called the human leukocyte antigen (HLA) system since in humans the antigens were first discovered on leukocytes.

The combinations of the many different allelic variants which are co-dominantly expressed means that the chance of two unrelated individuals having a completely identical set of alleles is extremely remote (less than the odds of winning the national lottery!). Thus the different allelic variants of the donor organ/tissue will be foreign to the recipient

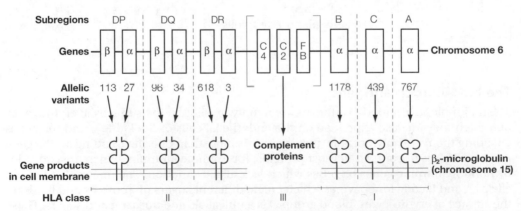

Figure 1. The human major histocompatibility locus. Class I and class II human leukocyte antigens (HLA) are encoded by three genes (A, B, and C) and six genes (DP, DQ, and DR), respectively. Each gene can be coded by many different alleles and this figure shows the multiple variants detected for each gene locus, the products of which, if different from self, are recognized as transplantation antigens. Thus, there are millions of different combinations of the different allelic products. The class III HLA locus encodes complement proteins. Other HLA genes have much less polymorphism and are involved in antigen processing and presentation and as inhibitory molecules (not shown; see Figure 5 in Section F2).

who does not have them and will therefore generate an immune response to them. Examples of alleles that might be expressed by donors/recipients are shown in Table 2 and the target of the recipient's immune system would be the products of the mismatched alleles.

**Table 2. Human leukocyte antigens (HLA): alleles of a hypothetical donor and recipient**

| Locus | Donor | Recipient | Alleles to which the recipient's immune system responds |
|-------|-------|-----------|---------------------------------------------------------|
| HLA-A | A2/A2* | A6/A2 | None** |
| HLA-B | B21/B26 | B23/B8 | B21, B26 |
| HLA-C | C5/C8 | C9/C4 | C5, C8 |
| HLA-DR | DR4/DR6 | DR8/DR3 | DR4, DR6 |
| HLA-DP | DP3/DP1 | DP2/DP1 | DP3 |
| HLA-DQ | DQ3/DQ3 | DQ4/DQ2 | DQ3 |

*Alleles defined by serotyping: **The recipient's immune system sees A2/A2 as self.

HLA alleles were first defined by serotyping, that is, identifying epitopes through antibodies produced to them following allografting or in the serum of multiparous mothers exposed to several fetuses with different HLA alleles. HLA typing has become more sophisticated, defining more precisely the different variants within the same allele, for example, HLA-B44 can be subdivided into HLA-B*4401, HLA-B*4402, HLA-B*4403, and so forth. This is important because even a single amino acid difference within an allelic product can induce an immune response to a transplant. Thus, molecular techniques, including gene sequencing, have defined multiple allelic forms that can potentially give rise to immune responses to transplants (Figure 1).

## Minor histocompatibility antigens

There are a number of minor transplantation antigens that include non-ABO blood group alloantigens, antigens associated with the sex chromosomes, for example the Y chromosome, and several others that have not been fully defined but are present in most tissues. These are usually "weaker" than the MHC antigens but can be targets for host responses against grafts or for grafts reacting against host tissues (Section M3).

# M3 Rejection mechanisms

---

**Key Notes**

| | |
|---|---|
| **Mechanisms of allograft rejection** | Rejection is mainly mediated by the adaptive immune system that recognizes the mismatched HLA allelic products expressed on donor tissues through its T- and B-cell receptors and is responsible for rejection. Both antibody-mediated and T-cell-mediated (CMI) rejection occur, depending on the source of tissue for the transplant (e.g., mainly CMI for skin, and both antibodies and CMI for kidney). The number of HLA mismatches between donor and recipient (i.e., transplantation antigens) usually determines the strength of rejection. |
| **Xenotransplant rejection** | Due to the inadequate supply of human donors, animals have been considered as an alternative source of organs/ tissues. The pig is deemed appropriate because the sizes of many of its internal organs are comparable to those of humans. Hyperacute rejection problems have arisen due to the presence in the pig of cell-surface sugars to which humans have natural hemagglutinins, similar to those against ABO antigens. |
| **Donor rejection of host tissues (graft-versus-host reactions)** | In addition to host rejecting graft tissue, T cells in bone marrow grafts are stimulated by mismatched host HLA leading to a graft-versus-host reaction. Care to avoid this response is required in using bone marrow as a source of stem cells in cases of anemia, metabolic diseases of the newborn, primary immunodeficiency, and tumors, especially leukemias. |
| **Related topics** | (E3) The cellular basis of the antibody response  (F1) The role of T cells in immune responses  (J1) Deficiencies in the immune system  (M4) Prevention of graft rejection |

---

## Mechanisms of allograft rejection

The immune system treats mismatched transplants in the same way as microbes. Thus, if a patient rejects a transplant through transplantation antigens, it will reject a second graft carrying the same or shared transplantation antigens much faster. This "second set" rejection is due to the sensitization by the first graft and a memory response on subsequent exposure. This is a property of the adaptive immune system.

Graft rejection is mediated by both cellular (T cell) and humoral immune mechanisms (antibodies). Furthermore, the number of mismatched alleles also determines the magnitude of the rejection response. The more mismatches, the larger the number of allo-antigens to which an immune response can be made. Thus in Section M2 in Table 2, the recipient's immune response could respond to eight different donor transplantation antigens. Although both T-cell-mediated responses and antibodies can be generated against the foreign antigens, the rejection of particular types of graft may be preferentially mediated more through antibodies than through T-cell-mediated immune (CMI) responses and *vice versa* (Table 1).

**Table 1. Main mechanisms of rejection of different kinds of grafts**

| Organ/tissue | Mechanism(s) |
| --- | --- |
| Blood | Antibodies (isohemagglutinins) |
| Kidney | Antibodies, CMI (T-cell) |
| Heart | Antibodies, CMI (T-cell) |
| Skin | CMI (T-cell) |
| Bone marrow | CMI (T-cell) |
| Cornea | Usually accepted unless vascularized, CMI (T-cell) |

There is evidence that the innate immune system, and particularly NK cells, also plays a role in both solid graft and stem cell graft rejection, as these cells have been shown to infiltrate cardiac grafts. They are activated by the absence of MHC molecules on target cells, and their inhibitory receptors are triggered by specific alleles of MHC class I molecules on cell surfaces. Thus, it is likely that they will damage some grafts resulting in their rejection.

There are three main types of allograft rejection patterns, as shown in Table 2 for the kidney, the best studied transplant.

● **Hyperacute rejection** occurs within a few minutes or hours and is believed to be mediated by pre-existing circulating antibody in the recipient to antigens of the donor. Unlike other transplants, the kidney has ABO-coded sugar antigens expressed on the endothelial cells of the blood vessels. Thus, if the donor has a different blood group from the recipient, the antibodies will result in a type II hypersensitivity reaction in the kidney graft (Section K3). Graft recipients might also have some memory responses to HLA through rejection of a previous graft. In addition, multiparous women recipients may have been sensitized to paternal HLA expressed by their child's cells. This could occur during pregnancy and at parturition when small amounts of blood of the newborn may get into the maternal circulation. Prior transfusion with blood containing some leukocytes of a recipient can also result in priming to HLA alleles.

● **Acute rejection** occurs within the first weeks or months following transplantation. The graft shows infiltrates of activated lymphocytes and monocytes. Antibody may be a factor in the process, but the effector mechanism is thought to be primarily through cytotoxic T cells or helper/delayed type hypersensitivity T cells (Section K5) and cytokines produced by monocytes/macrophages.

● **Chronic rejection** is the gradual loss of function of the grafted organ occurring over months to years. The lesion often shows infiltration with large numbers of mononuclear cells, predominantly T cells. The mechanism of rejection is not clear, but following transplantation, memory (and primary) responses that generate antibody and cellular immunity to HLA may take some time, especially because the patient will be immunosuppressed to improve graft "take" (Section M4) and there might be only a

**Table 2. Kidney graft rejection**

| Type of rejection | Time to rejection | Cause |
|---|---|---|
| Hyperacute | Within hours | Preformed antibodies (anti-ABO and/or anti-HLA) |
| Acute | Weeks to months | Cell-mediated (CD8$^+$, CD4$^+$ T cells) |
| Chronic | Months to years | Cell-mediated (CD8$^+$ T cells), responses to minor transplantation antigens |

limited number of mismatched alleles. Furthermore, minor transplantation antigens may eventually produce a sufficiently large immune response to result in rejection.

## Xenotransplant rejection

The inadequate supply of donor organs/tissues has led to consideration of animals as donors. In particular, the pig appears to be a suitable source of transplantable tissues because the sizes of many of its internal organs are comparable to those of humans. However, a major unforeseen problem is that pig cells have sugars which are not found on human cells and to which humans have serum IgM hemagglutinating antibodies (similar to the ABO isohemagglutinins, Table 1 in Section M2). Thus, pig organs will be rejected through a hyperacute mechanism due to preformed hemagglutinins which activate complement resulting in lysis of the grafted cells. Strategies planned to prevent this include:

● Trying to inactivate the gene encoding the glycosyltransferase responsible for the sugar residues.

● Introducing genes into the pig which code for molecules that inhibit the lytic component of complement activation (Sections D8 and G1).

Even if these strategies are successful, there is still the problem of the MHC molecules expressed by pig tissues. Moreover, animal sources of grafts have additional problems, including ethical issues and the possibility of transferring unknown viruses (zoonoses) that, in the long term, could enter the germ line.

## Donor rejection of host tissues (graft-versus-host reactions)

Rejection of transplanted organs and tissues is the result of the immune system of the recipient recognizing and responding to the donor's HLA (host-versus-graft response). However, in the case of bone marrow transplants there is an additional problem in that the graft (the bone marrow containing stem cells) will contain viable lymphocytes. In particular, T cells in the graft may recognize mismatched HLA alleles and produce a graft-versus-host reaction (GVH; Table 3).

**Table 3. Host-versus-graft and graft-versus-host reactions**

| Host-versus-graft reaction | Graft-versus-host reaction (GVH) |
|---|---|
| Response to donor HLA by host immune system | Response to recipient HLA by donor's T cells |

This often results in skin rashes and gastrointestinal problems and can be life-threatening. Like host-versus-graft reactions, the more HLA mismatches, the stronger the GVH reaction. These responses can also be seen against minor transplantation antigens. The pathology is probably mediated by inflammatory cytokines released from the donor T cells. Bone marrow stem cells are given for a number of clinical conditions to provide functional genes. These conditions include some primary immunodeficiency diseases, anemias, tumors, and metabolic diseases (Table 4). Bone marrow grafts are also being used to treat a variety of solid tumors, for example breast cancer, and some autoimmune diseases, for example rheumatoid arthritis, following heavy chemotherapy/irradiation to remove the tumor and lymphoid cells, respectively. As an alternative to bone marrow, stem cells for treatment of a number of diseases can be prepared from the blood of donors, by driving them into the circulation with cytokines (Section B2), or from umbilical cord blood, followed by further purification of these cells.

**Table 4. Some clinical conditions for which bone marrow (stem cell) grafts are given**

| Anemias | Metabolic diseases | Immunodeficiency diseases | Tumors |
|---|---|---|---|
| Fanconi's anemia | Gaucher's disease | Reticular dysgenesis | Acute lymphoblastic leukemia |
| | Thalassemias | SCID* | Acute myeloid leukemia |
| Aplastic anemia | Osteopetrosis | Chronic granulomatous disease | Chronic myeloid leukemia |
| | | Wiskott–Aldrich syndrome | Chronic lymphocytic leukemia |

*Severe combined immunodeficiency.

# M4 Prevention of graft rejection

## Key Notes

| | |
|---|---|
| **Familial grafting** | Due to the inheritance pattern of the HLA genes, transplantation within families greatly reduces HLA mismatches. Transplants from parents to siblings have at least a 50% match of HLA alleles, whilst sibling-to-sibling grafts have a 25% chance of having identical HLA alleles. |
| **Tissue typing** | Typing of the HLA of both transplant donor and recipient can be done by antibodies or "typing" cells. Molecular-genetics-based techniques are also now widely used. |
| **Cross-matching** | Cross-matching is used to test for preformed antibodies in the recipient directed to donor tissues. This is measured by mixing serum from the recipient with blood lymphocytes from the donor. |
| **Immuno-suppression** | Suppression of the immune system by drugs is usually necessary to aid in maintenance of the graft. The drugs used include corticosteroids, antiproliferative drugs (e.g., azathioprine), and inhibitors of T-cell activation (e.g., cyclosporine). |
| **The special case of the "fetal transplant"** | The fetus is an allograft, and yet in most cases it is not rejected. There are probably several mechanisms involved in suppression of the rejection process, including lack of expression of conventional HLA on the trophoblast, complement inhibitory proteins expressed by the trophoblast, and immunosuppressive molecules produced in the placenta. |
| **Related topics** | (D8) Antibody functions      (M3) Rejection mechanisms |

## Familial grafting

Transplantation within families significantly reduces allele mismatches because of the inheritance patterns of HLA (Figure 1). In general, there is little crossover within the locus and the whole locus is usually inherited *en bloc*. Thus, if parents donate grafts to their children there is an equal to or greater than (due to chance) 50% match of the HLA alleles. If siblings (brothers and sisters) donate to each other there is a one in four chance of a complete match. Thus, if you need a transplant, make sure you come from a family with lots of brothers and sisters! Other tissue antigens that trigger far less vigorous rejection responses (minor histocompatibility antigens) are encoded outside the MHC locus and include male-specific antigens. In fact, mismatches of minor transplantation antigens can be important in determining the fate of grafts between HLA-matched donor and recipient, especially as it relates to chronic rejection over a longer period of time.

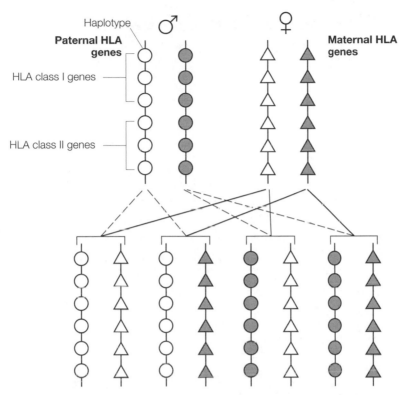

Figure 1. Inheritance of HLA genes. Each individual receives one set of HLA genes from each parent (i.e., they receive one haplotype from each parent). Because of their position on the chromosomes, alleles are inherited *en bloc*. Grafts from parents to siblings and *vice versa* have at least 50% of matched alleles while sibling-to-sibling grafts have a one in four chance of a complete match.

## Tissue typing

If a familial donor is not available, then the extent of the mismatches between alleles must be determined by tissue typing, in order to best match donor and recipient. In this context, one of the most useful assays involves cytotoxic antibodies (usually monoclonal antibodies) to individual HLAs. The principal of the antibody method depends on the surface expression of the HLA. Donor and recipient blood for typing are enriched for B cells (they express both class I and II HLA) and specific cytotoxic antibodies are added. Binding of the antibody to a surface HLA in the presence of complement results in the direct killing of the B cells (Figure 2). These can be microscopically scored. Using a panel of antibodies, each of which is directed to a different HLA allele, it is possible to HLA-type for the majority of alleles.

Many HLA-typing laboratories are now turning to identification of the inherited HLA genes via molecular-genetics-based tests that utilize the restriction fragment length polymorphism (RFLP) or polymerase chain reaction (PCR) amplification techniques. These technologies determine the nucleotide sequence of the HLA genes in question and give unequivocal results. In addition to its use in tissue typing for transplants, this technology has been particularly important in identifying minor polymorphisms within the HLA class II genes, which might be associated with susceptibility to particular kinds of diseases (Section L2).

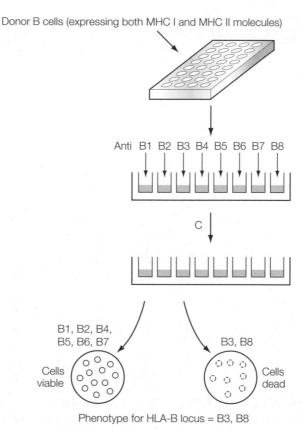

Donor B cells (expressing both MHC I and MHC II molecules)

Anti  B1  B2  B3  B4  B5  B6  B7  B8

C

B1, B2, B4,
B5, B6, B7                                B3, B8

Cells                                              Cells
viable                                             dead

Phenotype for HLA-B locus = B3, B8

Figure 2. Tissue typing. B cells obtained from the blood of the donor/recipient to be typed are placed in microplates and antibodies to the different MHC allelic products added together with complement. These include antibodies to HLA-A, -B, and -C loci and some D antigens. Only antibodies to B1 to B8 are shown here to illustrate the concept. Following incubation at 37°C, lysis (cell death) of the B cells occurs in those wells where antibodies have attached to the B cells. Thus, in this example, death of the B cells indicated that the donor was heterozygous for the B locus—B3 and B8.

Typing can also be done using the "mixed lymphocyte" reaction also called "mixed lymphocyte response," which primarily identifies HLA-D class II antigens. In this case, "typing cells" (usually cell lines carrying specific homozygous HLA-D allelic products) are treated with a drug to inhibit their proliferation. They are then mixed with the potential recipient's blood lymphocytes and cultured for 3–5 days. If the recipient's T cells do not carry the typing cell's HLA, they will proliferate in response to "foreign" HLA because they will not have been eliminated by negative selection in the thymus (Section G2). By using panels of typing cells, it is possible to determine the HLA type of the donor and recipient.

Matching of HLA for liver transplants does not appear to be of major advantage, probably due to the weak expression of HLA by hepatic cells.

## Cross-matching

Cross-matching is used to check that there are no preformed antibodies to donor HLA in the recipient. Blood lymphocytes from the donor are mixed with serum from the recipient (Figure 3). Anti-donor antibodies are detected by complement-mediated lysis of the cells or by using fluorescent staining and flow cytometry. The presence of such antibodies is contraindicatory to the use of the tissues from that donor. Cross-matching for blood groups is also important for renal transplants (Section M3).

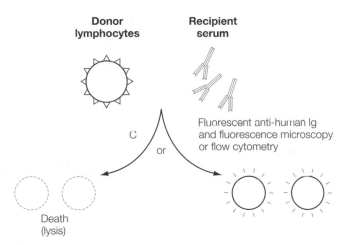

**Figure 3.** Cross-matching. Serum from the potential recipient is mixed with donor lymphocytes and is evaluated for lysis (Figure 2) in the presence of complement, or stained with fluorescent antibodies to human immunoglobulin (Section D7) and assayed by fluorescence microscopy or flow cytometry. Dead cells or positive fluorescence signifies the presence of anti-donor antibodies which could lead to a hyperacute rejection of the graft. This is contraindicatory to the use of this donor/recipient combination. This assay identifies HLA antibodies in the recipient serum. Cross-matching for blood groups is also carried out for renal transplants.

## Immunosuppression

In the vast majority of cases, there will be some HLA allelic mismatches and some donor minor histocompatibility antigen mismatches; therefore the immune system of the recipient has to be suppressed to avoid rejection. The mainstay drug treatment is a mixture of corticosteroids, antiproliferative drugs, and cyclosporine (ciclosporin) (a fungal nonapeptide). The mechanisms of immunosuppression by these and other drugs used are shown in Table 1. Not surprisingly, a major problem with these drugs is that by inhibiting the immune response against the graft they can also lead to increased susceptibility to infection (Section J1). In fact, infection and rejection are the main reasons for the failure of kidney grafts to be maintained. Drugs that inhibit T-cell activation, other than cyclosporine A, are also used to prevent graft rejection. In addition, some monoclonal antibodies (mAbs) are now licensed for treatment of transplant patients.

In order to achieve the successful acceptance of an allograft, it is necessary to achieve "transplantation tolerance." A number of experimental models and clinical experiences have shown that regulatory T cells (Tregs) are important to the successful outcome of grafting. Clinical studies using different treatment regimens appear to preferentially induce Tregs. This is a growing research area in transplantation that has as its goal not only the prevention of host-versus-graft reactions but GVH reactions as well.

## Table 1.  Drugs used to suppress graft rejection

| Drug | Mechanism(s) of immunosuppression |
| --- | --- |
| **Corticosteroids** | |
| Prednisone | Blocking of migration of neutrophils: Inhibition of IL-1, IL-6, and IL-2 production, lysis of immune cells |
| **Anti-proliferative drugs** | |
| Azathioprine | Inhibits DNA synthesis |
| Methotrexate | Inhibits DNA synthesis |
| Cyclophosphamide | Cross-links DNA resulting in cell death |
| **Inhibitors of T-cell activation** | |
| Cyclosporine (ciclosporin)<br>Tacrolimus (FK506)<br>Sirolimus (rapamycin) | These drugs inhibit IL-2 production and/or responses to IL-2 |
| **Immunotherapeutics (mAbs)** | |
| Basiliximab<br>Daclizumab | These antibodies target CD25 and block the binding of IL-2 to effector cells |

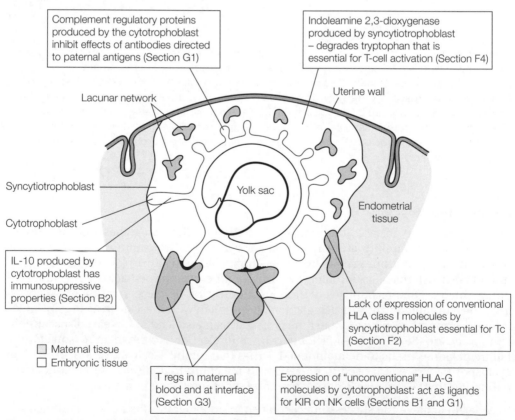

Figure 4. Mechanisms for preventing the rejection of an embryonic/fetal allograft. Both fetal and maternal components are important. During pregnancy, there is a bias towards a Th2 response mediated through estrogen and progesterone, and this together with the effects of other hormones such as human choriogonadotropin are thought to contribute to the maintenance of the fetal allograft (Section O2).

Currently, to reduce the risk of GVH reactions, monoclonal antibodies against CD52 are commonly added to bone marrow enriched for stem cells, to remove lymphocytes and other unwanted leukocytes. It is interesting to note that for stem cell treatment of some tumors (following their ablation by irradiation and chemotherapy), it is useful to have some T cells present that will recognize mismatched HLA molecules on the tumor cells and kill them (graft-versus-leukemia effect).

## The special case of the "fetal transplant"

The fetus is a chimera carrying HLA alleles from both parents. It is therefore effectively a semi-allograft in close apposition to maternal tissues. The main potential mechanisms for prevention of rejection are shown (Figure 4) for a recently implanted embryo (day 14), but also play an important role throughout gestation.

# N1 Origin and host defense against tumors

---

**Key Note**

| | |
|---|---|
| **Origin and host defense against tumors** | While the etiology of many human tumors is still unknown, it is clear that radiation as well as a variety of viruses and chemical carcinogens can induce tumors and that the genetics of an individual can play a significant role in susceptibility to tumor development. Immune responses in tumor-bearing patients can develop, or be induced to develop, against antigens associated with these tumors. These responses may be important in tumor regression. Promising therapeutic approaches based on these findings have been developed which are efficacious in the treatment of at least some tumors. Vaccines are also beginning to become available for therapy of tumors. |
| **Related topics** | (N5) Cytokine and cellular immunotherapy of tumors     (N6) Immunotherapy of tumors with antibodies |

---

## Origin and host defense against tumors

The origin and host response to tumors is currently the focus of extensive basic and clinical research. With regard to origin, a large number of environmental factors have been shown to be carcinogenic and/or mutagenic in animals. Several tumors in humans have, in fact, been associated with exposure to certain substances (asbestos with meso-theliomas in shipyard workers, hydrocarbons with scrotal cancer in chimney sweeps, and smoking in lung cancer). Viruses are also known to induce tumors in animals and humans. In humans, the Epstein–Barr DNA virus is involved in Burkitt's lymphoma and nasopharyngeal carcinoma, the hepatitis B virus in liver cancer, and the papillomavirus (HPV) in cervical carcinomas. The human T-lymphotropic virus 1 (HTLV-1, also known as human T-cell leukemia virus 1) is involved in certain forms of lymphocytic leukemia and human herpesvirus-8 (HHV-8) is associated with Kaposi's sarcoma.

Certainly, genetics can play an important role in the development of cancer. Although many factors are involved, proto-oncogenes (genes that encode proteins that regulate cell growth and differentiation) are particularly important. If mutated or expressed at high levels these normal genes can become oncogenes, which can then lead to disruption of normal growth and in some instances induction of a tumor. Some individuals inherit germ-line mutations that make them more susceptible to the development of a tumor.

In many instances host immune responses develop against tumors and in some instances may be protective. Based on the development of a clearer understanding of tumor immunology, numerous immunotherapeutic approaches have been explored for the treatment of cancer. Although the results from the use of monoclonal antibodies (mAbs), cellular therapy, and cytokines were initially less promising than hoped, much has been learned about how best to use them. Promising therapeutic approaches have been developed that are efficacious in the treatment of tumors, and many other effective immunotherapeutic approaches will also soon be available for the treatment of cancer.

# N2 Tumor antigens

---

## Key Notes

| | |
|---|---|
| **Introduction** | Tumor cells can be distinguished from normal cells by quantitative and qualitative differences in their antigens. Tumor-specific antigens (TSA) are unique to tumor cells but are rare, whereas tumor-associated antigens (TAA) are also on normal cells. Tumor antigens can be classified based on their origin or nature. |
| **Antigens associated with virally induced tumors** | Oncogenic DNA and RNA viruses code for viral antigens which are expressed by the tumor, and which are shared by all tumors induced by the same virus. |
| **Oncofetal antigens** | Antigens such as carcinoembryonic antigen (CEA) and alpha-fetoprotein (AFP) are highly associated with gastrointestinal (GI)-derived tumors and hepatomas, respectively. They are not unique to tumor cells since they are also found in normal cells during embryonic development and at low levels in normal human serum. |
| **Differentiation antigens** | Since many tumors result from the expansion of a single cell, a tumor will express the normal antigens characteristic of the type and differentiation stage of the cell that became malignant. This has permitted a clearer understanding of tumors as well as their classification and prognosis. |
| **Tumor-specific antigens and aberrantly expressed normal proteins** | Random mutagenesis of DNA, induced by chemical carcinogens or irradiation, may lead to tumors expressing tumor-specific antigens. These antigens are unique to the individual tumor and may result from mutated self proteins, translocated oncogenes, or products of oncogene fusion proteins. In addition, some tumors over or aberrantly express certain normal proteins to which host immune responses may be directed. |
| **Related topics** | (A4) Antigens         (D7) Immunoassay<br>(D5) Monoclonal antibodies |

---

## Introduction

A number of properties distinguish tumor from normal cells, including their invasiveness, loss of growth contact inhibition, and lack of response to regulation. In addition, there is considerable evidence for quantitative and qualitative differences in antigens associated with normal versus tumor cells. These antigens can be divided into tumor-specific antigens (TSA)—those unique to tumor cells—and tumor-associated antigens (TAA)—those also found on some normal cells. Another classification system is based on the origin or nature of the antigens and includes viral, chemical, oncofetal, and differentiation antigens.

## Antigens associated with virally induced tumors

In animal models, oncogenic DNA viruses (Table 1) code for both cell-surface and nuclear antigens that become expressed by the tumor. RNA tumor viruses induce expression of tumor cell-surface antigens that are viral proteins. These antigens are shared by all tumors induced by the same virus.

### Table 1. Virally induced/associated tumors

| Tumor viruses | Human tumor |
| --- | --- |
| **RNA** | |
| Human T-lymphotropic virus 1/human T-cell leukemia virus 1 (HTLV) | Adult T-cell leukemia/lymphoma |
| **DNA** | |
| Epstein–Barr (EBV) | B cell and Hodgkins lymphomas, nasopharyngeal carcinoma |
| Human papillomavirus (HPV) | Cervical carcinomas |
| Hepatitis B or C virus (HBV, HCV) | Hepatocellular carcinoma |
| Herpesvirus-8 (HHV-8) | Kaposi's sarcoma |
| Merkel cell polyomavirus | Merkel cell carcinoma |

## Oncofetal antigens

Although highly associated with some tumors, both on their cell surface and in the serum, oncofetal antigens are not unique to tumor cells as they are found at very low levels in normal human serum (Table 2), and are also present during embryonic development.

### Table 2. Examples of oncofetal antigens

| Antigen | MW (kDa) | Nature | Associated tumor |
| --- | --- | --- | --- |
| Carcinoembryonic antigen (CEA) | 180 | Glycoprotein | Gastrointestinal, breast |
| Alpha-fetoprotein (AFP) | 70 | Glycoprotein | Hepatomas |

Carcinoembryonic antigen (CEA) and alpha-fetoprotein (AFP) are two such antigens. CEA is expressed both on the cells and in the extracellular fluids of many patients with gastrointestinal-derived tumors including colon carcinoma, and pancreatic, liver, gall bladder, and breast tumors. CEA is also expressed by the gut, liver, and pancreas of the human fetus (2–6 months). AFP is found in secretions of yolk sac and fetal liver epithelium as well as in the serum of patients with hepatomas (liver tumors). Thus, these oncofetal antigens are not specific to tumors. Nor is their presence in the serum, even at high concentration, diagnostic of cancer, because high levels can result from non-neoplastic diseases, including chronic inflammation of the bowel or cirrhosis of the liver. However, the quantitation of these molecules in the serum can be used to evaluate the tumor burden and effectiveness of drug treatment (Section N4).

## Differentiation antigens

Some normal cellular antigens are expressed at specific stages of cell differentiation. These differentiation antigens can also be found on tumor cells and can be detected using monoclonal antibodies (Section D5). Moreover, since most tumors result from the expansion of a single cell arrested at some point in its differentiation, monoclonal antibodies to differentiation antigens are used to determine the approximate stage of differentiation at which the malignant event occurred. Using this approach, for example, it has been found that most T-cell leukemias are derived from early thymocytes or prothymocytes. Moreover, this also permits subclassification of similar tumors into groups that, on further study, often provides important information for prognosis and therapy (Section N4). Similar approaches have been applied to B-cell tumors and other malignant states (Table 3).

### Table 3. Differentiation antigens on lymphoid and myeloid malignancies

| Acute leukemias | | Chronic leukemias | |
| --- | --- | --- | --- |
| Disease | Markers | Disease | Markers |
| Common ALL | CD10 (CALLA), CD19, TdT (n) | B-CLL | CD19, CD20, CD5 |
| Null ALL | CD19, TdT (n) | HCL | CD19, CD20, TRAP |
| Pre-B cell ALL | CD19, IgM (m; cyt) | PLL | CD19, CD20, |
| T cell ALL | CD7, CD3 (cyt), TdT (n) | MF/S | CD3, CD4 |
| Myeloid leukemia | CD13, CD33, myeloperoxidase | T-CLL | CD3, CD8 |

ALL, acute lymphoblastic leukemia; CALLA, common ALL antigen; CLL, chronic lymphocytic leukemia; T-CLL, T-cell CLL; HCL, hairy cell leukemia; PLL, prolymphocytic leukemia; TdT, terminal deoxynucleotidyl transferase; n, nuclear; cyt, cytoplasmic; m, membrane; MF/S, mycosis fungoides/Sézary syndrome.

Melanoma, a cancer of melanocytes (pigment-producing cells in the skin), has also been a very informative model for understanding cancer immunity. Melanoma cells typically express a family of differentiation antigens that are also expressed by mature healthy melanocytes. These proteins are called melanosomal membrane proteins, and are normally involved in the biosynthesis of melanin. Although patients with melanoma often have antibody or T-cell responses against these differentiation antigens, they may not be sufficient to control growth of the tumor.

## Tumor-specific antigens and aberrantly expressed normal proteins

Random mutagenesis of DNA, induced by chemical carcinogens or irradiation, may lead to tumors expressing unique tumor-specific antigens. These antigens may result from mutated self proteins, translocated oncogenes, or products of oncogene fusion proteins. Because they result from random DNA mutagenesis, tumors induced by carcinogens or irradiation express antigens unique to the individual tumor. These unique antigens may be immunogenic and may induce development of host T- and/or B-cell responses.

Some tumors overexpress certain normal proteins that may also be overexpressed in other patients with similar tumors. In other cases, tumors produce proteins (e.g., tyrosine kinase, gp100) that would not normally be express to the same extent, or not at the differentiation stages of the normal cells from which the tumor was derived. These aberrantly expressed, but normal, proteins might be expected to be nonimmunogenic, but they are often able to induce immune responses in the host, perhaps because their aberrant expression occurs at a time when tolerance to the protein has waned.

# N3 Immune responses to tumors

---

**Key Notes**

**Immune surveillance**

That the immune system surveys constantly for neoplastic cells and destroys them is indicated by the observation of increased incidence of tumors of lymphoid or epithelial cells in immunodeficient animals and humans. In addition, a less-specific tumor surveillance system, perhaps NK cells, may also search for and eliminate certain types of tumor cells early in their development.

**Effector mechanisms**

Specific anti-tumor immunity appears to develop in tumor-bearing patients in much the same way as it does to pathogens or foreign antigens. Both TSAs and TAAs associated with tumor cells appear to be processed and presented in association with MHC class I molecules, making them potential targets for cytotoxic T cells. NK cells kill tumor cells not expressing MHC class I. Antibody-coated tumor cells can be killed by complement activation, macrophage- and PMN-mediated phagocytosis, ADCC, and induction of apoptosis.

**Tumor escape**

Mechanisms by which tumor cells may escape killing by the immune system include (i) induction of tolerance to tumor antigens; (ii) development of tumor antigen-negative variants; (iii) modulation of tumor antigen expression; (iv) tumor suppression of anti-tumor immunity; (v) poor immunogenicity of the tumor (e.g., due to lack of expression of MHC class I); and (vi) tumor-mediated Fas ligand–induced effector cell apoptosis.

**Related topics**

(B1) Cells of the innate immune system
(E3) The cellular basis of the antibody response
(F5) Clonal expansion and development of memory and effector function

(G2) Central tolerance
(G3) Peripheral tolerance
(J1) Deficiencies in the immune system

---

## Immune surveillance

Recent studies have supported the long-suspected belief that the immune system surveys for antigens associated with a newly developing tumor, and destroys the cells bearing them. Evidence supportive of this possibility comes from the observation of increased tumor incidence in immunodeficient animals or humans. However, congenitally athymic mice do not have high tumor rates, suggesting that the T-cell system may not be the major player in surveillance of certain tumors. Moreover, congenitally immunodeficient and immunosuppressed patients have high rates of tumors only of lymphoid or epithelial cells. Thus, a less-specific tumor surveillance system, perhaps NK cells, may predominantly search for and eliminate certain types of tumor cells early in their development. The best evidence for a surveillance mechanism involving T cells comes from experimental mouse models with virus-induced tumors, but here the response is essentially directed to viral antigens and not tumor antigens.

## Effector mechanisms

If a tumor evades the surveillance system, it might then be recognized by the specific immune systems. In models of chemically and virally induced tumors, the TAAs are immunogenic and trigger specific cellular and antibody responses against the tumor. This immunity may be protective and can be passively transferred with immune cells. In tumor-bearing patients as well, it is possible to demonstrate anti-tumor antibody, which may mediate some tumor cell lysis.

It is likely that anti-tumor immune responses develop in tumor-bearing patients in much the same way as they do to pathogens or foreign antigens. Thus, anti-tumor antibodies and T cells are generated and, along with nonspecific immune defense mechanisms, play a role in tumor immunity. More specifically, it is likely that both TSAs and TAAs are associated with tumor cells and, after their intracellular synthesis, are processed and presented in association with MHC class I molecules, making them potential targets for cytotoxic T cells. Overall, the potential effector mechanisms that may be involved in human tumor cell lysis *in vivo* are the same as those used in microbial immunity (Table 1).

## Table 1. Potential tumor immune effector mechanisms

Killing by specific cytotoxic T cells recognizing TAA or TSA peptides associated with MHC class I

Antibody induction of apoptosis

Killing mediated by antibody and complement

Antibody-dependent cellular cytotoxicity (ADCC) mediated by macrophages, PMNs, and NK cells

Phagocytosis by activated macrophages

NK-cell-mediated killing of tumor cells that lack MHC class I or those expressing ligands to which NK-cell receptors (e.g., NKG2D) bind (Section B1)

## Tumor escape

If tumors possess immunogenic antigens which eventually stimulate specific immune responses, how do they escape rejection? The various possibilities which may explain tumor cell escape from the immune system include:

- *Tolerance to tumor antigens.* This might happen if the antigen is a TAA and thus also expressed by normal cells and/or if the antigen is presented in a form or under conditions such that T cells are rendered unresponsive to it.

- *Selection for tumor-antigen-negative variants.* If antigens associated with tumor cells are able to elicit strong effective immune responses, tumor cells bearing these antigens would be rapidly eliminated, and only those tumor cells lacking, or with decreased amounts of, these antigens would survive.

- *Modulation of tumor antigen expression.* Binding of antibody to antigens on the surface of tumor cells may result in rapid internalization of antigen and its loss from the cell surface, permitting the tumor cell to escape, temporarily, from further detection by antibody and thus from FcR-bearing effector cells.

- *The tumor may immunosuppress the patient.* Tumors may release molecules such as TGFβ or IL-10 and/or may induce the development of T- or myeloid-regulatory/suppressor cells that have immunosuppressive properties (Section B2).

- *The tumor may have low immunogenicity.* Tumor cells having little or no MHC class I on their surface are able to avoid recognition by cytolytic T cells. Although these tumor cells are more susceptible to NK cells, NK cells do not have memory and thus there may be insufficient NK cells to deal with a large tumor burden.

- *Tumor cells sometimes express Fas ligand (FasL).* When FasL on the tumor interacts with Fas on T cells, T-cell apoptosis may result (Section F5).

# N4 Immunodiagnosis

---

**Key Notes**

| | |
|---|---|
| **Classification** | Monoclonal antibodies (mAbs) to antigens associated with a particular differentiation state can sometimes be used to classify the origin of the tumor and its stage in normal cell differentiation. This information permits prediction of the likelihood of success of current therapy. |
| **Monitoring** | mAbs can sometimes be used to determine the amount and rate of change of TAA (oncofetal antigens, PSA, CA-125, CA-19-9) in the serum of a patient as a measure of tumor progression and duration of remission. Using cytological analysis, mAbs to certain TAA (e.g., cytokeratin, MUC-1) can also be used to search for micrometastases. |
| **Imaging** | Radioconjugated mAbs specific for an appropriate TAA can sometimes be used to locate and image metastases in a tumor-bearing patient. |
| **Related topics** | (A5) Hemopoiesis – development of blood cells      (D5) Monoclonal antibodies     (D7) Immunoassay |

---

## Classification

Although numerous monoclonal antibodies (mAbs) have been developed against antigens associated with tumor cells, few of these antibodies are tumor-specific. Thus, binding of these mAbs to tissues from a patient will not necessarily indicate the presence or location of a tumor. However, because tumors often appear to be monoclonal in origin (develop from a single cell that has undergone a malignant event) and to have characteristics of the cell of origin, mAbs to antigens associated with particular differentiation states can be used to classify the origin of the tumor and the stage in normal cell differentiation most similar to that of the tumor cell (Section N2). One of the most prominent uses of this approach is in the subgrouping of leukemias (Figure 1).

In particular, mAbs have defined a large number of markers associated with lymphoid and myeloid cell populations, and with different stages of their differentiation. Information obtained using panels of such mAbs permits subclassification of some types of tumors. As a result, it is possible to develop patterns of tumor cell progression and responsiveness to therapy for tumors subclassified in this way. Thus, it becomes possible to predict, for a particular tumor subtype, whether or not current therapy will be effective, and if it is not, the need to pursue a different therapeutic approach.

## Monitoring

mAbs to TAAs can sometimes be used to monitor the progression of tumor growth in a patient. Oncofetal antigens, because of their presence in serum, are useful for this purpose. That is, because they are normally only present at very low levels in normal human serum, the presence of large amounts of CEA and AFP may indicate a gastrointestinal or

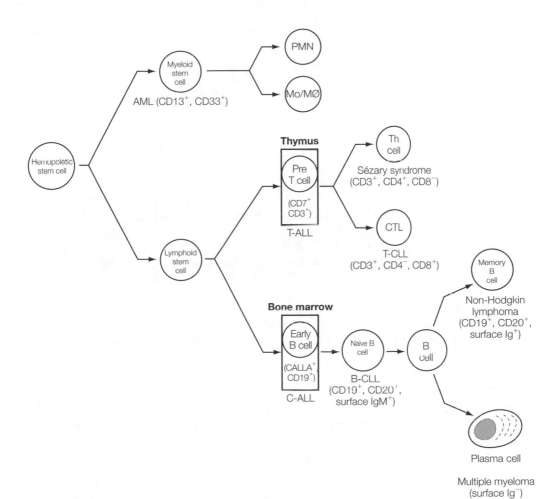

Figure 1. Subgrouping of leukemias. Each myeloid or lymphoid tumor expresses a set of markers (molecules) typical of normal myeloid or lymphoid cells at a particular stage of their differentiation. AML, acute myeloid leukemia; T-ALL, thymic acute lymphoblastic leukemia; T-CLL, T-cell chronic lymphocytic leukemia; c-ALL, common acute lymphoblastic leukemia; B-CLL, B-cell chronic lymphocytic leukemia; PMN, polymorphonuclear cell; Mo/MØ, monocyte/macrophage; CTL, cytotoxic T lymphocyte.

liver tumor, respectively. However, since conditions other than tumors elevate the level of these molecules in the serum, their levels are most useful in tumor-bearing patients whose serum level of the oncofetal protein is known. Relapses or duration of remission can then be followed by monitoring the rate of change of the amount of oncofetal antigen in the serum (Figure 2).

In addition, quantitation of other TAAs is used to monitor tumor presence and growth in patients with other tumor types. The serum levels of the mucins CA-125 and CA-19-9 (both high-molecular-weight proteoglycans) in a patient with ovarian cancer are useful in following the status and progression of this tumor. Prostate-specific antigen (PSA) is similarly useful in prostate cancer.

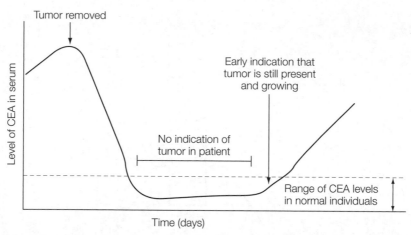

Figure 2. Monitoring serum levels of CEA in a cancer patient with a CEA-expressing tumor.

Using cytological analysis, mAbs to certain TAAs (e.g., cytokeratin) can be used to search for micrometastases in bone marrow or lymph nodes. Similarly, another mucin, MUC-1, is expressed on breast carcinomas in a pattern different from that on normal breast epithelium.

## Imaging

By linking a radioisotope (e.g., [131]I) to a mAb specific for an appropriate TAA (e.g., CEA), and intravenously injecting this construct into a tumor-bearing patient, it is possible to image tumor metastases using scintigraphy.

# N5 Cytokine and cellular immunotherapy of tumors

## Key Notes

| | |
|---|---|
| **Immunostimulation and cytokines** | Nonspecific immunostimulants induce cytokine-producing immune responses that activate effector cells, but have limited ability to mediate tumor cell lysis. To be effective in tumor therapy cytokines need to be used in concert with specific immunotherapy. |
| **Lymphokine-activated killer (LAK) cells** | Lymphocytes from a tumor-bearing patient are expanded and activated with IL-2. These cytotoxic LAK cells, primarily NK cells, are infused back into the patient with or without more IL-2. Although some tumor regression occurs, toxicity limits use of this approach. |
| **Tumor-infiltrating lymphocytes (TILs)** | CD8$^+$ lymphocytes isolated from the patient's tumor, at least some of which are tumor-specific, are expanded and activated with IL-2. These TILs are infused into the patient with or without IL-2. TILs can induce tumor regression, especially in patients with renal cell carcinoma, but toxicity due to IL-2 is significant. |
| **Macrophage-activated killer (MAK) cells** | Monocytes isolated from the blood of tumor-bearing patients are cultured with cytokines *in vitro* to activate these cells for enhanced cytotoxicity before reinjection into the patient. |
| **Related topics** | (B1) Cells of the innate immune system<br>(B2) Molecules of the innate immune system<br>(F5) Clonal expansion and development of memory and effector function |

## Immunostimulation and cytokines

Initially, immunotherapy in humans utilized nonspecific immunostimulants such as BCG and *Cryptosporidium parvum*, which resulted in some tumor cell killing but overall did little to reduce the tumor cell burden. These results probably reflect the development of strong immune responses against antigens associated with these microbes, including production of cytokines capable of activating immune effector cells. The resulting activated cells (e.g., macrophages) then mediated increased tumor cell lysis.

When recombinant cytokines became available, they were tried, but again with limited success. Thus, although cytokines are critical to the development of specific immune responses, when used alone they primarily enhance nonspecific activation of immune cells (Section B2). To be effective anti-tumor agents they will need to be used in concert with induction of more specific immune responses to the tumor. Even so, IFNα remains a standard treatment for patients with melanoma.

## Lymphokine-activated killer (LAK) cells

This approach is based on the fact that many tumor-bearing patients have lymphocytes reactive to their tumor. Patient blood lymphocytes are expanded and activated *in vitro* using IL-2 and then injected back into the patient with or without more IL-2 (Figure 1). These LAK cells are primarily NK cells and thus do not have the selective specificity of T cells. Rather, they react with and kill tumor cells that express little or no cell surface MHC class I molecules, or that express ligands (e.g., MICA or MICB) to which certain NK-cell receptors (e.g., NKG2D) can bind (Table 1). Although some tumor regression occurs with this approach, significant toxicity is evident if high doses of IL-2 are used. It is important to note that, because of significant MHC differences in individuals, this kind of cellular therapy, as with TILs and MAKs, is patient-specific. Thus, a new cell preparation must be developed for each patient from their own cells.

## Tumor-infiltrating lymphocytes (TILs)

As with LAK cells, TILs are obtained from tumor-bearing patients, expanded and activated with IL-2 (Figure 1). In this case, however, lymphocytes, primarily CD8+ T cells, are

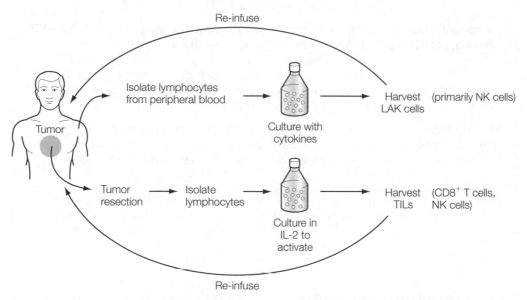

Figure 1. Therapy with LAK cells and TILs.

## Table 1. Natural killer (NK) cells

| | |
|---|---|
| 1. Activated for anti-tumor activity by IFNγ and/or IL-2 | |
| 2. Specificity: | Have cell-surface receptors for the Fc portion of Ig (FcγRIII)<br>Have cell-surface receptors for certain self molecules<br>Have cell-surface receptors for ligands expressed primarily by tumor cells |
| 3. Mediate killing of: | Antibody-coated, virus-infected, or tumor cells<br>Virus-infected or tumor cells with little or no MHC class I<br>Tumor cells expressing ligands to which NK-cell receptors (e.g., NKG2D) bind |

isolated from patient tumor samples, at least some of which are thought to be specific for tumor antigens. They are also infused back into the patient with or without more IL-2. TIL therapy induces tumor regression in some patients, especially in patients with renal cell carcinoma. Again, there is significant toxicity if high doses of IL-2 are used to maintain the active status of the TIL cells *in vivo*.

## Macrophage-activated killer (MAK) cells

Another immunotherapeutic approach involves the use of cytokines and activated macrophages. Monocytes are isolated from the peripheral blood of tumor-bearing patients and cultured *in vitro* with cytokines (e.g., IFNγ) to activate these cells for enhanced cytotoxicity. They are then injected into the patient. Although these cells are highly cytotoxic and phagocytic, they are relatively nonspecific, and may require co-injection with antibody to TAAs to be most effective.

# N6 Immunotherapy of tumors with antibodies

---

**Key Notes**

| | |
|---|---|
| **Specificity of monoclonal antibodies (mAbs) to tumors** | The vast majority of mAbs prepared against human tumors are not tumor-specific. The TSAs which have been identified include (i) idiotypes of antibody on a B-cell tumor and (ii) a mutant form of epidermal growth factor receptor (EGF-R). Additional TSAs have been identified by analysis of translocations associated with tumor development. |
| **Tumor therapy with antibodies alone** | Monoclonal Abs kill tumor cells by apoptosis, complement activation, ADCC, or phagocytosis. More than 10 mAbs have been approved for therapy of tumors, including mAbs for treatment of breast cancer, B-cell tumors, myeloid leukemia, and colorectal cancer. Thus, mAbs can be effective if they react with antigens highly expressed on the tumor, are used to treat minimal disease, and are used in patients whose immune system is functional. |
| **Tumor therapy with immunotoxins (ITs)** | ITs are mAbs to TAAs that are linked to a toxin or radionuclide. mAbs coupled to toxin inhibit critical cellular processes when they are internalized. Radionuclide-coupled mAbs mediate killing by DNA damage from decay and release of high-energy particles. |
| **Tumor therapy with bispecific antibodies (BsAbs)** | BsAbs are engineered molecules that have two different covalently linked specificities, one against a TAA, the other to a trigger molecule on a killer cell. *In vivo*, BsAbs bind to immune effector cells, arming them to seek out and kill tumor cells. |
| **Related topics** | (A5) Hemopoiesis – development of blood cells (D4) Allotypes and idiotypes     (D5) Monoclonal antibodies (M1) The transplantation problem (Q3) Antibody-mediated immunotherapy |

---

## Specificity of monoclonal antibodies (mAbs) to tumors

Considerable effort has been expended on the development of monoclonal antibodies (mAbs) to tumor-specific antigens (TSAs), since it was thought that only mAbs specific for tumor cells would be useful in the diagnosis and treatment of tumors. However, few if any mAbs prepared against human tumors have been found to be truly tumor-specific. Examples of antigens which could be considered to be TSAs include:

- The idiotype of the antibody on a B-cell tumor (e.g., CLL). The first successful use of a mAb in tumor therapy involved treatment of a patient with anti-idiotype antibody prepared specifically against the patient's tumor. This approach could also be used to treat T-cell tumors based on their expression of their unique binding site.

● A mutant form of the epidermal growth factor receptor (EGF-R) which has a deletion of an extracellular domain. That this molecule is antigenic and uniquely expressed on tumors, including glioblastomas, is the basis for approaches using an antibody-based therapeutic agent and for a vaccine to induce a CTL response.

● As information becomes available on the mutations and translocations associated with various tumors, unique gene products are being identified that serve as TSAs.

## Tumor therapy with antibodies alone

Although mAbs can cause tumor cell lysis through complement activation, by targeting NK-cell, monocyte, and macrophage ADCC or phagocytosis, or by inducing apoptosis, the use of mAbs to treat human tumors was not initially successful. To some extent this resulted from (i) the lack of specificity of the mAb utilized; (ii) the presence of soluble forms of the antigen in the serum that effectively interfered with the interaction of antibody with the tumor cell; (iii) modulation and loss of the antibody–antigen complexes from the tumor-cell surface before antibody-mediated killing could occur; (iv) outgrowth of (selection for) tumor cells not expressing the antigen. Moreover, since most mAbs used in therapy were originally of mouse origin, they did not interface well with human effector molecules (complement) and cells (NK cells, macrophages, and PMNs), and being foreign, induced a human anti-mouse antibody (HAMA) response that compromised the effectiveness of the mAb. Nonetheless, the clearer appreciation that developed on how to use mAbs more effectively in cancer therapy has resulted in a renaissance in antibody therapy (Figure 1).

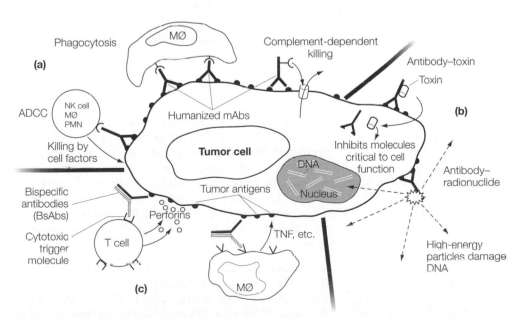

Figure 1. Antibody-based tumor therapy: (a) antibody alone; (b) antibody–toxin/radionuclide constructions; (c) bispecific antibodies.

Most trials now use human or humanized mAbs (Section D5) and many have been approved by the US FDA for treatment of different cancers (Table 1), including those for therapy of breast cancer, B-cell tumors, myeloid leukemia, and colorectal cancer. Other mAbs still in clinical trials are also demonstrating efficacy as indicated by tumor regression in some patients. Thus, there is growing optimism that mAbs will be very useful tumor therapeutic agents, especially if (i) they are human or humanized to permit long-term use; (ii) the antigen to which they react is expressed at a high level on the tumor; (iii) the mAb is used to treat minimal disease; and (iv) the patient's immune system is functional.

### Table 1. Monoclonal antibodies approved for cancer therapy

| Name | Type | Date approved | Specificity | Tumor |
|------|------|---------------|-------------|-------|
| **Antibodies alone** | | | | |
| Trastuzumab | Humanized | 1998 | HER2/neu | Breast, lung, ovarian, pancreatic cancer |
| Rituximab | Chimeric | 1997 | CD20 | B-cell lymphoma |
| Alemtuzumab | Humanized | 2001 | CD52 | B-cell CLL |
| Cetuximab | Chimeric | 2004 | EGFR | Colorectal and head and neck cancer |
| Panitumumab | Human | 2006 | EGFR | Metastatic colorectal cancer |
| Bevacizumab | Humanized | 2004 | VEGF | Colon, breast, kidney |
| Ipilimumab | Human | 2010 | CTLA-4 | Melanoma |
| **Antibody–toxin/radionuclide conjugates** | | | | |
| Gemtuzumab ozogamicin | Humanized | 2000 | CD33 | AML |
| Ibritumomab tiuxetan | Chimeric | 2002 | CD20 | B-cell |
| Tositumomab | Murine | 2003 | CD20 | Lymphoma (NHL) |

It is important to note that not all of the mAbs approved, or in clinical trials, react directly with the tumor, as some react with molecules on cells that in one way or another are important to tumor growth. In particular, bevacizumab reacts with vascular endothelial growth factor (VEGF), a molecule that is responsible for inducing the creation of new blood vessels. Although this is important in early life and when there is inadequate circulation, tumors are able to use VEGF to create blood vessels critical to their growth. Bevacizumab prevents this from happening and the ability of the tumor to grow is decreased. Although not a cure, when it is used with other therapies it is often important for changing the course of the disease.

Another mAb approved for therapy of cancer that does not react directly with the tumor is ipilimumab. This human mAb reacts with CTLA-4, a co-stimulatory and regulatory molecule expressed on activated T cells. CTLA-4 interacts with CD28 on APCs and in so doing causes the T cell to shut down. In the case of tumors, a CTL anti-tumor response would be inhibited. Ipilimumab binds to CTLA-4 and prevents it from interacting with CD28 and thus stops the T cell from shutting down.

## Tumor therapy with immunotoxins (ITs)

Many studies, including clinical trials, have used mAbs to which toxins or radionuclides have been coupled (Figure 1). Thus, when injected into a patient, ITs would not need to activate patient effector mechanisms. Rather, ITs would seek out and bind to tumor cell antigens, and mediate their own lethal hit. Toxins such as ricin are very potent inhibitors of critical intracellular processes, with a single molecule able to kill a cell. It is essential, however, that the targeting mAbs react with a TAA that is internalized on binding of IT. Gemtuzumab ozogamicin, an IT in which a humanized mAb to CD33 has been conjugated to the toxin calicheamicin, has been approved for treatment of patients with acute myeloid leukemia.

Radionuclide-coupled mAbs mediate killing by DNA damage from decay and release of high-energy particles. This kind of IT will kill bystander cells (those nearby), which may result in killing of normal cells, but also of adjacent tumor cells that may not express the targeted antigen. The development of useful ITs has taken more time than initially anticipated due to toxic side effects, but many of the problems have been solved and some ITs have been approved for therapy (Table 1), others are in late-stage clinical trials. Note that the same anti-CD20 mAb approved for therapy of B-cell lymphoma has also been coupled to a radionuclide, yttrium, and the resulting immunotoxin (ibritumomab tiuxetan) has been approved for therapy of B-cell lymphoma.

## Tumor therapy with bispecific antibodies (BsAbs)

Directing or redirecting immune effector cells is also being explored as a way to enhance the ability of a patient's own immune system to reject their tumor. BsAbs, consisting of the binding sites of two different covalently linked mAbs, have been engineered as anti-tumor therapeutics. One specificity of this BsAb is to a TAA (e.g., HER2/neu), the other to a trigger molecule on a killer cell (e.g., CD64 on macrophages). When injected into a tumor-bearing patient, the BsAb binds to the immune effector cell thereby arming it to seek out, and kill, tumor cells. Several BsAbs have shown promise and are in clinical trials for therapy of cancer.

# N7 Tumor vaccines

## Key Notes

| | |
|---|---|
| **Prophylactic versus therapeutic vaccines** | The development of vaccines for cancer therapy is being actively pursued. Prophylactic vaccines induce immunity to viruses associated with tumor development. Other approaches are designed to enhance/induce effective immunity in tumor-bearing patients. |
| **Immunization with tumors, tumor antigens, and tumor antigen DNA** | Killed or irradiated patient tumor cells or appropriate TAAs and their peptides, as well as DNA of TAAs, are being tested for their ability to induce patient anti-tumor immunity. |
| **Immunization with transfected tumors** | Transfecting tumor cells with co-stimulatory molecules enhances their immunogenicity and ability to induce a CTL response. Tumor cells transfected with cytokine genes attract, expand, and activate immune cells reactive to tumor antigens. |
| **Immunization with APCs loaded with TAA** | Since immature DCs are best able to ingest antigen and mature DCs are best at presenting antigen, considerable effort is directed at determining optimal conditions for loading and maturing DCs *ex vivo* so they induce strong CTL anti-tumor responses when reintroduced into the patient. It has been demonstrated that this approach can work. |
| **Related topics** | (F5) Clonal expansion and development of memory and effector function     (I3) Antigen preparations    (I4) Vaccines to pathogens and tumors    (Q4) Cellular immunotherapy |

## Prophylactic versus therapeutic vaccines

Numerous approaches are being used to develop vaccines for use in the treatment of cancer (Table 1). Prophylactic approaches focus on the use of vaccines that induce immunity to viruses or microbes known to be associated with the development of a tumor (Sections N2 and Q4). Hepatitis B vaccines prevent infection by this virus and reduce the incidence of liver cancer. Human papillomavirus (HPV) vaccines reduce the development of cervical carcinoma. Most other tumor vaccine approaches are designed to enhance or induce effective tumor immunity in patients who have already developed cancer.

## Immunization with tumors, tumor antigens, and tumor antigen DNA

A variety of approaches are being explored for inducing or enhancing a patient's immunity to their tumor. These include injecting killed or irradiated tumor cells from the patient, an approach that has had little success. The identification of appropriate TAAs (those expressed at low levels on normal cells and high levels on tumors) and their potentially

**Table 1. Main types of cancer vaccine approaches**

| | |
|---|---|
| Whole tumor cell | Transfected to produce GM-CSF, irradiated to prevent division, injected as a vaccine |
| Protein antigen | Taken-up, processed, presented by APCs |
| Peptide epitopes | Occupy empty MHC molecules on APCs |
| DNA transfection | Transfect APCs or other cells with DNA that will produce/secrete antigen for uptake by APCs. |
| Dendritic cells | Load with peptides, proteins, or whole tumor lysate; traffic to lymph nodes and expand CD8$^+$ T cells |

immunogenic peptides has resulted in their use in attempts to focus the patient's immune system to respond to antigens that are primarily tumor associated. As with immunization using whole cells, these antigens would most likely induce a T-helper cell rather than a more desirable CTL response, as they would enter APCs by the exogenous pathway and be presented on MHC class II (Section F2). However, it is now clear that the APC presenting antigen to the CTL must first be conditioned by interaction with a T-helper cell before it can effectively induce a CTL response (Section F5). Moreover, antigens entering by the exogenous pathway may in some instances be presented on MHC class I molecules (**cross-presentation**) and initiate a CTL response.

Yet another actively pursued approach involves immunizing with DNA encoding the TAA or peptide, either alone or in an appropriate expression vector. This DNA introduced into a cell would be integrated, expressed, and translated into proteins in the cytosol, some of which would be degraded to peptides for loading onto MHC class I molecules and thus potential induction of a CTL response.

## Immunization with transfected tumors

Since most kinds of tumor cells do not express the co-stimulatory molecules (e.g., CD80 and CD86) important to the induction of an immune response, studies have been carried out in animal models to determine if transfecting tumor cells with these molecules would enhance their immunogenicity. In fact, B7-transfected tumor cells induce a strong CTL response against the tumor. Furthermore, these CTLs are sometimes able to lyse parent tumor cells not expressing B7, because, once activated, CTLs do not need B7 co-stimulatory signals to kill.

Another approach involves transfecting tumor cells from a patient with a cytokine gene, as certain cytokines expressed by the tumor may attract, expand, and activate cells of the immune system and induce or enhance immunity to tumor antigens. In experimental models, tumor cells transfected with cytokine genes (e.g., IL-2, IFNγ, GM-CSF) are able to induce immunity to the tumor resulting in its regression or rejection. IL-2 may, for example, enhance the development of cytotoxic cells to TAAs from their precursors (e.g., in melanoma).

## Immunization with APCs loaded with TAA

A very active area of tumor vaccine research involves loading patient dendritic cells (DCs) *in vitro* with TAA and re-injection of these cells into the patient. This approach has the benefit that potential APCs can be isolated from a patient's peripheral blood and manipulated such that their antigen-presenting capabilities are optimal. In particular, monocytes readily obtained from the peripheral blood of a patient can be induced with cytokines to differentiate into immature DCs (Figure 1). Since immature DCs are best able to ingest antigen and mature DCs are best at presenting antigen, loading of mono-cyte-derived immature DCs followed by cytokine-induced differentiation of these cells to mature DCs is more readily accomplished *in vitro* than *in vivo*. These mature, loaded APCs are then reintroduced into the patient, fully able to stimulate T cells. Many research groups are currently trying to define the optimal conditions for obtaining immature DCs, for loading and maturing them, and for their reintroduction into the patient.

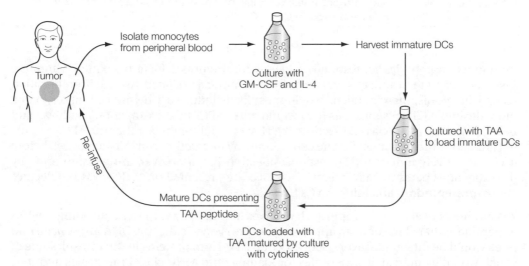

Figure 1. Overview of dendritic cell vaccines

The most successful of these approaches, and perhaps a breakthrough in the develop-ment of tumor vaccines for therapy, was sipuleucel-T, a therapy pioneered by Dendreon. In particular, patient DC precursors were incubated with a fusion protein consisting of prostatic acid phosphatase (a molecule produced by prostate cancer cells) linked to GM-CSF (an activator of human plasmacytoid pre-DCs) and infused these cells back into the patient. In the pivotal phase III clinical trial it was found that patients with prostate can-cer, when treated with sipuleucel-T, lived an average of 4.1 months longer than patients treated with the control. In April of 2010, the FDA approved sipuleucel-T for use in the treatment of advanced prostate cancer—the first therapeutic tumor vaccine approved in the US.

# O1 Overview

---

**Key Note**

**Overview**
Gender-associated differences in immunological function have been recognized for some time. These are seen at the level of the genes inherited on sex chromosomes, the different anatomical structures of females versus males, and the influence of sex hormones on immunity. Females show a more robust cell- and antibody-mediated response to many infections compared with males and generally tend to have higher levels of serum IgG and IgM. The anatomical distribution of lymphoid tissue is the same in both sexes with the exception of the reproductive tracts. In both the male and female, immune cells and molecules are distributed throughout the reproductive tract and in the female appear to be influenced by hormones associated with menstruation and pregnancy. Since most cells of the immune system have receptors for sex hormones it is not surprising that these hormones have important effects on the immune system. For example, estrogens increase levels of Th2 cells and B-cell synthesis of immunoglobulins while testosterone suppresses B-cell activity. From observations of changes of disease activity of some autoimmune diseases during pregnancy and in experimental models, it seems likely that hormones are involved in the gender-based increase in the frequency of autoimmune diseases in females compared with males.

**Related topics**
(C3) Mucosa-associated lymphoid tissues
(D2) Antibody classes
(E3) The cellular basis of the antibody response
(E4) Antibody responses in different tissues

---

## Overview

Before the advent of antibiotics there was an increase in prevalence of infectious disease (morbidity) and mortality in male children compared with females and while the mortality has decreased, morbidity is still greater in male children. In recent years a better understanding of the effect of gender on the immune system has thrown some light on the basis of these findings. The effects are seen at the level of the genes inherited on sex chromosomes, the different reproductive anatomy of females versus males, and the influence of sex hormones on immunity. At the genetic level, several of the genes important for immune responses (e.g., CD40L) have been localized on the X chromosome and several X-linked primary immunodeficiency diseases have been described that primarily affect males (Section J2). However, in addition to the role of the X chromosome, females generally have a higher absolute number of CD4⁺ T cells in their circulation and show a more robust cell- and antibody-mediated response to many infections compared with males. For example, there is a tendency for levels of serum IgG and IgM antibodies to be higher throughout life in females compared with males. Furthermore, increased specific primary and secondary responses to microbial infections such as *Escherichia coli*, *Brucella*, measles, rubella, and hepatitis B have been reported in females. This could contribute to the decreased susceptibility to infection and general trend in heightened immune response to microbial infections compared with males (Table 1).

**Table 1. Gender-associated immunological differences**

|  | Female | Male |
| --- | --- | --- |
| Susceptibility to infection | < | > |
| Immune response to infection | > | < |
| Frequency of autoimmune disease | High | Low |

The anatomical distribution of lymphoid tissue is the same in both sexes with the exception of the reproductive tracts. Although both tracts are lined by epithelial cells that secrete antimicrobial proteins (Section B2), organized lymphoid tissue is distributed throughout the female reproductive tract and appears to be influenced by hormones associated with menstruation and pregnancy. The male, in contrast, has fewer lymphocytes distributed throughout his reproductive tract. The lymphoid tissue associated with the reproductive tracts is part of the mucosa-associated immune system (Section C3).

Perhaps as a consequence of their tendency to heightened immune response, females have a much higher frequency of autoimmune diseases pre- and post-puberty compared with their male counterparts. These autoimmune diseases include juvenile arthritis, Hashimoto's thyroiditis, systemic lupus erythematosus (SLE), primary biliary cirrhosis, and rheumatoid arthritis (Section L2). In certain strains of mice (New Zealand Black), female mice are much more susceptible to the human equivalent of SLE than their male counterparts. Furthermore, male mice that have had their testes removed develop the same incidence as females, while removal of the ovaries has a protective effect in females. This provides strong evidence for the role of sex hormones on the adverse immune response in autoimmune diseases. In addition, they have been shown to have pivotal and different immunological functions in the developmental phase of the immune system and in its functional activity. In experimental systems it has been shown that estrogens are associated with increased B-cell synthesis of immunoglobulins while testosterone is associated with suppression of B-cell activity. Sex hormones appear to have major influences on immune cell types and function during the menstrual cycle and pregnancy. For example, during the cycle, NK cells increase in numbers in the secretory phase compared with the proliferative phase and in pregnancy these cells are the major immune cell population in the endometrium following implantation of the blastocyst.

# O2 Immune cells and molecules associated with the reproductive tracts

| Key Notes | |
|---|---|
| **The female reproductive tract** | The female reproductive tract (FRT) contains all of the cells associated with an appropriate immune response. These are distributed throughout the reproductive tract in the vagina, cervix, endometrium, and the epithelial layers of the fallopian tubes. The epithelial cells lining the tract express pattern recognition receptors (PRRs) that react with microbes and induce the production of antimicrobial peptides by these cells. Protective antibodies of both IgA and IgG classes are also present in the lumen of the tract. |
| **Immunological changes during the menstrual cycle** | There are both numerical and functional changes in NK cells and in T-cell subsets in the endometrium during the menstrual cycle, which are thought to be related to the marked fluctuations in estradiol and progesterone during the cycle. |
| **Immune-associated changes during pregnancy** | Pregnancy is considered an immune privileged state that permits acceptance of the fetal allograft as a result of the influence of pregnancy-associated hormonal regulation. Immunological changes include a shift towards a Th2-type cytokine profile. |
| **The role of the lactating breast in immune defense** | Plasma cells associated with the acini of the lactating breast produce IgA that, as part of the colostrum and breast milk, is important for protection of the newborn. In addition, colostrum and breast milk contain antimicrobial proteins. |
| **The male reproductive tract** | The lower male reproductive tract (penile urethra) resembles other mucosal sites in that it has abundant CD4$^+$ and CD8$^+$ T lymphocytes in the lamina propria (although CD8$^+$ T cells predominate). Macrophages and dendritic cells are found in the distal tip of the urethra, especially the prepuce. Occasional lymphocytes, mainly CD8$^+$ T cells, are found among the epithelial cells in the vas deferens and epididymis. As in the FRT, epithelial cells lining the tract express PRRs. IgA antibodies derived from plasma cells in the urethra and prostate glands are present in the seminal fluid, which also contains IgG and prostaglandins. |

| Related topics | (B1) Cells of the innate immune system | (C3) Mucosa-associated lymphoid tissues |
|---|---|---|
| | (B2) Molecules of the innate immune system | (D2) Antibody classes |
| | | (D8) Antibody functions |
| | (C1) Lymphocytes | (E4) Antibody responses in different tissues |

## The female reproductive tract

The female reproductive tract (FRT) is composed of the vagina and cervix (lower FRT) and, uterus (containing the endometrium), ovaries, and fallopian tubes (upper FRT; Figure 1). The endometrium is a multifunctional part of the FRT that provides protection against microbial invasion but also an environment for normal reproductive physiology (i.e., preparation for implantation) and, during pregnancy, a "tolerant" environment for the fetus, which is a semi-allograft (Section M4).

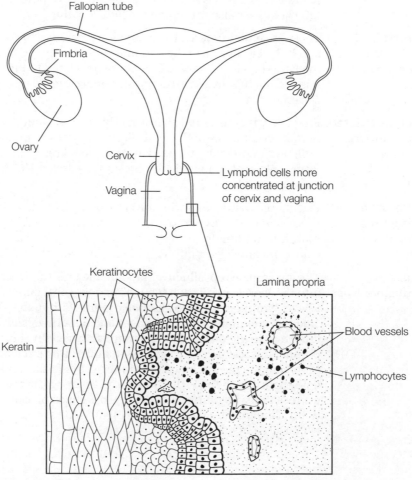

Figure 1. The female reproductive tract.

## Innate immunity in the FRT

The lower FRT—the vagina and cervix—is the main portal for microbial infection. It is nonsterile and home to a number of commensal microbes, including lactobacilli that help to protect against pathogens. The epithelial lining of the complete FRT contains PRRs, including the TLRs 1–9. There is also evidence for other PRRs, including NODs (Section B2). In addition, vaginal fluids contain a number of antimicrobial proteins that appear to be made in tissues of both the upper and lower FRTs. These include human β defensins and enzymes that have antimicrobial activity. Macrophages and NK cells are found in the tissues of the vagina, cervix, and endometrium (Table 1). Although their function in the lower FRT, where there is a high microbial load, is likely to be that of protection, in the endometrium they might also produce cytokines involved in tissue remodeling during the cycle. NKT cells and γδ T cells (in the epithelial layer of the endometrium) are also present in the FRT and may have an immunoregulatory function.

**Table 1. Cells of the immune system associated with the female reproductive tract**

| | |
|---|---|
| **Innate** | |
| NK cells | Vagina, cervix, endometrium |
| Macrophages | Vagina, cervix, endometrium |
| Dendritic cells | Vagina, cervix, endometrium |
| **Adaptive** | |
| CD4⁺, CD8⁺ T cells | Vagina, cervix, endometrium, fallopian tubes |
| γδ T cells | Endometrium |
| Tregs | Endometrium |
| B cells | Endometrium |
| Plasma cells | Vagina, cervix, endometrium, fallopian tubes |

## Adaptive immunity in the FRT

Both CD4⁺ and CD8⁺ lymphocytes, and MHC class II-positive dendritic cells, are distributed throughout the lower reproductive tract in both epithelial and subepithelial layers of the cervix and the vagina. Unlike the ileum of the gastrointestinal tract, epithelial M cells that can transport antigens into the subepithelial layers have not been found in the FRT (Section C3). The majority of lymphocytes appear to be located at the junction of the cervix (ectocervix) and vagina, that is, the site of entry of microbes into the FRT). The cells in the epithelial layer of the cervix are mainly CD8⁺ T cells whereas those in the subepithelial layer are mainly CD4⁺ T cells. The CD4⁺ T cells are of both the Th1 and Th2 type although their relative proportion changes during the menstrual cycle. Interestingly, recent data have suggested that there are regulatory T cells in the FRT, especially in the endometrium, that may play a role in providing an environment for tolerance of a fertilized egg and subsequently the fetus (Section M4). Some B lymphocytes are found in the FRT although they tend to localize to the endometrium. However, plasma cells (especially those producing IgA), although distributed mainly in the lamina propria of the cervix and vagina, have also been found in the fallopian tubes and endometrium (Table 1).

IgG and IgA are both normally found in secretions of the cervix and vagina. IgA is derived locally from IgA plasma cells while IgG is thought to be mainly derived from the serum, although some local IgG production does take place. IgA, and some IgM, is transported across the mucosal surfaces of the tract via poly-Ig receptors (pIgR) on specialized epithelial cells (Section E4). It is less certain how IgG is transported, although it is likely that there are IgG Fc receptors on specialized epithelial cells that can carry out this activity. Interestingly, the levels of IgG are usually higher than those of IgA in the nonpregnant female reproductive tract, but vary during the cycle. In contrast, at other mucosal sites, for example intestine, secretory IgA is predominant (Section D2).

The FRT, like other mucosal surfaces, is constantly open to the outside world and therefore has direct exposure to pathogens (Section C3). Vaginal immunization gives rise to both IgA and IgG responses to the immunizing antigen in both the vagina and cervix, confirming the ability of the lower reproductive tract to respond to foreign antigens. Independent of immunization, post-menopausal women tend to have higher concentrations of IgA and IgG in the cervix than nonimmunized pre-menopausal women. Since the vagina is an efficient site for immunization, it could be expected that sexually active women would mount potent responses to sperm antigens and other proteins associated with the male ejaculate. For the most part, however, this appears not to be the case, as females are usually unresponsive to sperm antigens, perhaps as the result of tolerance induction (Sections G2 and G3). In addition, the high concentrations of prostaglandins (which have potent immunosuppressive properties) in seminal fluid may inhibit immune responses. However, antibodies to sperm antigens are produced in some women, which can lead to infertility.

## Immunological changes during the menstrual cycle

As described earlier, NK cells, macrophages, CD4+ T cells, CD8+ T cells, and B cells are distributed throughout the endometrium. The number of NK cells increases from the late proliferative phase (LP), also called the follicular phase, to the late secretory phase (LS), also called the luteal phase, at which time they constitute the largest number of lymphocytes (Figure 2).

T-cell numbers appear to remain relatively unchanged during the cycle, although the ratio of CD8+ to CD4+ T cells is lower at the LS phase compared with the LP phase, a finding consistent with a higher cytotoxic cell function during the proliferative phase.

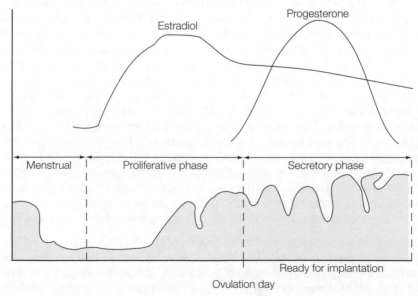

Figure 2. Endometrial thickening during the menstrual cycle.

Measured as percentage of lymphocytes, NK cells increase from 20% to 80%, whereas T cells decrease from 50% to less than 10% between the LP and LS phases of the menstrual cycle. In contrast, NKT cells that are able to respond to glycolipids and produce a range of cytokines (Section B1) are increased in the late secretory phase of the menstrual cycle. Few classical MHC class II-positive dendritic cells are present in the uterus. Changes in cell populations are summarized in Table 2. Organized lymphoid aggregates consisting of a core of B cells surrounded by more numerous CD8$^+$ T cells and an outer halo of monocytes/macrophages are also found in the endometrium. The number and size of these aggregates increase during the menstrual cycle and their absence in post-menopausal women suggests that they are hormonally influenced. The function of these aggregates is at present unknown. The phagocytic activity of mononuclear cells decreases during the early phase of the cycle compared with the later phase, whereas T-cell responses to *Candida* antigens are reduced during the secretory phase. Interestingly, this appears to be the time of increased susceptibility to *Candida albicans* infections. Taken together, the marked fluctuations in estradiol or progesterone during the menstrual cycle may influence not only the accumulation of different populations of immune cells but also their response to microbial antigens. In addition to changes in lymphocyte populations, there is also evidence for an alteration in expression of TLRs during cycle which might also contribute to changes in susceptibility to infection.

**Table 2. Changes in lymphocyte populations during the menstrual cycle**

|  | Proliferative phase | Secretory phase |
| --- | --- | --- |
| NK cells | Lower | Higher |
| T-cell subsets | Mostly Th2 | Mostly Th1 |
| NKT cells | Lower | Higher |
| Tregs (peaks at ovulation) | Higher | Lower |

## Immune-associated changes during pregnancy

As mentioned earlier, the endometrium not only protects against microbial invasion but, during pregnancy, also creates a tolerogenic environment for the fetus. It is believed that in conjunction with hormonal changes, for example increased levels of estrogen and progesterone, the immune-associated changes during pregnancy are responsible for this environment. Following implantation of the blastocyst, there is an increase in estradiol and progesterone, the latter being essential for maintaining the pregnant state. This prepares the uterus for reception and development of the fertilized ovum and induces changes in the endometrium resulting in a modified mucosa (decidua) (Figure 3). Changes occur in immune cell populations in the endometrium and in the overall immune system, providing an "immune privileged state" for the maintenance of the fetal "transplant" (Section M4). During the first trimester of pregnancy, endometrial NK cells (now called decidual NK cells) account for 60–80% of the immune cells and these rapidly decrease after the second trimester. The remainder of the immune cell population consists of macrophages, CD3$^+$ T cells (of which CD4$^+$ and CD8$^+$ are expressed in equal numbers), γδ T cells, NKT cells, Tregs (Sections B1 and C1), and a small percentage of B cells.

During the early phases of pregnancy, Th2 lymphocytes are the dominant T-lymphocyte subset in the decidua. Th2-type cytokines, such as IL-4 and IL-6, induce the release of human choriogonadotropin (HCG) from trophoblasts, which in turn stimulates the production of progesterone from the corpus luteum. The decidua (part of the placenta) itself secretes IL-6, IL-10, IL-13, and TGFβ, which decrease the secretion of Th1-type cytokines.

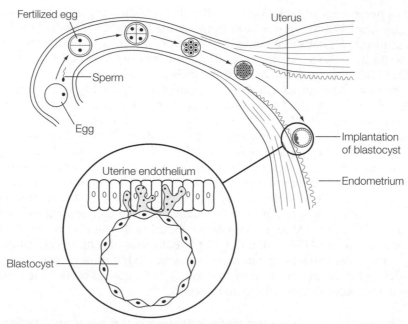

Figure 3. Fertilization and implantation of the blastocyst.

Moreover, trophoblasts are a source of IL-4 and IL-10. Thus Th2-type T cells and placenta-derived Th2 cytokines may contribute to the maintenance of pregnancy by modulating the immune (e.g., Th1-cell function; Section G5) and endocrine systems. Interestingly, in animal models of pregnancy, Th1-type cytokines such as IFN$\gamma$ have been shown to be associated with spontaneous abortion. During pregnancy the increase in Tregs and in the ratio of Th2 to Th1 is reflected in the circulation and throughout the body, contributing to a reduced immune state required for maintenance of the semi-allograft fetus (Section M4).

The decreased responsiveness of the maternal immune system aimed at allowing the maintenance of the fetus may to some extent increase the risk of infection to the mother and baby. Certainly, infections during pregnancy can be very serious and lead to preterm deliveries. However, PRRs expressed by trophoblasts that form the fetal–maternal interface may help to prevent transmission of microbial pathogens across the placenta.

## The role of the lactating breast in immune defense

Hormonal changes during pregnancy prepare the female breast for the supply of milk to ensure the healthy development of the newborn child. The World Health Organization has emphasized the importance of breast feeding in the prevention of respiratory and gastrointestinal infections and has predicted that increasing breast feeding by 40% would reduce respiratory and gastrointestinal death by approximately 50% worldwide in children under the age of 18 months.

### Localization of immune cells in the lactating breast

The female breast is composed of fatty and connective tissue, milk ducts, and lobules (Figure 4). The lobules contain specialized epithelial cells for producing milk and are surrounded by stroma or lamina propria containing capillaries, lymphatics, mononuclear cells, and granulocytes. The initial secretions from the breast (colostrum) contain between

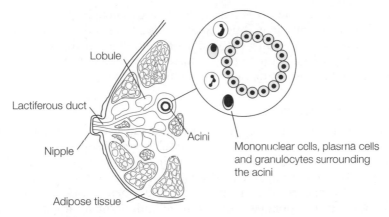

Figure 4. The lactating breast.

$10^6$ and $10^7$ leukocytes per ml that reduce to between $10^4$ and $10^5$ per ml in mature milk. Colostrum and breast milk contain lactoferrin and also a variety of cytokines and growth factors. These include TGFβ, colony-stimulating factors, IL-1, IL-6, TNFα, and epidermal growth factor. While the antimicrobial role of lactoferrin is well documented, it is not clear how these components contribute to protection in the hostile environment of the infants' intestine, although they may have a role in maintaining the integrity of the neonatal intestinal tract.

### Antibodies in colostrum and breast milk

The lactating breast is an important site of mucosal immunity. In humans, approximately 80% of plasma cells in the lactating breast produce IgA. This immunoglobulin, and other factors such as cytokines and growth and nutritional factors, provide help to the newborn in thriving and preventing infection. Secretory IgA is produced by plasma cells derived from B cells that, under hormonal influences, originally homed to the breast from other mucosal surfaces—major portals of microbial entry (Section C3). Lymphocytes entering the breast tissue contain homing molecules to allow them to enter mucosal tissue (Section C4). As in the intestinal wall, IgA dimers produced by plasma cells in the mammary gland are bound by the polymeric immunoglobulin receptors located on the basolateral surface membranes of epithelial cells. The antibody receptor complex is then internalized and transported to the apical surface of the cell. Since these secretory IgA antibodies are derived from B cells that have originated from other mucosal surfaces, such as the respiratory or gastrointestinal tracts, they usually have specificity against microbial antigens found in these tissues.

The concentration of IgA is greatest in colostrum and postpartum falls rapidly in the breast milk to the equivalent of serum IgA concentrations. While IgA in the serum is mainly IgA1—85% compared with 15% IgA2—mammary gland IgA contains more IgA2 than IgA1, a ratio similar to that for intestinal IgA (Section D2). The importance of IgA2 in the secretions may be related to its increased resistance to degradation by proteases produced by microbial pathogens (*Pseudomonas* sp., *Neisseria* sp., *Haemophilus influenzae*, and *Streptococcus pneumoniae*, etc.) found in the mucosa. The mechanisms by which ingested IgA functions to protect against infection presumably involve aggregation and blocking of attachment and entry of microbes (Section D8). This passive IgA mediated immunity is especially important immediately after birth because, as described previously, production of IgA by the infant only begins at birth (Section C5).

Many of the IgA antibodies found in breast milk are also directed against antigens derived from the mother's diet. Thus, in addition to protecting against microbial infection these antibodies may also protect against absorption of certain food antigens in early development, perhaps decreasing allergy to these substances. Small amounts of IgM and even smaller amounts of IgG are also found in breast milk.

## The male reproductive tract

The male reproductive tract (Figure 5) is composed of the testes, epididymis, and vas deferens, (upper reproductive tract) and seminal vesicles, prostate gland, and urethra (lower reproductive tract). Like those of the FRT, the epithelial cells lining the tracts have PRRs, including the TLRs 1–9 that mediate protection against most microbes (Section B3). In addition, the epithelial surface is bathed in antimicrobial proteins including human β-defensins and some chemokines that also have antimicrobial properties. Only occasional lymphocytes, mainly CD8$^+$ cells, and macrophages are found in the upper reproductive tract among the epithelial cells in the vas deferens and epididymis. The urethra is the first point of contact of microbes entering the lower male reproductive tract.

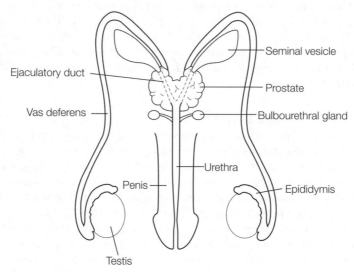

Figure 5. The male reproductive tract.

As in other mucosal sites, there are abundant CD4$^+$ and CD8$^+$ T lymphocytes in the lamina propria and epithelium (although CD8$^+$ T cells predominate). These are mainly memory T cells and many of them carry mucosa-associated homing molecules (Section C4). Macrophages are also frequent within the lamina propria but dendritic cells tend to be most plentiful in the distal tip of the urethra and particularly in the prepuce of uncircumcised males. These HLA class II-positive antigen-presenting cells are probably the first cells to come across sexually transmitted viruses, including HIV (Sections H2 and J3). IgA plasma cells are found in the prostate gland and the lamina propria (especially of the urethra) and home there from other mucosal surfaces. As in the gut, epithelial cells with poly-Ig receptors transport the IgA into the lumen (Section D2). Thus, the urethral mucosal tissue is an extremely important site of immunological protection against ascending microbial infections. The seminal fluid contains antimicrobial proteins produced by the epithelial cells of the tract, a high concentration of prostaglandins, and IgA and IgG (see above).

# O3 Functional effects of sex hormones on the immune system

---

## Key Notes

| | |
|---|---|
| **Effects of estrogen and progesterone on immune function** | Estrogens (and progesterone) have many effects on immune function. Estrogens increase immunoglobulin synthesis, cause a shift from Th1- to Th2-type cytokine production, inhibit NK-cell activity, and enhance dendritic cell differentiation. |
| **The effect of testosterone on immune function** | Testosterone appears to have a contrary effect to estrogens on some immunological functions in that it inhibits immunoglobulin synthesis by B cells and is associated with protection against some autoimmune diseases. |
| **Gender-associated autoimmunity: the role of sex hormones** | Most autoimmune diseases are more frequent in females than in males, suggesting that sex hormones play a pivotal role in the development and/or maintenance of these diseases. |

| **Related topics** | (E2) B-cell activation | (I2) Factors contributing to the development of autoimmune disease |
|---|---|---|
| | (G5) Regulation by T cells and antibody | |
| | (G6) Neuroendocrine regulation of immune responses | |

---

## Effects of estrogen and progesterone on immune function

Progesterone and estrogen (estradiol, estriol, and estrone) are the main female sex hormones. Estradiol is produced by the ovaries and progesterone by the corpus luteum and the uterine endometrium. Experimental studies *in vitro* and *in vivo* have shown several diverse effects of estrogen (and progesterone) on the immune system (Table 1). NK cells, dendritic cells, monocytes/macrophages and T and B cells all have estrogen receptors. Estrogen suppresses the cytotoxic activity of NK cells and downregulates IFNγ production. In contrast, it enhances the differentiation of IL-12-producing dendritic cells and antibody production, probably as a result of its effect on B cells and/or plasma cells, as well as enhancement of a Th2 cytokine response. Consistent with this finding, the anti-estrogen drug tamoxifen causes an increase in the production of IL-2 and IFNγ in animal models, while reducing the levels of the Th2-associated cytokine IL-10. Similar effects are seen with progesterone, suggesting that both of these hormones can regulate immune responses (Section G5). That estrogens can directly activate some lymphocyte populations is indicated by their ability to stimulate lymphocyte proliferation *in vitro* and modulate expression of lymphocyte surface molecules. In experimental models, estrogen has been shown to influence development of T and B cells. Thymic stromal cells and immature thymocytes express estrogen receptors and their absence in knockout mice results in decreased thymic size. Estrogen also inhibits the synthesis of thymosin $\alpha_1$, a thymic hormone, produced by thymic epithelial cells. Since thymosin $\alpha_1$ has a role in the maintenance of thymic homeostasis, the inhibition of this hormone may have a role in involution of the thymus.

In addition to preparing the uterine endometrium for implantation, progesterone has some effects on immune cells similar to those of estrogens (Table 1). Thus, the decreased NK-cell function and biased Th2 response throughout pregnancy mediated by these two hormones is likely to reduce the potential "rejection" of the fetal allograft mediated through a Th1-type mechanism (Section M4).

## Table 1.  Effects of estrogen on the immune system

Enhances differentiation of dendritic cells

Decreases IL-1β and IL-6 production by macrophages

Decreases NK-cell cytotoxicity*

Enhances Th-cell expression of chemokine receptors involved in homing

Induces Th2-type responses*

Increases Treg cell numbers

Enhances IgM and IgG synthesis by B cells

Downregulates IFNγ production

Prevents Fas-dependent apoptosis of Th2 cells

* Progesterone also has these effects.

## The effect of testosterone on immune function

Testosterone is the main male-associated sex hormone and is produced mainly by the testes. Receptors for testosterone on cells associated with the immune system appear to have a distribution similar to estrogen receptors. Testosterone, like estrogen, is able to dramatically induce thymic regression. In animal models, removal of the testes (orchidectomy) or removal of the ovaries (oophorectomy) causes thymic enlargement (hypertrophy). Although the mechanisms whereby testosterone and estrogen cause the thymus to regress are not fully understood it is likely that receptors for estrogen and testosterone on thymocytes and thymic epithelial cells are involved.

Testosterone has also been shown to induce Th2 cells to produce IL-10 (Table 2). This would tend to suppress Th1-cell responses, leading to decreased cell-mediated immunity. Unlike the effects of estrogen, T- and B-cell proliferation and production of

## Table 2.  Effect of testosterone on the immune system

Decreases expression of some TLRs on macrophages

Induces production of IL-10 from Th cells

Inhibits T- and B-cell proliferation and cytokine production

Inhibits B-cell production of immunoglobulins

Promotes Fas-dependent apoptosis of Th2 cells

immunoglobulins is inhibited (and not enhanced) by testosterone. In addition, one of the mechanisms by which lymphocytes are regulated is through Fas-mediated cell death through apoptosis (Section G3). Testosterone promotes, while estrogen prevents, Fas-dependent apoptosis of Th2 cells by reducing the expression of the apoptosis-suppressing mitochondrial proteins. Interestingly, there is now evidence that lymphocytes can produce small amounts of testosterone but the significance of this is currently unclear.

## Gender-associated autoimmunity: the role of sex hormones

Autoimmune diseases are disproportionately more common in females than in males (Table 3), with a few exceptions; for example, ankylosing spondylitis is found mainly in males. That autoimmune diseases are more frequent in females than males and occur more frequently post-puberty suggests a major immunological role for sex hormones in these conditions. Moreover, post-menopausal women have less clinical disease activity than their pre-menopausal counterparts, perhaps as a result of the decreased estrogen levels in post-menopausal women and/or their increased IL-2 and IFNγ levels. In addition, hormone replacement therapy reverses the decreased clinical disease activity in post-menopausal women. Clinical activity in SLE is also known to cycle with menses. Furthermore, pregnancy can ameliorate certain autoimmune diseases; for example, rheumatoid arthritis symptoms are markedly reduced during pregnancy. Thus, hormones altered during pregnancy can have a major influence on the immune system.

In animal models of diabetes such as the nonobese diabetic (NOD) mouse, diabetes develops spontaneously but the severity of the disease is much more common in the female of the species. This severity can be reduced by removal of reproductive organs in female mice or treatment with testosterone. Sjögren's syndrome, which is primarily a female autoimmune disease, is also found in hypogonadal males. In mouse models of Sjögren's syndrome, treatment with testosterone delays disease progression. Overall, the greater predisposition of females to autoimmune diseases appears related to the effects of the different sex hormones.

## Table 3. Incidence of autoimmune diseases in females compared with males

| Hashimoto's disease | 30:1 |
|---|---|
| Sjögren's syndrome | 9:1 |
| Systemic lupus erythematosus | 9:1 |
| Primary biliary cirrhosis | 9:1 |
| Antiphospholipid syndrome | 9:1 |
| Chronic active hepatitis | 8:1 |
| Mixed connective tissue disease | 8:1 |
| Graves' disease | 6:1 |
| Rheumatoid arthritis | 3:1 |
| Scleroderma | 3:1 |
| Type 1 diabetes | 2:1 |
| Multiple sclerosis | 2:1 |
| Myasthenia gravis | 2:1 |
| Celiac disease | 2:1 |

# P1 Overview

**Key Note**

**Overview**

It is now clear that as humans age, defects in both innate and adaptive immune systems occur, and the overall effectiveness of their immune system declines (termed *immunosenescence*). T-cell function is reduced; the affinity of antibodies decreases and the response to vaccination is diminished. The incidence of microbial infections increases, with greater morbidity and mortality. Moreover, there is an increased predisposition to malignancies.

**Related topics**

| | |
|---|---|
| (B1) Cells of the innate immune system | (G5) Regulation by T cells and antibody |
| (C1) Lymphocytes | (G6) Neuroendocrine regulation of immune responses |
| (D1) Antibody structure | |
| (F1) The role of T cells in immune responses | |

## Overview

A decline in immune competence is well recognized in the elderly and is due to changes in both the innate and adaptive immune systems. Key immunological manifestations of immunosenescence include poor responsiveness to new emerging pathogens, such as West Nile virus and severe acute respiratory syndrome (SARS), and the reduced efficacy of vaccination. The incidence of microbial disease is increased compared with the young (Table 1); for example, urinary tract infections, septicemia (bacteria in the bloodstream), and periodontal disease are much more common in the aged. Morbidity associated with gastroenteritis caused by *Salmonella* and other enteric bacteria such as *Escherichia coli* O157 is greater. Pneumonia, resulting from infections with influenza, rhinovirus, and cytomegalovirus (CMV), is much more frequent and morbidity and mortality greater. There is a 10-fold increase in the incidence of TB in the elderly. Furthermore, some studies indicate that the major cause of death in individuals older than 80 years is infections.

**Table 1. Microbial infections associated with aging**

| | |
|---|---|
| Tuberculosis | Shingles |
| Pneumonia | Influenza |
| Urinary tract infections | Cytomegalovirus (CMV) |
| Bacterial dysentery | Gram-negative sepsis |
| Periodontal disease | |

Malignancies are seen much more frequently in older people and while many of these may be related to inappropriate DNA translational events, a defective immune system is also responsible, since there is an association between immune deficiency and increased malignancy (Section J3).

Aged individuals show a decline in many aspects of protective immunity, including a tendency to produce lower-affinity antibodies, a poor outcome from vaccination, and a loss of delayed-type hypersensitivity to antigens previously encountered in life (Section K5). While not all studies agree, there is reliable data indicating that defects develop in innate immunity as well as in the adaptive immune response. Although an increase in NK-cell numbers is a feature of aging, their cytotoxic function is diminished. The function of both NKT and $\gamma\delta$ T cells is also decreased (Sections B1 and C1). IL-6 and IL-10 production by monocytes is increased with aging as well as the pro-inflammatory cytokines IL-1$\beta$ and TNF$\alpha$. MHC molecules are expressed at lower density on a variety of cells, and fewer T cells expressing CD28, important for T-cell signaling, are found in the elderly (Section F1).

Hemopoiesis is impaired, with fewer progenitor cells produced. Thymic involution is significant in the elderly, with fewer naive T cells entering the vascular pool and hence secondary lymphoid organs. Activation-induced cell death (AICD) and apoptosis are increased. The incidence of shingles, a good indicator of immunosenescence, suggests decreased functional activity in the T, NK, and NKT cell compartments. Decreased T-cell function can readily be shown by DTH tests to recall antigens (Section J4). However, other factors such as physical well-being, nutritional status, stress, gender, and previous vaccination history must all be considered. Individuals who are malnourished are prone to vaccination failure. Autoimmunity, more common in females (Section L2), is also more frequent as aging progresses. Autoantibodies are increased although not necessarily associated with pathology. Interestingly, systemic lupus erythematosus (SLE) becomes milder in patients as they age, as does primary Sjögren's syndrome, suggesting that immunosenescence may also have benefits in some clinical situations. Furthermore, acute rejection of kidney, heart, or liver grafts is less in the elderly and the incidence of asthma and specific IgE responses to allergens decreases.

Age-related changes in hormonal and neurotransmitter function may also have an impact on immune function and may determine morbidity, mortality, and longevity.

# P2 Developmental changes in primary lymphoid tissue and lymphocytes with age

## Key Notes

| | |
|---|---|
| **Hemopoiesis and aging** | With increasing age, the number of progenitor cells produced decreases. This is seen in the bone marrow and in the thymus, organs that give rise to the specialized cells of the immune system. |
| **Thymic involution and aging** | One of the major immunological events associated with aging is the involution of the thymus. This begins at puberty and continues to senescence and as a consequence results in a decreased output of thymic T cells to the vascular pool. |
| **Lymphopenia, apoptosis, and aging** | Lymphopenia is associated with aging. While the number of mature naive T cells entering the vascular pool is reduced, many studies have shown that T-cell apoptosis (programmed cell death) is increased (Section F5). This appears to be associated with increased expression of CD95 (Fas) and its ligand (CD95L) on T cells and T-cell subsets as well as the presence of soluble CD95 in the sera of the aged. Thus, in the aged, increased T-cell susceptibility to death may contribute significantly to immunodeficiency and decreased response to vaccination. |
| **Related topics** | (A5) Hemopoiesis – development of blood cells<br>(C2) Lymphoid organs and tissues<br>(F3) Shaping the T-cell repertoire<br>(G2) Central tolerance<br>(G3) Peripheral tolerance |

## Hemopoiesis and aging

Aging has a profound influence on the differentiation of hemopoietic stem cells (HSCs). This is supported by studies showing that aged HSCs are less able to support erythropoiesis and lymphopoiesis than those from the young. These findings help to explain age-associated anemia and the decline in the renewal and function of T and B cells in aged individuals.

The proliferative capacity of cells of the bone marrow peaks during middle age and decreases thereafter. This is associated with a decrease in colony-stimulating factors (Section B2), a decrease in the number of progenitor cells (including those for T and B cells) produced, and an increase in apoptosis. There is also a decreased capacity to deal with severe bleeding in the elderly, suggesting a decline in the function of the coagulation system. Other factors that may have a role in hemopoietic senescence include changes in microenvironment, hormones, stromal cells, and cytokines responsible for stem cell growth and differentiation. These changes probably also contribute to immunosenescence in that there would be fewer progenitor cells committed to the maturation processes of the immune system, leading to fewer naive lymphocytes entering the vascular pool and secondary lymphoid tissues. However, even though there are fewer progenitor cells in the marrow of aged individuals, they are still functional, as indicated by the finding that there are sufficient numbers of these cells to achieve hemopoiesis in autograft treatments for multiple myeloma and other malignant blood disorders.

## Thymic involution and aging

One of the most obvious effects of aging on the immune system is the involution of the thymus. After an initial increase in thymus size up to puberty there is a progressive decrease to near total atrophy in the elderly, where much lymphoid tissue is replaced by fat. There is an approximate 3% decrease per year in the size of the thymus at puberty until middle age and an approximate 1% decrease per year thereafter (Figure 1). As a result of this, the number of T cells entering the vascular pool and secondary lymphoid organs and tissues decreases with age. Even so, data from studies on the rearrangement of genes encoding the T-cell receptor indicate that, although reduced, there is continuing thymic function in the elderly.

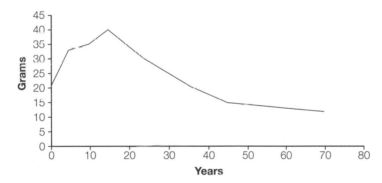

Figure 1. Average thymus weight change with age.

That changes occur in the thymus during aging is also indicated by observations on the recovery of CD4+ T cells following chemotherapy or therapy with anti-CD4 antibodies. The CD4+ T cells return to normal levels faster in younger individuals than adults and are mainly of the naive phenotype (CD45RA), compared with those with a memory phenotype (CD45R0) in the elderly.

Although thymic atrophy is probably under the control of many factors, including T-cell cytokines, thymic hormones, and products of both the nervous and endocrine systems, it is thought to result primarily from failure of the thymic microenvironment to support lymphopoiesis. Significant cytokine changes occur in the thymus during aging, including increases in IL-6 and macrophage colony-stimulating factor (M-CSF), and decreases in IL-2, IL-10, and IL-13. Other important thymic cytokines such as IL-7 and IL-15 remain stable during the aging process. There is also an age-associated increase in acetylcholinesterase-positive structures in the human thymus. In rodent models, older mice have increased noradrenergic sympathetic nerves and a 15-fold increase in the concentration of noradrenaline (norepinephrine), which indicates a role for the nervous system in thymic function especially in old age (Section P4).

## Lymphopenia, apoptosis, and aging

The role of TNFα in apoptosis mediated through Fas (CD95) and Fas ligand (CD95L) inter-actions (Section F5) has been studied extensively in the aged. T cells in the elderly have increased CD95 and CD95L expressed on their surfaces. Moreover, there is an increase in the level of soluble CD95L in the sera of aged individuals. TNFα, a cytokine produced mainly by activated macrophages, is also increased in the aged. Moreover, human T cells and T-cell subsets display increased sensitivity to apoptosis mediated through CD95/95L and an increased sensitivity to TNFα-mediated killing. This may be fundamental to the immunodeficiency, decreased responses to vaccination, and immunosenescence seen in aged individuals (Table 1).

**Table 1. Factors that influence immunosenescence**

| | |
|---|---|
| Hemopoietic failure | Thymic involution |
| Endocrine senescence | Neural senescence |
| Apoptosis | |

# P3 The effects of aging on the innate and adaptive immune systems

---

**Key Notes**

| | |
|---|---|
| **Innate immunity in aging** | Neutrophils, monocytes, macrophages, dendritic cells, and NKT cells are the major components of the innate immune system (Section B1). They function immediately after birth and continue to do so throughout life. Even so, there is some loss of functionality associated with the aging process in all compartments of the innate immune system, including cytokine production and the expression of pattern recognition receptors (e.g., TLRs), all or some of which may contribute to immunosenescence. |
| **The adaptive immune system in aging** | Both T- and B-cell defects have been reported during aging. T-cell abnormalities include increased numbers of memory cells, decreased naive cells, reduced IL-2 and IL-2R expression, and poor responses to T-cell stimulants. B-cell abnormalities have also been shown and include a lack of CD40 expression and decreased antibody affinities and specificities. Autoantibodies are increased and there is an increase in monoclonal gammopathies. Of particular importance, there is a diminished ability to produce appropriate T- and B-cell responses to microbes and vaccines. |
| **Related topics** | (B1) Cells of the innate immune system · (F1) The role of T cells in immune responses |
| | (B2) Molecules of the innate immune system · (F2) T-cell recognition of antigen |
| | (D1) Antibody structure · (F4) T-cell activation |
| | (E2) B-cell activation |

---

## Innate immunity in aging

Innate immunity functions immediately after birth and persists throughout life, although some functional changes occur with aging. Defects during aging have been described in neutrophils, monocytes, macrophages, dendritic cells, NK cells, and NKT cells, the major components of the innate immune system.

Pattern recognition receptors (PRRs), for example Toll-like receptors (TLRs), play an important role in defense against infectious diseases as they are expressed on a variety of different cell types, including macrophages and dendritic cells (DCs) (Section B3), and are responsible for directing these cells to microbes. Studies have shown that signaling through TLR-1, 2, 7, and 9 is reduced in DCs and may be responsible for the decreased Th1 responses seen in the elderly. Taken together, these data suggest that, in the aged, an inappropriate microenvironment for antigen presentation and T-cell proliferation contributes to poorer responses against microbial infections (Table 1).

**Table 1. Major changes in the innate immune system with aging**

| |
| --- |
| Increased IL-10 production |
| Decreased TLR-1, 2, 7, 9 expression |
| Decreased superoxide production |
| Decreased numbers of DCs in skin epidermis |
| Decrease in cytotoxicity |
| Decrease in IL-2, IL-12, and TNFα |

## Neutrophils

Although there are no changes in the number of mature neutrophils in the blood between aged and young individuals, a decrease in chemotaxis and phagocytosis has been reported. Other neutrophil abnormalities include reduced superoxide production in response to staphylococcal infections (Section B1) and reduced ability to respond to survival factors such as GM-CSF and G-CSF (Section B2). There is also a marked reduction with aging in the expression of CD16, an IgG Fc receptor on these cells that mediates phagocytosis of antibody-coated microbes (Section B1).

## Monocytes, macrophages, dendritic cells

After leaving the bone marrow and entering the blood, monocytes differentiate continuously into tissue macrophages (Section B1). Monocytes, macrophages, and DCs are highly plastic and the tissue microenvironment in which they reside as well as the aging process can modify their number and function. Monocytes from older adults frequently show signs of activation and secrete increased levels of IL-10 compared with monocytes from young adults. In the elderly, macrophages produce less superoxide, a molecule associated with intracellular killing, and are less able to migrate in response to a chemotactic signal (Section B1). Similar findings have been found with DCs in the elderly. Moreover, DCs (Langerhans cells) are reduced in number in the skin of the aged (Section B1).

Accessory cell function is critical for T-cell activation and is achieved by presenting antigens in association with class I and class II MHC molecules. Compared with younger controls, MHC class I molecules (Section M2) are less well transcribed (based on mRNA levels) in monocytes from the elderly, which may give rise to less-efficient antigen presentation and poorer immune responses, especially those involving cytotoxic CD8+ T cells (Section F5).

## NK and NKT cells

NK and NKT cells are very important in host defense against microbial infection and tumors (Section B1), as both are cytotoxic and can destroy tumor cells and cells infected with microbes. Although often increased in number in the elderly, NK cells have impaired cytotoxicity that potentially represents a risk factor for microbial infections and malignancy in the aged and may contribute to their increased susceptibility to some age-related diseases. Age-related declines in perforin, cytotoxic granules, and granzymes

(Section F5) have been reported, and are significantly greater in men than women, which may help to account for the greater longevity of women.

The cytokines IL-2, IL-12, and TNFα that are involved in the activation of NK cells are reduced during aging. Thus, on a cell-to-cell basis, NK cells in the elderly release only 25% of the IFNγ released by NK cells from young adults. In mouse models of aging, similar results have been found. Note that NK cells from spleen and lymph nodes of older animals have a profound loss of function, but whether this is true in humans has yet to be ascertained. Diet may also have a major effect on NK-cell activity, as indicated by the finding that dietary zinc can restore NK-cell function and increase IFNγ production in the aged.

NKT cells, which have characteristics of both NK and T cells (Section B1), increase in number with age, but produce less IFNγ. Moreover, like NK cells, cytotoxicity of NKT cells also decreases with age.

## The adaptive immune system in aging

T and B cells are the main cell types of the adaptive immune system and although the numbers of both cell types are relatively stable in the aged compared with the young, they are to some extent phenotypically and functionally impaired in the aged.

### T cells

The primary changes in T-cell phenotypes with aging are a marked decrease in naive T cells (CD45RA⁺) and an increase in memory T cells (CD45RO). In contrast, most T cells in young adults resemble "naive" cells, and in newborn blood nearly 100% are naive. CD4⁺ T cells expressing CTLA-4 (CD152) are also increased with aging and there is an increase of soluble CTLA-4 in the sera. While the pronounced increase in the number of memory cells is seen in both the CD4⁺ and CD8⁺ T-cell populations it is more pronounced in CD8⁺ T cells. This may, in part, be a result of the longer half-life of CD8⁺ memory cells that are found in older people and may account for the skewing of the CD4⁺/CD8⁺ ratio in favor of CD8⁺ cells.

Another marker thought to be important in evaluating T-cell function is CD28, a receptor critical to T-cell activation. Most T cells are CD28⁺ at birth, but with aging, the ratio of CD28⁺ to CD28⁻ cells decreases, particularly in the CD8⁺ subset. This decrease with age in CD28 would be expected to decrease the ability of aging individuals to respond aggressively to pathogens. It is interesting, however, that nonagenarians, who have very successfully aged, have a similar decrease in CD28⁺ cells. Even so, their CD8⁺ T-cell functions remain intact, indicating that a decrease in CD28 positivity is not necessarily an indicator of decreased immune function. Rather, there must be other factors that compensate for decreased CD28 expression with aging, that permit these cells to function appropriately, and that perhaps facilitate rejuvenation of the T-cell population, that is, development of more naive T cells.

Various explanations have been proposed for the decreased number of naive T cells and increased memory cells in the aged, including thymic involution, persistent antigen challenge, and replicative senescence. In the case of thymic involution there would be fewer new thymic emigrants (naive T cells) entering the vascular pool, and thus fewer naive T cells in the tissues and circulation.

Constant exposure to foreign antigens over time, as well as chronic stimulation by endogenous viruses such as those of the herpes family (e.g., Epstein–Barr virus (EBV), varicella-zoster virus (VZV), and CMV)), may persistently stimulate memory cells. This may happen to the extent that the immunological burden of the resulting virus-specific CD8⁺ T-cell population may make it difficult for generation and/or housing of naive T cells. For example, in industrial countries, CMV has a prevalence of 60% increasing to 90% in the aged. T-cell responses to the virus are broad and target many of its antigenic epitopes. This results in large numbers of CD8⁺ memory T cells, so many that the number of CMV antigen-specific memory T cells may represent 50% of the entire CD8⁺ memory T-cell pool. Over time (with aging) this would result in the filling of the immunological space with memory cells.

With regard to replication senescence, except for cancer cells and some stem cells, most cells cannot divide indefinitely, as errors may occur with each replication. Although these are usually fixed promptly, with aging, repair mechanisms become less able to accurately, and in a timely way, repair mistakes. The innate and adaptive immune systems suffer these difficulties along with all other systems.

Functionally, T cells from aged individuals show poor proliferative responses to phytohemagglutinin (PHA), anti-CD3, or IL-2. They also show poor delayed-type hypersensitivity (DTH) responses (Section K5). They produce significantly lower levels of IL-2 and express fewer IL-2 receptors. If insufficient amounts of IL-2 are produced or if T cells cannot respond effectively to IL-2, T-cell function will be impaired. Impairments in the activation of transcription factors associated with IL-2 synthesis have also been described in the elderly (Table 2).

**Table 2. Major T-cell changes associated with aging**

| |
| --- |
| Increase in memory T cells (CD45RO) |
| Decrease in naive T cells (CD45RA) |
| Increase in CD4 cells expressing CTLA-4 |
| Decrease in T cells expressing CD28 |
| Decreased proliferative responses to mitogens and antigens |
| Decreased DTH responses |
| Decreased IL-2 production |
| Decreased IL-2R expression |

### Th1 versus Th2 responses

Studies regarding Th1 and Th2 responses in aging are unclear. Some studies suggest an imbalance, with a decline in Th1 and an increase in Th2 activity accompanied by increased synthesis of IL-4 and IL-10. However, other studies indicate that the microenvironment in which CD4⁺ T cells develop in the elderly permits production of more cells committed to Th1 than in younger subjects. Moreover, animal studies have generally provided evidence of enhanced Th1 over Th2 responses.

### B cells

Abnormalities associated with aging have also been identified in B cells. Two B-cell groups (B1 and B2) can be distinguished based on their requirements for T-cell help (Section E2).

B1 B cells predominate in peritoneum and mucosal tissues, and represent a self-renewing pool that is established early in life. B1 B cells are characterized by production of polyspecific, low-affinity "natural" antibodies, primarily of the IgM class, in response to microbial antigens, although they may also be self-reactive. They are also distinguished by their expression of CD5, a molecule that binds to another B-cell molecule CD72, which thus permits B-cell–B-cell interactions.

B2 B cells are the primary B-cell group involved in production of high-affinity, high-specificity antibodies of the various Ig classes. Their activation requires T-cell help, which results in affinity maturation. Many newly formed immature B2 cells migrate from the bone marrow to the spleen (as transitional B cells) where they mature. Some seed into the splenic follicular zone, others migrate via the bloodstream into other secondary lymphoid tissues such as lymph nodes and mucosal tissues. Entry into the spleen for maturation purposes ensures a constant turnover of splenic B2 cells and the maintenance of the B2-cell repertoire.

Marginal zone B cells are not easily categorized, as they recirculate and have somatic mutations in their B-cell receptor, and may best be viewed as IgM memory B cells that develop diversity independent of T cells.

The developmental pathways of these B-cell types vary with age. In prenatal and neonatal life, the B1 pathway predominates, whereas after birth the B2 pathway becomes dominant. However, with age, the number of immature B2 cells migrating from the bone marrow to the spleen is reduced. As B2-cell production slows, the relative ratio of B1 to B2 cells increases. The consequences of increased proportional representation of B1 versus B2 are unclear but may contribute to some repertoire differences and the increased incidence of autoantibodies found in aging.

Mature B2 cells can be divided into different subsets on the basis of their expression of certain phenotypic markers. Memory B cells are CD27+ while naive B cells are CD27− IgD+. In the aged, double-negative (CD27−IgD−) cells are significantly increased. These double-negative cells do not seem to act as antigen-presenting cells, nor do they express significant levels of CD40, the molecule that interacts with CD154 on T lymphocytes. Hence, these expanded cells may be late memory or exhausted cells that have downmodulated the expression of CD27, and filled the immunologic space in the elderly. Interestingly, B2 naive lymphocytes are increased in healthy centenarians, when compared with control subjects, and would appear to give a survival advantage.

Although the amount of antibody in the circulation does not appear to decrease with age, there are significant changes. There is a decrease in antibody affinity and specificity with a concomitant decline in vaccination efficacy. Antibody diversity in the aged also significantly deviates from normal and correlates with health status; the more restricted the antibody repertoire the greater the association with fragility. In addition, antibody clonality is more common and monoclonal gammopathies are increased. Aging is also associated with increased autoantibody responses, although these are of low specificity and of unknown pathological significance.

Some of the abnormal antibody responses described in aging may be related to dysfunctional CD4+ T-cell support for B-cell activation and/or proliferation, for example through defective T-cell signaling via CD86–CD28 and/or CD154–CD40, or reduced or abnormal cytokine responses. However, T independent polysaccharide responses, which are crucial for antibacterial protection, also appear to be lacking later in life.

It is also interesting that there is evidence for a decrease in IgE-mediated allergies (Section K2) with age and an increase in salivary IgG and IgA levels, perhaps reflecting changes in mucosal immunity. Moreover, gastrointestinal immunosenescence is associated with deficits in differentiation and homing of IgA plasmablasts to the lamina propria and the initiation and regulation of local antibody production.

B-cell-associated changes in aging include:

- Decreased numbers of immature B cells migrating from bone marrow
- Relative increase in B1 cells
- Increased double-negative (IgD⁻,CD27⁻) B1 cells
- Lack of CD40 expression
- Decreased antibody affinity, specificity, and diversity
- Diminished ability to produce antibodies to vaccines and to novel antigens
- An increase in monoclonal gammopathies
- An increase in autoantibody production especially rheumatoid factor, anti-dsDNA, anti-histones, and anti-cardiolipin antibodies (Section L)
- Decreased incidence of IgE mediated allergies

# P4 The effect of the neuroendocrine system on the immune system in aging

| Key Note | | | |
|---|---|---|---|
| **Immunosenescence and the neuroendocrine–immune network** | Cells of the innate and adaptive immune systems can respond to signals from the neuroendocrine system resulting in cellular activation and cytokine production. Defects in this system as a result of senescence may have a significant impact on the immune system as it ages. | | |
| **Related topics** | (F2) T-cell recognition of antigen | (G6) | Neuroendocrine regulation of immune responses |
| | (H2) Immunity to different organisms | | |

## Immunosenescence and the neuroendocrine–immune network

The immune, endocrine, and nervous systems influence each other through cytokines, hormones, and neurotransmitters. During aging, changes occur in all areas of these interactive systems.

## The endocrine system

During aging there is a decrease in dehydroepiandrosterone (DHEA), dehydroepiandrosterone sulfate (DHEAS), progesterone, and aldosterone, as well as in growth hormone, melatonin, and the sex hormones estrogen and testosterone. Although the extent to which decreases in these hormones influence immunosenescence is not clear, circumstantial evidence indicates that deficiencies in at least some of these molecules contribute to significant changes in immune defense. As examples of the role of these hormones: (1) The levels of cortisol, a powerful anti-inflammatory stress-related hormone, are increased in aging; (2) estrogen stimulates immune function and testosterone is inhibitory (Section O); (3) in mouse models, melatonin improves the DTH response, prevents thymic involution, and increases antibody responses; and (4) hormonal treatments (T3, T4, melatonin, growth hormone) restore NK-cell cytotoxicity and IL-2 and IFNγ production in aged mice.

Leptin is another hormone known to affect the pro- and anti-inflammatory cytokine responses. While commonly associated with obesity, increased leptin levels are known to affect the Th1/Th2 balance by increasing the pro-inflammatory cytokine response. Leptin levels are increased by estrogen and decreased by testosterone. Since these hormones decrease with age, changes in leptin levels may contribute to the T-cell cytokine status.

## The nervous system

The nervous system in humans consists of the central nervous system (CNS) and the peripheral nervous system (PNS) and includes billions of nerve cells (neurons) connected through nerve fibers (axons) that communicate with other cells via synapses—membrane-to-membrane junctions that allow chemical or electrical signaling. Nerve fibers that release neurotransmitters are widely distributed in tissues throughout the body, including those of the endocrine system and the primary and secondary lymphoid organs. Receptors for neurotransmitters are expressed on cells of both the innate and adaptive immune systems and include receptors for neuropeptide Y (NPY), somatostatin (SOM), vasoactive intestinal peptide (VIP), substance P, dopamine, serotonin, norepinephrine (noradrenaline), and many others. These receptors on different immune cell types interact with neurotransmitters and modulate immunological changes such as cellular adhesion and trafficking, cytokine production, mast cell degranulation and histamine release, as well as other inflammatory responses (Section G6).

The expression of dopamine D3 receptors on lymphocytes is also reduced during aging. Since dopamine induces T-cell adherence to fibronectin, which is important in T-cell trafficking, the lower expression of dopamine receptors would compromise T-cell trafficking and thus immune defense against pathogens.

Changes to the nervous system occur during aging. In particular, the sympathetic response, which aids in the control of the body's internal organs and in mobilizing the body's resources under stress, is increased, which may in turn alter its influence on the immune system. Among other events, there is an increase in norepinephrine in the circulation, which would be expected to inhibit IL-2, IL-12, and IFNγ synthesis, and stimulate IL-6 and IL-10 production (Figure 1) resulting in induction of Th2-type responses at the expense of Th1 responses. Thus, the Th1/Th2 balance in the elderly might decrease and compromise immunity while enhancing asthma and IgE-mediated allergies. For the most part this does not seem to be the case. Moreover, other studies suggest that the microenvironment in which CD4⁺ T cells develop in the elderly permits production of more cells committed to Th1 than Th2, and more Th1 than that in younger subjects. Thus, although the nervous and endocrine systems play a role in modulating immune defense, the interactions are complex and are likely to be severely complicated by many other different factors (including the variability of the elderly populations that are studied) that also influence the nature and effectiveness of the immune response to an antigen.

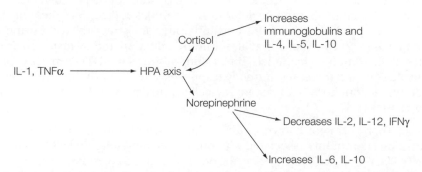

Figure 1. Interplay between the immune, endocrine, and nervous systems. HPA, hypothalamic–pituitary–adrenal.

Taken together these findings show the interplay between the nervous, endocrine, and immune systems and the complex associated changes with aging (Table 1) that eventually lead to immunosenescence.

**Table 1. Changes in molecules of the neuroendocrine system with aging**

| | |
|---|---|
| Cortisol | Increased |
| Norepinephrine | Increased |
| VIP | Increased |
| Melatonin | Decreased |
| DHEA, DHEAS | Decreased |
| Growth hormone | Decreased |
| Estrogen | Decreased |
| Testosterone | Decreased |
| Substance P | Decreased |
| Serotonin levels | Decreased |

VIP, vasoactive intestinal peptide; DHEA, dehydroepiandrosterone; DHEAS, dehydroepiandrosterone sulfate.

# Q1 Cytokine-, antibody-, and cell-mediated immunotherapy

---

**Key Notes**

| | |
|---|---|
| **Cytokine-mediated immunotherapy** | Cytokines have been known to have an important role in immunity since the 1960s and were viewed as promising agents for therapy of tumors and other diseases. When these cytokines became readily available they were therapeutically tested in a variety of diseases with disappointing results. Currently only a few cytokines are approved for therapy, others are still being studied. |
| **Antibody-mediated immunotherapy** | In 1952 Ogden Bruton diagnosed a boy with a history of infection as deficient in antibodies and successfully treated him with human Ig, an approach still used for therapy. In 1975 Kohler and Milstein developed an approach to create mAbs that are now widely used for immunotherapy. Although initially of mouse origin, they are made more human by replacing mouse H- and L-chain C regions with human C regions to create a **chimeric** mAb in which only the V regions are mouse. **Humanized** mAbs are created by inserting the hypervariable regions of the mouse mAb into a human mAb, creating a mAb with only the binding regions of the mouse. Fully human mAbs are developed in mice whose antibody genes are replaced by human antibody genes, or from Fv libraries using phage display technology. |
| **Cell-mediated immunotherapy** | The exponential growth of our understanding of the cells and molecules critical to immune defense over the last 20 years has led to the development of realistic approaches to effective cellular immunotherapy. In relation to tumors, the current non-antibody-mediated immunotherapeutic approaches include those focused on direct cellular therapy of the tumor, prophylactic vaccines to microbes capable of inducing a tumor, and therapeutic cellular vaccines. |
| **Clinical trials** | Immunotherapeutics are thoroughly tested before approval for human use. This requires preclinical testing, *in vitro* and in animals, followed by five phases of clinical trials. Phase 0 trials determine how the body processes the drug and how it works in the body. Phase I trials determine the toxicity with escalating doses. Phase II trials test safety and efficacy. Phase III "double-blind" trials determine drug effectiveness, after which successful drugs are approved for patient use. Phase IV trials evaluate long-term safety. |

| | | | |
|---|---|---|---|
| **Related topics** | (D5) Monoclonal antibodies | (N5) Cytokine and cellular immunotherapy of tumors | |
| | (J2) Primary/congenital (inherited) immunodeficiency | (N6) Immunotherapy of tumors with antibodies | |
| | (L5) Diagnosis and treatment of autoimmune disease | | |

## Cytokine-mediated immunotherapy

Cytokines, soluble extracellular proteins or glycoproteins, involved in the development and regulation of immune responses have been known to exist since the 1960s. They were thought to be promising agents for therapy of tumors and other diseases, and were initially prepared *in vitro* or from serum, although the quantities available were too small to do definitive clinical trials. When cytokines became readily available, through recombinant gene technology, they were tested in a variety of diseases with disappointing results. Nonetheless, several cytokines have been approved for therapy and many others are still being studied as promising candidates for immunotherapy.

## Antibody-mediated immunotherapy

One of the earliest events in the development of antibody-mediated immunotherapy occurred in 1952, when Colonel Ogden Bruton noted the absence of immunoglobulin (Ig) in a boy with a history of pneumonia and other bacterial pulmonary infections. He was the first physician to provide specific immunotherapy for this X-linked disorder (now called Bruton's agammaglobulinemia, Section J2) by giving injections of human IgG, which successfully resolved the problem. Intravenous immunoglobulin, IVIG, as a similar but far more sophisticated drug is now called, is still used to treat immunodeficiency diseases, as well as many other disorders.

Another major breakthrough in this area occurred in 1975 when Kohler and Milstein at Cambridge (Section D5) described an approach to the development of monoclonal antibodies (mAbs), for which they received the Nobel Prize in 1984. Thus, by the early 1980s many immunologists around the world were making mAbs to almost everything, with a special emphasis on cancer, as it was imagined that patients could be readily cured with these "magic bullets."

Although most therapeutic mAbs have been developed in mice, their immunogenicity in humans and their inability to interact well with human effector cells has required that these mAbs be made "more human." Such difficulties also stimulated new approaches to the development of mAbs (Section D5). Mouse mAbs can be made more human by genetically replacing the C regions of the H and L chains of the mouse mAb with the C-region genes of human antibodies to create a **chimeric** mAb, where only the V regions are mouse. **Humanized** mAbs are created by genetically inserting the hypervariable regions of the mouse mAb into a human mAb, creating a mAb that is 95% human with only the binding regions being mouse. Fully human mAbs are developed in mice whose antibody genes have been replaced by human antibody genes, so that when the mouse is immunized it makes fully human antibodies against the antigen. Another way to develop human mAbs involves preparing mRNA for the VH and VL regions from a large number of human B cells. From this mRNA, cDNA for each

H-chain V region is prepared and joined to the cDNA for each L-chain V region to create Fv libraries (Section D5). Overall, these libraries encode a vast number of different antigen-combining sites from which binding sites specific for different antigens can be selected and linked genetically to human L- and H-chain C regions to create a human mAb.

In 2008, an International Working Group appointed by the World Health Organization (WHO) developed a standardized nomenclature for mAbs. The common stem for mAbs is -**mab** at the end of the name and denotes any product containing an immunoglobulin variable domain that binds to a defined target. The species of the immunoglobulin on which the mAb is based is given as follows: **a**, rat; **axo**, rat/mouse; **e**, hamster; **i**, primate; **u**, human; **xi**, chimeric; **zu**, humanized. The **tu** signifies that the target is a tumor. For example, rituximab is a chimeric anti-tumor mAb and trastuzumab is a humanized anti-tumor mAb, and so on. The prefixes "ri" and "tras" are merely unique identifiers for the antibody.

## Cell-mediated immunotherapy

Cellular immunotherapy did not truly develop until the critical cells and molecules necessary for protection from, or therapy of, viral infections and tumors were identified. Moreover, there needed to be a far better understanding of how these components worked together to mediate killing of tumors and virally infected cells. Over the past 30 years, as a result of contributions by many laboratories, much has been learned about these cells and molecules and how they work, and thus how best to design effective prophylactic and therapeutic approaches to successful cellular immunotherapy. In relation to tumors, the current non-antibody-mediated immunotherapeutic approaches include those focused on (1) direct cellular therapy of the tumor (i.e., the use of cytotoxic cells of the immune system to kill tumor cells); (2) prophylactic vaccines (i.e., the use of vaccines against microbes associated with the development of a tumor); and (3) therapeutic vaccines (i.e., the use of vaccines in patients who already have a tumor).

## Clinical trials

Like all new drugs, immunotherapeutics are required to be thoroughly tested before they are approved for human use. This is achieved through clinical trials, of which there are five main phases. Before this, preclinical testing, *in vitro* and in animals, would have to been carried out to determine the potential effectiveness and toxicity of the study drug. Phase 0 clinical trials may or may not be required, but if they are, single sub-therapeutic doses of the drug are given to a small number of subjects (10–15) to determine how the body processes the drug and how it works in the body. For Phase I trials, the subjects involved would depend on the diseases targeted by the drug. That is, if the drug targeted cancer patients, the subjects involved would be patients with late-stage cancer. If, however, the drug was targeted to other nonlethal diseases, it could involve a small group of healthy volunteers (12–40). This phase, usually taking up to 6 months, is designed to test the toxicity of the drug with escalating doses. In Phase II trials, a larger number of actual patients (20–300) who have the condition needing treatment are given the drug. This is designed to test the efficacy of the drug and safety issues, and may take up to 2 years. Phase III trials involve larger patient groups (300–3000) and are carried out in multiple centers. Patients are usually randomized into two groups, one of which receives a placebo, the other the test drug, with neither the patients nor the physician knowing which patients are receiving the drug. These "double-blind" trials are designed to determine how effective the drug is, in comparison with current "gold standard" treatment. Successful drugs would be submitted for approval for patient use after this Phase. Phase IV trials involve determining the further long-term safety issues of the drug and the potential effectiveness of the drug for therapy of other conditions.

# Q2 Cytokine-mediated immunotherapy

## Key Notes

| | |
|---|---|
| **Cytokine-mediated immunotherapy** | The therapeutic potential of recombinant cytokines has been extensively explored for many diseases, with limited therapeutic success. Even so, IL-2 and some of the interferons (primarily INFα) have been approved for therapy of viral infections and some tumors. Other cytokines (IL-17, IL-15, IL-21, and IL-37) are in late stages of development. |
| **Interferons** | Interferons are master regulators of inflammatory responses and efforts focused on their use in treating diseases have resulted in their approval for tumor and pathogen therapy. |
| **Recombinant IFNs** | Several recombinant interferons, including IFNα-2a, 2b, β-1b, β-1a, α-n3, and γ-1b, have been approved for therapy of tumors as well as for hepatitis B and C. In addition some of these have been approved for treatment of other diseases, including multiple sclerosis and chronic granulomatous disease. |
| **Natural human α-interferon** | This human leukocyte cytokine preparation containing several forms of natural IFNα is approved for therapy of metastatic renal cell carcinoma and malignant melanoma. |
| **PEGylated IFNα-2a and α-2b** | Covalent conjugation of recombinant IFNα-2a and IFNα-2b, respectively, with polyethylene glycol (PEG) increases the half-life of these cytokines. They are approved for therapy of patients with chronic hepatitis B and C. |
| **Interleukin-2** | Interleukin-2 is an important growth factor for T cells and is approved for the treatment of malignant melanoma. It is also the only drug approved for treatment of metastatic renal cell cancer (RCC). |
| **Related topics** | (B2) Molecules of the innate immune system     (F5) Clonal expansion and development of memory and effector function |

## Cytokine-mediated immunotherapy

When recombinant cytokines became available, they were used in clinical trials for therapy of a wide range of diseases, but with limited success. Thus, although cytokines are critical to the development of specific immune responses, when used alone they primarily enhance nonspecific activation of immune cells (Section B2). To be effective against tumors or other diseases, it is clear that, in most cases, cytokines will need to be used in concert with induction of more specific immune responses to the tumor or disease. Even so, IL-2 and some of the interferons (primarily IFNα), have been approved for therapy (Table 1). Moreover, IFNα remains a standard treatment for patients with melanoma. Other cytokines, including IL-17, IL-15, IL-21, and IL-37 are in late stages of development.

**Table 1. Cytokines approved for therapy of tumors and viral infections**

| Name | Cytokine | Diseases |
|---|---|---|
| Interferons | | |
| Recombinant IFN | IFNα-2a, 2b, β-1b, β-1a, α-n3, and γ-1b | HCL, KS, CML, melanoma, hepatitis B and C |
| NHu α-interferon | IFN-α1, α2, α8, α10, α14, and α21 | RCC, melanoma |
| PEGylated IFNα-2a | IFNα-2a | Hepatitis B and C |
| PEG-IFNα-2b | IFNα-2b | Hepatitis B and C |
| Interleukin-2 | IL-2 | RCC |

HCL, hairy cell leukemia; KS, AIDS-related Kaposi's sarcoma; CML, chronic myeloid leukemia; RCC, renal cell cancer; MS, multiple sclerosis; HPV, human papillomavirus; CGD, chronic granulomatous disease.

## Interferons

Interferons are master regulators of inflammatory responses and thus have the potential to mediate enhanced immune defense against pathogens as well as tumors. To that end, considerable effort has focused on their use in treating various diseases, some of which have been approved for therapy.

## Recombinant IFNs

Recombinant IFNs include IFNα-2a, 2b, β-1b, β-1a, α-n3, and γ-1b. Several recombinant interferons have been approved for therapy of tumors, including IFNα-2a for hairy cell leukemia, AIDS-related Kaposi's sarcoma, and chronic myeloid leukemia. IFNα-2b is also approved for the treatment of hairy cell leukemia and Kaposi's sarcoma as well as malignant melanoma and for chronic hepatitis B and C, when used in combination with ribavirin (a pro-drug that when metabolized interferes with viral replication). IFNβ-1b and IFNβ-1a are approved for treatment of multiple sclerosis. IFNα-n3 is used for treatment of genital and perianal warts caused by human papillomavirus, and IFNγ-1B for treatment of chronic granulomatous disease (CGD), and severe, malignant osteopetrosis.

## Natural human α-interferon

Natural human α-interferon (Multiferon) is a human IFNα mixture, purified from activated human blood leukocytes, that contains several forms of natural IFNα, including IFN-α1, α2, α8, α10, α14, and α21. It has been approved for treatment of renal cell carcinoma and melanoma. These IFNαs have an inflammatory affect that tends to induce Th1 immune responses. In particular, they enhance B-cell proliferation, activate NK cells, induce dendritic cells to initiate immune responses, and increase HLA class I and II expression, thus increasing CD8+ CTL-mediated killing of tumor or virus-infected cells. Overall, it inhibits tumor cell growth and is approved for treatment of metastatic renal carcinoma and malignant melanoma.

## PEGylated IFNα-2a and α-2b

PEGylated IFNα-2a is a covalent conjugate of recombinant IFNα-2a and polyethylene glycol (PEG), which is used to increase IFNα-2a half-life. It is approved for therapy of patients with chronic hepatitis C (including those who also have HIV), and for treatment of chronic hepatitis B, usually in combination with ribavirin. It does however, have significant side effects, including lymphopenia. PEG-interferon α-2b, like Pegasys, is a covalent conjugate, in this case, of IFNα2b and PEG, and is used as an antiviral agent. IFNα2b induces production of proteins that mediate antiviral, antiproliferative, anticancer, and immune-modulating effects. Again, PEG extends the duration of therapeutic effects.

## Interleukin-2

Interleukin-2 is approved for the treatment of malignant melanoma and is the only drug approved in the US for the treatment of metastatic renal cell cancer (RCC). It is also in clinical trials for therapy of chronic viral infections, and as a booster (adjuvant) for vaccines. IL-2 increases CD4$^+$ T-cell numbers by expanding CD4$^+$ cells (Section F5) and prolonging their half-life. However, there are significant side effects with high doses, including low blood pressure, irregular heart rhythms, fluid accumulation in the lungs, and fever. Of note, many immunosuppressive drugs (e.g., corticosteroids) used to treat autoimmunity and organ transplant rejection work by inhibiting IL-2 production or by blocking IL-2R signaling, thus preventing T-cell expansion and function.

# Q3 Antibody-mediated immunotherapy

## Key Notes

**Immunotherapy of cancer**

Most mAbs used in tumor therapy are now human or humanized and target tumor-associated antigens that are also expressed on normal cells. However, as mutations and translocations associated with tumors are identified, unique gene products (tumor-specific antigens, TSAs) can then be targeted. More than 10 mAbs have been approved for treatment of cancer, including those for therapy of breast cancer, B-cell tumors, myeloid leukemia, and colorectal cancer. Although most mAbs approved for therapy target tumor antigens and mediate killing of tumor cells, some target antigens not on tumor cells. Bevacizumab targets VEGF and in so doing decreases the blood supply to the tumor and thus its viability. Another mAb, ipilimumab, blocks CTLA-4, a molecule on activated T cells that transmits an inhibitory signal and prevents T cells from shutting down. Other molecules like CTLA-4 exist as immunologic **checkpoints** that, when targeted by mAbs, may also enhance immune defense.

**Tumor therapy with antibodies alone**

Examples of mAbs approved for treatment on their own include those with specificity for cell-surface molecules on tumor and normal cells, for example, HER2/neu on some breast carcinoma cells (trastuzumab), CD20 on B-cell lymphomas and B-cell chronic lymphocytic leukemia (rituximab), and the EGFR on colorectal cancers (cetuximab). Monoclonal antibodies to non-tumor-associated molecules include bevacizumab (targeting VEGF) and ipilimumab (targeting CTLA-4).

**Tumor therapy with immunotoxins (ITs)/radionuclides**

Some cancer therapeutics have used mAbs to which toxins or radionuclides have been coupled that, when injected into patients, seek out and bind to tumor cell antigens, and mediate their own lethal hit. Approved mAb conjugates include gemtuzumab ozogamicin, which targets CD33 in acute myeloid leukemia, and two that target CD20: ibritumomab tiuxetan and tositumomab.

**Immunotherapy of autoimmune and other inflammatory diseases**

A number of mAbs and fusion proteins have been approved for treating autoimmune and other inflammatory diseases through targeting: lymphocytes– B-cell CD20 (rituximab), T-cell CTLA-4 (abatacept); cytokines– including TNFα (infliximab), IL-1β (canakinumab), and IL-12 and IL-23 (ustekinumab); and cytokine receptors, such as IL-6R (tocilizumab). Other mAbs target VEGF (ranibizumab) and

|  |  |  |  |
|---|---|---|---|
|  | C5 (eculizumab). A mAb targeting IgE (omalizumab) is approved for treating allergic asthma. |  |  |
| **Immunotherapy of other disorders** | Monoclonal antibodies have been developed for therapy of other disorders, including infection, heart attacks, and graft rejection. In particular, palivizumab protects against RSV infection of premature infants. Abciximab, a Fab fragment of a mAb that targets the platelet glycoprotein (GP) IIb/IIIa receptor, decreases the chance of heart attack in patients undergoing surgery to unblock heart arteries. Basiliximab, a mAb to CD25 prevents organ transplant rejection, as does daclizumab, a mAb also specific for CD25. Muromonab-CD3, a mAb that binds to the CD3ε chain of the T-cell-receptor–CD3 complex, also blocks transplant rejection. |  |  |
| **Related topics** | (B2) Molecules of the innate immune system | (L5) Diagnosis and treatment of autoimmune disease | |
|  | (D5) Monoclonal antibodies | (N6) Immunotherapy of tumors with antibodies | |

## Immunotherapy of cancer

Although it was clear early on that mAbs can mediate killing of tumor or other cells through a variety of mechanisms—including complement activation, targeting NK cells, monocytes, and macrophages to mediate ADCC and/or phagocytosis, or by inducing apoptosis—the use of mAbs to treat human tumors or other diseases was not initially successful.

Many clinical trials were carried out, but with little success—so little that by the early 1990s most of the big pharmaceutical companies had given up on mAb therapy. To some extent this resulted from (Section N6) (1) the lack of specificity of the mAb utilized; (2) the presence of soluble forms of the antigen in the serum that effectively interfered with the interaction of antibody with the tumor cell; (3) modulation and loss of the antibody–antigen complexes from the tumor cell surface before antibody-mediated killing could occur; and (4) outgrowth of (selection for) tumor cells not expressing the antigen. Perhaps most important, mAbs used in therapy during the 1980s were of mouse origin and did not interface well with human effector molecules (complement) and cells (NK cells, macrophages, PMNs), and, being foreign, induced a human anti-mouse antibody (HAMA) response that compromised the effectiveness of the mAb.

Nonetheless, some companies continued to pursue mAbs, and one of them (IDEC Pharmaceuticals) developed an antibody (rituximab) specific for a well-defined molecule (CD20) expressed on B-cell leukemias, and even though CD20 was also expressed on normal B cells, rituximab was found to be effective in treating these tumors without significant side effects. It was approved for therapy in 1997 and became a blockbuster therapeutic that dramatically changed the field and the thoughts on mAbs. Certainly, the clearer appreciation that developed on how to use mAbs more effectively in cancer therapy was also important in the renaissance in antibody therapy that ensued. Thus, it is quite clear that mAbs are, and will continue to be, very useful tumor therapeutic agents, especially if (1) they are human or humanized to permit long-term use (Section D5); (2) the antigen to which they react is expressed at a high level on the tumor; (3) the mAb is used to treat minimal disease; and (4) the patient's immune system is functional.

It is of note that when Kohler and Milstein asked the Cambridge technology transfer office to submit for a patent on this technology, they were refused and told that their technology did not appear to have much value! There are currently over 20 mAbs approved for therapy and the sales for just one of these, rituximab, have already totaled over 10 billion dollars.

## Tumor therapy with antibodies alone

Few mAbs prepared against human tumors have been found to be truly tumor-specific (Section N6), although it is likely that, as mutations and translocations associated with various tumors are examined, unique gene products (TSAs) will be identified. Even so, it became clear from the anti-CD20 rituximab results that, to be effective, a mAb need not be tumor-specific. Most mAbs used in tumor therapy are now human or humanized (Section D5) and target antigens that are also expressed on normal cells. More than 10 of these have been approved for treatment of cancer (Table 1), including those for therapy of breast cancer, B-cell tumors, myeloid leukemia, and colorectal cancer. Other mAbs still in clinical trials are also demonstrating efficacy as indicated by tumor regression in some patients.

It is important to note that although most antibodies approved for therapy of tumors target tumor cell-surface antigens, some target antigens not found on tumor cells. That is, some mAbs approved for tumor therapy, or in clinical trials, react with molecules

## Table 1. Antibodies approved for cancer therapy

| Name | Type | Target | Tumors | Function |
|------|------|--------|--------|----------|
| **Antibodies alone** | | | | |
| Trastuzumab | Humanized | HER2/neu | Breast, lung, ovarian, pancreatic cancer | Tumor cell killing |
| Rituximab | Chimeric | CD20 | B-cell lymphoma | Tumor cell killing |
| Ofatumumab | Human | CD20 | B-cell CLL | Inhibits early-stage B-cell activation |
| Alemtuzumab | Humanized | CD52 | B-cell CLL | Tumor cell killing |
| Cetuximab | Chimeric | EGFR | Colorectal, head and neck cancer | Tumor cell killing |
| Panitumumab | Human | EGFR | Metastatic colorectal cancer | Tumor cell killing |
| Bevacizumab | Humanized | VEGF | Colon, breast, kidney | Shuts down blood supply to the tumor |
| Ipilimumab | Human | CTLA-4 | Melanoma | Stops CTL shutdown |
| **Antibody–toxin/radionuclide conjugates** | | | | |
| Gemtuzumab ozogamicin | Humanized | CD33 | AML | Toxin killing |
| Ibritumomab tiuxetan | Chimeric | CD20 | B cell | Radionuclide killing |
| Tositumomab | Murine | CD20 | Lymphoma (NHL) | Radionuclide killing |

on non-tumor cells that, in one way or another, are important to tumor growth. In particular, bevacizumab reacts with vascular endothelial growth factor (VEGF), a molecule that is responsible for inducing the development of new blood vessels. Although this is important in early life and when there is inadequate circulation, tumors are able to use VEGF to create blood vessels critical to their growth. Bevacizumab prevents this from happening and the ability of the tumor to grow is decreased. Although not a cure, when it is used with other therapies it may often change the course of the disease.

Another mAb approved for therapy of cancer that does not react directly with the tumor is ipilimumab. This human mAb reacts with CTLA-4, a co-stimulatory and regulatory molecule expressed on activated T cells. Like CD28 on T cells, CTLA-4 binds to the co-stimulatory molecules CD80 and CD86 on APCs, but, unlike CD28, when it binds to these co-stimulatory molecules it transmits an inhibitory signal, causing T cells to shut down. In the case of tumors, a CTL anti-tumor response would be inhibited. Thus, treatment with ipilimumab, which binds to and blocks CTLA-4, prevents CTLA-4 interacting with CD80 and CD86 and thus stops the T cell from shutting down. This mAb has been approved for therapy of human metastatic melanoma and will likely be used to treat prostate and other cancers.

Viewed from a larger perspective, CTLA-4 is just one of a group of molecules that are immunologic **checkpoints**, in this case of T-cell activation. It is now clear that, by using mAbs to these checkpoints, the development of a T-cell response can be either negatively or positively influenced, and thus have potentially dramatic effects on autoimmune disease or protection against tumors and infections.

## Tumor therapy with immunotoxins (ITs)/radionuclides

Many cancer trials have used mAbs to which toxins or radionuclides have been coupled (Section N6), and several such constructs have been approved for tumor therapy (Table 1). When injected into patients, these ITs seek out and bind to tumor-cell antigens, and mediate their own lethal hit. Toxins can be quite potent, with a single molecule able to kill a cell. It is essential, however, that the targeting mAbs react with a TAA that is internalized on binding of the IT, and that killing of normal cells be minimized. Gemtuzumab ozogamicin, a humanized mAb to CD33 conjugated to the toxin calicheamicin, has been approved for treatment of patients with acute myeloid leukemia.

Radionuclide-coupled mAbs mediate killing by DNA damage from decay and release of high-energy particles. These ITs kill bystander cells (those nearby), which may result in the killing of both normal cells and adjacent tumor cells that may not express the targeted antigen.

The development of useful ITs has taken more time than initially anticipated due to toxic side effects, but several ITs have been approved for therapy (Table 1), others are in clinical trials. Of note, the same anti-CD20 mAb approved for therapy of B-cell lymphoma has been coupled to a radionuclide, yttrium, and the resulting immunotoxin (ibritumomab tiuxetan) has been approved for therapy of B-cell lymphoma.

Chemotherapy targeted through mAbs is also being developed. Here, a prodrug is coupled to the mAbs and injected into the cancer patient. This is followed by administration of the "inactive" cytotoxic/cytostatic drug systemically. The prodrug converts the inactive drug to an "active" form locally at the tumor site, thereby achieving a high concentration of the drug targeting the tumor.

## Immunotherapy of autoimmune and other inflammatory diseases

mAbs that specifically target immune cells associated with a disease, especially T or B cells, can eliminate lymphocytes responsible for the disease. However, it is important to avoid elimination of too many immune cells as so doing may lead to secondary immunodeficiency (Section J3). Monoclonal antibodies approved for treatment of autoimmune and other inflammatory diseases are shown in Table 2. Although induction of specific immune tolerance to the particular autoantigen would be ideal, more than one autoantigen is often involved and induction of tolerance is very difficult to achieve during an ongoing immune response (Section G3). However, in celiac disease, since the inducer of the disease is known (gluten), its removal from the diet leads a return to clinical normality and the cessation of autoantibody production to the target enzyme transglutaminase.

**Intravenous immunoglobulin (IVIG)** is an IgG solution prepared from the blood of multiple donors that is administered intravenously. It is used primarily to treat immune deficiencies, inflammatory and autoimmune diseases, and acute infection. Because it provides passive immunity it lasts between 2 weeks and 3 months, after which another injection is required. In particular, it is approved for therapy of primary immune deficiencies, and acquired compromised immune conditions (secondary immune deficiencies), characterized by low antibody levels.

**Approved indications for therapeutic use of IVIG include:**

- Hematopoietic stem cell, allogeneic bone marrow, and kidney transplant recipients
- Primary immunodeficiencies
- B-cell chronic lymphocytic leukemia (B-CLL)
- Immune-mediated thrombocytopenia (ITP)
- Pediatric HIV
- Kawasaki disease
- Chronic inflammatory demyelinating polyneuropathy (CIDP)

**Natalizumab** is a humanized mAb against the integrin $\alpha_4$ adhesion molecule. It is approved for treatment of multiple sclerosis (MS) and Crohn's disease and works by inhibiting the attachment and passage of inflammatory immune cells through tight junctions between endothelial cells lining the intestines and blood–brain barrier (extravasation, Sections B4 and C4). Natalizumab decreases relapse and improves the quality of life for patients with MS, and similarly helps patients with Crohn's disease. However, its use has been associated with the development of progressive multifocal leukoencephalopathy (PML, a fatal viral disease involving inflammation of the white matter of the brain). Initially withdrawn from the market, natalizumab was returned in 2006. Although there have been 31 cases of PML through January of 2010, its clinical benefits outweigh its risks.

**Alefacept** is a fusion protein consisting of LFA3 fused to the Fc region of a human IgG1 antibody. It is approved for treatment of psoriasis and appears to work by binding to CD2 and inducing T-cell apoptosis and inhibition of CD4 and CD8 activation, thus reducing T-cell-stimulated proliferation of keratinocytes, the basis for psoriasis symptoms. Alefacept is also being studied for treatment of T-cell lymphomas.

**Rituximab**, as described above, is a mouse–human chimeric IgG1 mAb that targets CD20, a molecule expressed on all B cells and on B-cell tumors. After approval for B-cell lymphomas, rituximab was found to be effective in therapy of antibody-mediated autoimmune

**Table 2. Antibodies approved for therapy of autoimmune and other inflammatory diseases**

| Name | Type | Target | Diseases | Function |
|---|---|---|---|---|
| IVIG | Pooled human IgG | Multiple | Multiple | Provide antibody-mediated passive immunity |
| Natalizumab | Humanized IgG4 | Integrin $\alpha_4$ $\alpha_4\beta_1$ subunit | MS, Crohn's disease | Blocks leukocyte adhesion |
| Alefacept | LFA3*–IgG1Fc fusion protein | CD2 | Plaque psoriasis | Blocks LFA3–CD2 interaction and lymphocyte activation |
| Rituximab | Chimeric IgG1 | CD20 | RA not responding to TNF inhibition | Induces apoptosis, ADCC and CDC |
| Ranibizumab | Fab fragment of bevacizumab | VEGF | "Wet" macular degeneration | Inhibits blood vessel development in the eye |
| Abatacept | CTLA4*–modFc | CD80/CD86 | RA, JIA | Blocks T-cell activation |
| Eculizumab | Humanized IgG2/IgG4 | C5 | PNH | Binds C5, blocks cleavage to C5a and C5b |
| Omalizumab | Humanized IgG1 | IgE | Persistent allergic asthma | Reduces release of allergic response mediators |
| Canakinumab | Human[†] IgG1 | IL-1β | Cryopyrin-associated periodic syndromes | Ligand binding/receptor antagonism |
| Tocilizumab | Humanized IgG1 | IL-6R | RA | Receptor binding/ligand blockade |
| Ustekinumab | Human[†] IgG1 | IL-12/IL-23 | Plaque psoriasis | Ligand binding/receptor antagonism |
| Infliximab | Chimeric IgG1 | TNFα | Crohn's disease, RA, psoriatic arthritis, plaque psoriasis | Blocks TNF activity |
| Golimumab | Human[†] IgG1 | TNFα | RA, psoriatic arthritis | Blocks TNF activity |
| Adalimumab | Human[‡] IgG1 | TNFα | RA, JIA, psoriatic arthritis, Crohn's disease, plaque psoriasis | Blocks TNF activity |
| Certolizumab pegol[§] | Humanized Fab | TNFα | Crohn's disease, RA | Blocks TNF activity |
| Etanercept | TNFR2*–Fc (IgG1) | TNFα | RA, JIA, psoriatic arthritis, plaque psoriasis | Blocks TNF activity |

MS, multiple sclerosis; RA, rheumatoid arthritis; JIA, juvenile idiopathic arthritis; PNH, paroxysmal nocturnal hemoglobinuria.
* Extracellular domain; † human Ab mice; ‡ phage-produced; § PEG conjugate.

diseases and is now approved for treatment of rheumatoid arthritis (RA) patients not responding to TNF inhibitors. As with tumors, it works by inducing B-cell apoptosis, by ADCC, and by complement-dependent killing. Several other anti-CD20 mAbs have also been developed and used for therapy of autoimmune disease, including **ofatumumab**, which inhibits early-stage B-lymphocyte activation and has shown potential in treating RA, relapsing remitting MS, follicular non-Hodgkin's lymphoma, and diffuse B-cell lymphoma.

**Ranibizumab** is a Fab fragment of the anti-angiogenic mAb bevacizumab and is approved for therapy of "wet" macular degeneration, a disease that results in an age-related loss of vision. Ranibizumab works by inhibiting VEGF-mediated generation of new vessels that contribute to the development of this disease.

**Abatacept** is a fusion protein consisting of the Fc region of an IgG1 human antibody fused to the extracellular domain of CTLA-4, a molecule that binds CD80/CD86 (Section G3). Abatacept blocks interaction of CD80/CD86 on APCs with CD28 on T cells, and thus the critical co-stimulatory second signal required for T-cell activation is not provided, shutting down T cells. It is approved for the treatment of RA after inadequate response to anti-TNFα.

**Belatacept**, a very slightly modified abatacept, has been tested in clinical trials of organ transplantation and has apparently demonstrated extended graft survival with limited toxicity; a request for approval has been submitted to the US FDA.

**Eculizumab** is a mAb that binds to and neutralizes complement component C5, which is critical to activation of the MAC (Section D8) and thus complement-mediated lysis. It has been approved for therapy of paroxysmal nocturnal hemoglobinuria (PNH), a rare inherited disease in which a gene (*PIGA*) that codes for a protein that protects red blood cells from complement-mediated lysis is missing, thus causing red-cell lysis whenever complement is activated.

**Omalizumab** is a humanized IgG1 mAb that inhibits the binding of IgE to the high-affinity IgE receptor (FcεRI) by binding to an epitope that is close to, and overlaps, the site on IgE that binds to its receptor on mast cells and basophils (Section K2). It is approved for patients with moderate to severe allergic asthma caused by hypersensitivity to environmental antigens. Note that in a small percent of patients (0.1–0.2%) omalizumab causes anaphylaxis (Section K2).

**Canakinumab** is a fully human mAb that blocks IL-1β (Section B2), a cytokine that has significant and multiple inflammatory properties. It has been approved for treatment of cryopyrin-associated periodic syndromes (CAPS). These are rare inherited auto-inflammatory syndromes, including familial cold auto-inflammatory syndrome (an inflammatory disorder triggered by cold exposure), Muckle–Wells syndrome (inflammation that results, in this case, in hearing loss, hives, and ameloidosis), and neonatal-onset multi-system inflammatory disease (acute inflammation in newborns).

**Tocilizumab** (also known as **atlizumab**), a humanized mAb against the IL-6 receptor (IL-6R), blocks binding of IL-6, a cytokine important in immune responses and implicated in the pathogenesis of autoimmune diseases as well as in tumors, particularly multiple myeloma and prostate cancer. Tocilizumab is approved for RA and for Castleman's disease, a rare benign tumor of B cells.

**Ustekinumab** is a human mAb specific for IL-12 and IL-23, cytokines that regulate the immune system and immune-mediated inflammatory disorders. It is approved for therapy of plaque psoriasis, a chronic autoimmune disease of the skin.

Monoclonal Abs that target **TNFα**, or soluble receptors for TNFα, are effective in suppressing the inflammatory process in RA and other inflammatory diseases and slowing their progression. Several of these have been approved for therapy, including the following:

**Infliximab** is a mouse–human chimeric mAb that targets TNFα. It was initially approved for treatment of Crohn's disease (inflammatory bowel disease) and is now approved for a variety of other autoimmune diseases including psoriasis, RA, ankylosing spondylitis (AS), and ulcerative colitis. Infliximab blocks the action of TNFα by preventing its binding to its receptor on the cell and in Crohn's disease causes programmed cell death of the TNFα-expressing activated T cells that mediate inflammation.

**Golimumab** is a human mAb approved for therapy of adults with moderately to severely active RA, psoriatic arthritis, and AS.

**Adalimumab** is a human mAb and like infliximab is approved for psoriasis, RA, AS, Crohn's disease, and juvenile idiopathic arthritis (JIA).

**Certolizumab pegol**, a PEGylated Fab′ fragment of a mAb directed against TNFα, has been approved for therapy of Crohn's disease and RA.

**Etanercept** also binds and inhibits the action of TNFα but is not a mAb. Rather it is a fusion protein consisting of the TNFα receptor fused to the Fc region of a human IgG1 antibody. Like anti-TNFα mAbs, it decreases excess inflammation mediated by TNFα. It is approved for a variety of autoimmune diseases including AS, RA, JIA, and psoriatic arthritis, and can potentially be used in other disorders mediated by excess TNFα.

## Immunotherapy of other disorders

A summary of the mAbs approved for use in other disorders is shown in Table 3.

**Palivizumab**, a humanized IgG mAb targeting an epitope in the F protein of respiratory syncytial virus (RSV), is approved for prevention of RSV infection, especially for those at high risk (e.g., premature infants). It significantly reduces risk of hospitalization due to RSV infection. This approved mAb is one of many antimicrobial mAbs that are in various stages of clinical trials. Target microbes include HIV, CMV, hepatitis C virus (HCV), rabies, *Clostridium difficile*, *Staphylococcus*, and *Bacillus anthracis*.

**Abciximab** is the Fab fragment of a chimeric human–mouse mAb that targets the human platelet glycoprotein (GP) IIb/IIIa receptor and inhibits platelet aggregation. It also binds to the vitronectin ($\alpha_v\beta_3$) receptor on platelets, blood vessel smooth muscle, and endothelial cells and decreases the chance of heart attack in patients undergoing surgery to unblock heart arteries.

**Basiliximab** is a chimeric mouse–human mAb to CD25, the α chain of the T-cell IL-2 receptor. It is used to prevent organ transplant rejection, especially kidney transplants, by blocking T-cell replication and activation of B cells that produce antibodies mediating transplant rejection.

**Daclizumab** is a humanized mAb also specific for CD25. Like basiliximab, it reduces the incidence and severity of acute rejection of kidney transplants without increasing infection.

**Muromonab-CD3** is a mouse IgG2a mAb that binds to the CD3 ε chain of the T-cell receptor CD3–complex. It reduces T-cell function, induces apoptosis of T cells, and blocks transplant rejection.

**Table 3. Monoclonal antibody therapy of other disorders**

| Name | Type | Target | Diseases | Function |
|------|------|--------|----------|----------|
| Palivizumab | Humanized IgG | F protein of RSV | RSV | Blocks RSV infection |
| Abciximab | Human–mouse Fab fragment | GPIIb/IIIa | Heart attack | Blocks platelet aggregation |
| Basiliximab | Human–mouse IgG | CD25 T-cell IL-2R | Graft rejection | Blocks T-cell activation and B-cell Ab production |
| Daclizumab | Humanized IgG | CD25 T-cell IL-2R | Graft rejection | Blocks T-cell activation and B-cell Ab production |
| Muromonab-CD3 | Mouse IgG2a | CD3 ε chain | Graft rejection | Inhibits and kills T cells |

RSV, respiratory syncytial virus.

# Q4 Cellular immunotherapy

---

### Key Notes

**Overview**

Current non-antibody-mediated immunotherapeutic approaches include those focused on (1) direct cellular therapy using cytotoxic cells to kill tumor cells; (2) prophylactic vaccines that target microbes associated with the development of a tumor; and (3) therapeutic vaccines to induce the patient's immune system to reject their tumor.

**Cellular therapeutic approaches**

The various cellular therapeutic approaches, including those involving lymphokine-activated killer cells, tumor-infiltrating lymphocytes, and macrophage-activated killer cells, have had limited success.

**Prophylactic vaccines**

Prophylactic vaccines have been developed that induce immunity to viruses known to be associated with the development of tumors. Hepatitis B vaccines prevent infection by this virus and reduce the incidence of liver cancer. Human papillomavirus (HPV) vaccines reduce the development of cervical carcinoma.

**Cellular immunotherapeutic vaccines**

Immunotherapeutic cellular approaches aimed at inducing immune responses to viruses or tumor antigens have been difficult to develop due in part to a number of factors: (1) most tumor-associated antigens are also expressed by some normal cells and thus immune tolerance has to be broken; (2) because of histo-incompatibility a new drug must be developed for each patient. Approaches that have had limited success include immunization with (a) tumor cells, their antigens, or peptides; (b) DNA encoding tumor antigens; and (c) tumor cells transfected with genes for cytokines and/or co-stimulatory molecules. Approaches that have had more success involve DCs loaded *in vitro* with killed tumor cells or their lysates, or tumor antigens, along with adjuvants and/or cytokines. The drug sipuleucel-T, which uses this approach for the therapy of prostate cancer, is the first therapeutic tumor vaccine approved in the US.

**Related topics**

## Overview

The development of realistic approaches to effective cellular immunotherapy has resulted from the exponential growth of our understanding of the cells and molecules critical to immune defense over the last 20 years. In relation to tumors, the current non-antibody-mediated immunotherapeutic approaches include those focused on (1) direct cellular therapy of the tumor (i.e., the use of cytotoxic cells of the immune system to kill tumor cells); (2) prophylactic vaccines (i.e., the use of vaccines against microbes associated with the development of a tumor); and (3) therapeutic vaccines (i.e., the use of vaccines in patients who already have a tumor, in an attempt to induce the patient's immune system to reject their tumor).

## Cellular therapeutic approaches

The various cellular therapeutic approaches include those involving (1) lymphokine-activated killer (LAK) cells; (2) tumor-infiltrating lymphocytes (TILs); and (3) macrophage-activated killer (MAK) cells. These therapeutic approaches are all patient-specific due to the requirement of MHC compatibility, and thus a new cell preparation must be developed for each patient from their own cells.

The use of LAK cells is based on the presence, in many tumor-bearing patients, of lymphocytes reactive to their tumor, and thus expansion of these cells *ex vivo* could permit larger numbers of these cells to be used in the patient's therapy. Blood lymphocytes obtained from the patient are expanded, activated with IL-2, and infused back into the patient with more IL-2. As LAK cells are primarily NK cells they do not have the selective specificity of T cells and only target tumor cells that either are without MHC class I or express ligands to which NK cells can bind. Although tumor regression occurs in some patients, there is significant toxicity using high doses of IL-2.

TIL therapy involves isolation of CD8+ lymphocytes from the patient's tumor, some of which are tumor-specific. These are expanded and activated with IL-2 and infused into the patient with or without IL-2. TIL therapy induces tumor regression in some patients, especially those with renal cell carcinoma, but again there is significant toxicity if high doses of IL-2 are used to maintain the active status of the TIL cells *in vivo*.

MAK cells are prepared from blood monocytes isolated from tumor-bearing patients by culture with cytokines (e.g., IFNγ) *in vitro*. This activates the cells for enhanced cytotoxicity, after which they are re-injected into the patient.

Overall these approaches have an effect on tumor burden but, for LAKs and TILs, the doses of IL-2 required are too high (they are quite toxic) to justify their use in most patients. Also, although MAK cells are highly cytotoxic and phagocytic, they are relatively nonspecific, and are likely to require co-injection with mAbs to tumor-associated antigens to be most effective.

## Prophylactic vaccines

Prophylactic vaccines have been developed and approved for prevention of a variety of infectious diseases. A number of these induce protective immunity to viruses, some of which are associated with tumor development. In this case, not only does the vaccine prevent the infectious agent from causing disease, but it also decreases the likelihood that the viral infection will lead to the development of cancer. Two approved prophylactic vaccines that induce immunity to viruses that cause tumors have been developed, and are quite effective (Sections N2 and N7). In particular, hepatitis B vaccines prevent infection by this virus and reduce the incidence of liver cancer, and the human papillomavirus (HPV) vaccine reduces the risk of cervical carcinoma (Table 1).

**Table 1. Virally induced/associated tumors**

| Tumor viruses | Human tumor | Vaccine approved |
|---|---|---|
| **RNA** | | |
| Human T-lymphotropic virus-1 (HTLV) | Adult T-cell leukemia | No |
| | T-cell lymphoma | No |
| **DNA** | | |
| Epstein–Barr virus (EBV) | B-cell lymphoma | No |
| | Hodgkin's lymphoma | No |
| | Nasopharyngeal carcinoma | No |
| Human papillomavirus (HPV) | Cervical carcinomas | **Yes** |
| Hepatitis B virus (HBV) | Hepatocellular carcinoma | **Yes** |
| Hepatitis C virus (HCV) | Hepatocellular carcinoma | No |
| Human herpesvirus-8 (HHV-8) | Kaposi's sarcoma | No |
| *Helicobacter pylori* | Gastric cancer | No |
| Merkel cell polyomavirus | Merkel cell carcinoma | No |

Although a significant achievement in reducing the development of these cancers, there are six other viruses (Section N2) and a bacterium that inhabits the stomach that are known to be associated with the development of tumors, and it is likely that others will be identified in the future. The development of prophylactic vaccines to these microbes would represent a monumental achievement, not only as a result of protection against these diseases, but, equally important, for significantly decreasing the risk of development of their associated tumors.

## Cellular immunotherapeutic vaccines

Other cellular immunotherapeutic approaches are designed to enhance/induce effective immunity in virally infected or tumor-bearing patients. Although these approaches may be directed toward killing virus-infected or tumor cells, they are most often focused on the induction, in a host already infected or with cancer, of an immune response sufficient to mediate cure. These have been far more difficult to develop because of (1) the difficulties of defining the appropriate "protective" antigens to be used for induction of the immune response; (2) the requirement that the cells involved in therapeutic immunity must traffic to and localize in specific areas of the lymph nodes (Section C4); and (3) the logistics associated with the preparation of the vaccine. In addition, since most tumor-associated antigens are also expressed on some normal cells and are therefore viewed as self by the patient's immune system, a successful vaccine will need to break immune tolerance to these antigens, without unacceptable anti-self side effects. Moreover, in virtually all cases, the therapeutic cells must be from the patient to be treated, as cells from other individuals would be rejected due to MHC incompatibility. Thus, a new drug must be developed for each patient.

A variety of approaches have been explored for the development of immunotherapeutic vaccines (Section N5), including immunization with (1) killed whole tumor cells; (2) tumor protein antigens; (3) tumor peptide epitopes; (4) DNA encoding tumor antigens; and (5) dendritic cells loaded with tumor antigens (Table 2). More specifically, killed or irradiated patient tumor cells transfected with co-stimulatory molecules are being explored in clinical trials as ways of inducing immunogenicity and an effective CTL response. Another approach has been to use tumor cells transfected with cytokine genes that attract, expand, and activate immune cells reactive to tumor-associated antigens. Although several companies and many researchers have put considerable effort into this area, with some positive results, it has not been as successful as initially hoped. Tumor-associated protein antigens and/or their peptides, as well as DNA for tumor-associated antigens, have also been explored as therapeutic vaccines, again with limited success thus far.

**Table 2. The main types of cancer vaccine approaches**

| | |
|---|---|
| Whole tumor cell | Tumor cells alone, or transfected to express co-stimulatory molecules or cytokines, including GM-CSF, are killed or irradiated and injected as a vaccine |
| Tumor protein antigen | Taken-up, processed, presented by APCs |
| Tumor peptide epitopes | Occupy empty MHC molecules on APCs |
| DNA encoding tumor antigen | Transfected into APCs or other cells that will produce/secrete antigen for uptake by APCs. |
| Dendritic cells | Loaded with peptides, proteins, whole tumor lysates, traffic to lymph nodes and expand CD8+ T cells |

More promising have been approaches that have focused on dendritic cells (DCs). Two DC-related, but rather different, approaches are currently being explored. One, a rather novel approach, involves the use of mAbs that specifically target human DCs, and which have been genetically linked to tumor-associated antigens. In this case, immunization is carried out by injecting the mAb carrying the tumor antigen(s) into the patient, after which the mAb seeks out and binds to DCs, is internalized, and delivers its cargo in such a way as to induce both T- and B-cell responses to the tumor antigens. This approach is in clinical trials for therapy of tumors and viral infections. The other, more classical, approach involves isolating DCs from patients and loading them with their own killed tumor cells, or their lysates, or with well-defined tumor-associated antigens, sometimes including adjuvants or cytokines. Since immature DCs are best able to ingest antigen and mature DCs are best at presenting antigen, considerable effort is directed at determining optimal conditions for loading and maturing DCs *ex vivo* so that they induce strong CTL anti-tumor responses when reintroduced into the patient. There are several reports of positive results from some of these trials.

The most successful of the "classical" approaches, and perhaps a breakthrough in the development of cellular tumor vaccines for therapy, resulted in the drug sipuleucel-T. Sipuleucel-T consists of dendritic cells that have been incubated with a fusion protein consisting of an enzyme specific to the prostate, prostatic acid phosphatase (PAP), linked to granulocyte–macrophage colony-stimulating factor (GM-CSF), an activator of human plasmacytoid pre-dendritic cells, as well as granulocytes and monocytes. In particular, DCs obtained from a patient are incubated with this fusion protein (PAP-GM-CSF), during which time there is receptor-mediated uptake by the DCs, and after which these cells are infused back into the patient.

In the pivotal phase III clinical trial it was found that patients with advanced prostate cancer, when treated with sipuleucel-T, lived an average of 4.1 months longer than patients treated with the control. In April of 2010, the FDA approved sipuleucel-T for use in the treatment of advanced prostate cancer—the first therapeutic tumor vaccine approved in the US.

Although the average increase in lifespan is not great, some patients live considerably longer. In addition, the patients used in the trial had late-stage prostate cancer, and it is likely that when used for earlier-stage patients the results will be more promising. More-over, in contrast to mAb therapy for tumors, which generally involves continued long-term treatment with the drug, sipuleucel-T is given only three times. That is, mAbs are a passive therapy whereas sipuleucel-T and drugs like it induce active immunity that continues to persist in the patient.

Although this is envisioned as a breakthrough for cellular therapeutics, it should be again noted that a new drug must be created for each patient, that the preparation of the drug is onerous and thus expensive, and that this drug will only be effective for treatment of prostate cancer. It does, however, pave the way for other similarly designed drugs for other tumors.

# Further reading

A large number of textbooks in immunology are now available that are good reference books for those interested in more detail. In addition, specific detailed information can often be obtained through the Web and through specialist journal databases, including Medline and PubMed.

## General textbooks

Abbas, A.K., Lichtman, A.H., Pillai, S. (2009) *Cellular and Molecular Immunology, Updated Edition*, 6th edn. W.B. Saunders, Philadelphia.

Delves, P.J., Martin, S.J., Burton, D.R., Roitt, I.M. (2011) *Roitt's Essential Immunology*, 12th edn. Wiley-Blackwell, Oxford.

Kindt, T.J., Goldsby, R.A., Osborne, B.A. (2006) *Kuby Immunology*, 6th edn. W.H. Freeman, San Francisco.

Male, D., Brostoff, J., Roth, D.B., Roitt, I.M. (eds) (2006) *Immunology*, 7th edn. Mosby, Philadelphia.

Murphy, K.M., Travers, P., Walport, M. (2007) *Janeway's Immunobiology*, 7th edn. Garland Science, New York.

Playfair, J.H.L., Chain, B.M. (2009) *Immunology at a Glance*, 9th edn. Wiley-Blackwell, Oxford.

# Abbreviations

| | |
|---|---|
| Ab | antibody |
| ADA | adenosine deaminase |
| ADCC | antibody-dependent cellular cytotoxicity |
| AFP | alpha-fetoprotein |
| Ag | antigen |
| AICD | activation-induced cell death |
| AID | activation-induced deaminase |
| AIDS | acquired immune deficiency syndrome |
| AIHA | autoimmune hemolytic anemia |
| AIRE | autoimmune regulator |
| ALPS | autoimmune lymphoproliferative syndrome |
| AMP | antimicrobial peptide |
| ANA | antinuclear antibody |
| ANCA | antibodies to neutrophil cytoplasmic antigen |
| APC | antigen presenting cell |
| APECED | autoimmune polyendocrinopathy–candidiasis–ectodermal syndrome |
| ARDS | acute respiratory distress syndrome |
| AT | ataxia telangiectasia |
| BALT | bronchus-associated lymphoid tissue |
| BCR | B-cell receptor |
| BsAb | bispecific antibody |
| Btk | Bruton's tyrosine kinase |
| CD | celiac disease |
| CEA | carcinoembryonic antigen |
| CGD | chronic granulomatous disease |
| CGRP | calcitonin gene-related peptide |
| CJD | Creutzfeldt–Jakob disease |
| CMI | cell-mediated immune (response) |
| CML | chronic myeloid leukemia |
| CMV | cytomegalovirus |
| CRD | carbohydrate recognition domain |
| CRH | corticotropin-releasing hormone |
| CRP | C-reactive protein |
| CSF | colony-stimulating factor |
| CTL | cytotoxic T lymphocyte |
| CVID | common variable immunodeficiency |
| DAB | 3,3′-diaminobenzidine |
| DAF | decay-accelerating factor |
| DAG | diacylglycerol |
| DC | dendritic cell |
| DHEA | dehydroepiandrosterone |
| DHEAS | dehydroepiandrosterone sulfate |
| DTH | delayed-type hypersensitivity |
| EAE | experimental allergic encephalomyelitis |
| EBV | Epstein–Barr virus |
| EGF-R | epidermal growth factor receptor |

| | |
|---|---|
| ELISA | enzyme-linked immunosorbent assay |
| ER | endoplasmic reticulum |
| FDC | follicular dendritic cell |
| FRT | female reproductive tract |
| Fv | antigen-combining site |
| GALT | gut-associated lymphoid tissue |
| GC | germinal center |
| G-CSF | granulocyte colony-stimulating factor |
| GEF | guanine nucleoside exchange factor |
| GI | gastrointestinal |
| GM-CSF | granulocyte–monocyte colony-stimulating factor |
| GOD | generation of diversity |
| GVH | graft versus host |
| HA | hemagglutinin |
| HAMA | human anti-mouse antibody |
| HBV | hepatitis B virus |
| HCG | human choriogonadotropin |
| HCL | hairy cell leukemia |
| HCV | hepatitis C virus |
| HDN | hemolytic disease of the newborn |
| HEV | high endothelial venules |
| HHV-8 | human herpesvirus-8 |
| HIV | human immunodeficiency virus |
| HLA | human leukocyte antigen |
| HPA | hypothalamus–pituitary–adrenal |
| HPV | human papillomavirus |
| HSC | hemopoietic stem cell |
| HSCT | HSC transplant |
| HSP | heat shock protein |
| HTLV | human T-lymphotropic virus, human T-cell leukemia virus |
| ICAM | intercellular adhesion molecule |
| IDC | interdigitating cell |
| IEL | intraepithelial lymphocyte |
| IFN | interferon |
| Ig | immunoglobulin |
| IHC | immunohistochemistry |
| IL | interleukin |
| IPEX | immune dysregulation polyendocrinopathy enteropathy X-linked syndrome |
| IT | immunotoxin |
| ITAM | immunoreceptor tyrosine-based activation motif |
| ITIM | immunoreceptor tyrosine-based inhibitory motif |
| ITP | immune thrombocytopenic purpura |
| IVIG | intravenous immunoglobulin |
| JIA | juvenile idiopathic arthritis |
| KAR | killer activation receptor |
| KIR | killer inhibitory receptor |
| KS | Kaposi's sarcoma |
| KSHV | Kaposi's sarcoma-associated herpesvirus (also known as human herpesvirus-8 [HHV-8]) |
| LAK | lymphokine-activated killer |

| | |
|---|---|
| LCMV | lymphocytic choriomeningitis virus |
| LFA | leukocyte function antigen |
| LH | Langerhans cell |
| LP | late proliferative |
| LPS | lipopolysaccharide |
| LS | late secretory |
| LSC | lymphoid stem cells |
| MØ | macrophage |
| mAb | monoclonal antibodies |
| MAC | membrane attack complex |
| MAK | macrophage-activated killer (cell) |
| MALT | mucosa-associated lymphoid tissue |
| MBL | mannose-binding lectin |
| MCP | membrane cofactor protein |
| MCP-1 | monocyte chemotactic protein |
| M-CSF | monocyte/macrophage colony-stimulating factor |
| MDP | muramyl dipeptide |
| MHC | major histocompatibility complex |
| MR | mannose receptor |
| MRSA | meticillin-resistant *Staphylococcus aureus* |
| MS | multiple sclerosis |
| MSC | myeloid suppressor cell |
| MTb | *Mycobacterium tuberculosis* |
| MZ | marginal zone |
| NALT | nasal-associated lymphoid tissue |
| NBT | nitroblue tetrazolium test |
| NFAT | nuclear factor of activated T cells |
| NK | natural killer |
| NLR | NOD-like receptor |
| NO | nitric oxide |
| NOD | nonobese diabetic |
| NPY | neuropeptide Y |
| NSAID | nonsteroidal anti-inflammatory drug |
| PAF | platelet-activating factor |
| PAGE | polyacrylamide gel electrophoresis |
| PALS | periarteriolar lymphoid sheath |
| PAMP | pathogen-associated molecular pattern |
| PAP | prostatic acid phosphatase |
| PCM | protein–calorie malnutrition |
| PCR | polymerase chain reaction |
| PEG | polyethylene glycol |
| pIgR | poly Ig receptor |
| PLGA | poly(lactide-co-glycolide) |
| PML | progressive multifocal leukoencephalopathy |
| PMN | polymorphonuclear cell |
| PNH | paroxysmal noctural hemoglobinuria |
| PNP | purine nucleoside phosphorylase |
| PPD | purified protein derivative |
| PRR | pattern recognition receptor |
| PS | phosphatidylserine |
| PSA | prostate-specific antigen |

| | |
|---|---|
| RA | rheumatoid arthritis |
| *RAG1* | recombinase activating gene 1 |
| RAST | radioallergosorbent test |
| RCC | renal cell cancer |
| RF | rheumatoid factor |
| RFLP | restriction fragment length polymorphism |
| RIA | radioimmunoassay |
| RIG-1 | retinoic-acid-inducible gene 1 |
| RLR | RIG-1-like receptor |
| RP | red pulp |
| RSV | respiratory syncytial virus |
| SAA | serum amyloid protein A |
| SARS | severe acute respiratory syndrome |
| SC | secretory component |
| SCF | stem cell factor |
| SCID | severe combined immunodeficiency |
| SDS | sodium dodecyl sulfate |
| SE | staphylococcal enterotoxin(s) |
| SLE | systemic lupus erythematosus |
| S-Lx | sialyl-Lewis X (molecule) |
| SOM | somatostatin |
| SR | scavenger receptor |
| SV | splenic vein |
| TAA | tumor-associated antigen |
| TB | tuberculosis |
| TBII | thyrotropin-binding inhibitory immunoglobulin |
| Tc | T-cytotoxic cell |
| TCR | T-cell antigen receptor |
| TGF | transforming growth factor |
| TGSI | thyroid growth-stimulating immunoglobulin |
| Th | T-helper cell |
| TIL | tumor-infiltrating lymphocyte |
| TLR | Toll-like receptor |
| TNF | tumor necrosis factor |
| Treg | regulatory T cell |
| TSA | tumor-specific antigen |
| TSH | thyroid-stimulating hormone |
| TSST | toxic shock syndrome toxin |
| VCAM | vascular cell adhesion molecule |
| VEGF | vascular endothelial growth facor |
| VIP | vasoactive intestinal peptide |
| WAS | Wiscott–Aldrich syndrome |

# Index

Page numbers in **bold** refer to tables; those in *italics* refer to figures